TRANSLATIONS

OF

MATHEMATICAL

MONOGRAPHS

Volume 32

TRIANGULAR AND JORDAN REPRESENTATIONS OF LINEAR OPERATORS

by

M. S. BRODSKIĬ

AMERICAN MATHEMATICAL SOCIETY

Providence, Rhode Island 02904

1971

ТРЕУГОЛЬНЫЕ И ЖОРДАНОВЫ
ПРЕДСТАВЛЕНИЯ ЛИНЕЙНЫХ ОПЕРАТОРОВ

(Серия „Современные проблемы математики")

М. С. БРОДСКИЙ

Издательство „Наука"
Главная Редакция
Физико-Математической Литературы

Москва 1969

Translated from the Russian by
J. M. Danskin

International Standard Book Number 0-8218-1582-2

Library of Congress Card Number 74-162998

Printed in the United States of America

PREFACE

In this book we present the foundations of the theory of triangular and Jordan representations of bounded linear operators in Hilbert space, a subject which has arisen in the last 10–15 years.

It is well known that for every selfadjoint matrix of finite order there exists a unitary transformation which carries it into diagonal form. Geometrically this means that a finite-dimensional Hilbert space, in which there is given a selfadjoint operator A, is representable in the form of the orthogonal sum of one-dimensional subspaces invariant relative to A. More than 60 years ago David Hilbert formulated the infinite-dimensional analog of this fact.

Any square matrix, according to Schur's theorem, can be reduced by means of a certain unitary transformation to triangular form. The geometrical picture here is to the effect that a linear operator A given in an n-dimensional space $\mathfrak{H}$ has a system of invariant subspaces, ordered by inclusion:

$$0 = \mathfrak{H}_0 \subset \mathfrak{H}_1 \subset \dots \subset \mathfrak{H}_{n-1} \subset \mathfrak{H}_n = \mathfrak{H} \qquad (\dim \mathfrak{H}_k = k).$$

Denote by P_k the orthoprojector onto $\mathfrak{H}_k$. Since $\Delta P_j A \Delta P_k = 0$ $(k < j$, $\Delta P_j = P_j - P_{j-1})$ and

$$A = \sum_{j,k=1}^{n} \Delta P_j A \Delta P_k = \sum_{j=1}^{n} \Delta P_j A \Delta P_j + \sum_{j<k} \Delta P_j A \Delta P_k,$$

we have the formula

$$A = \sum_{j=1}^{n} \lambda_j \Delta P_j + 2i \sum_{j=1}^{n} P_{j-1} A_I \Delta P_j \tag{1}$$

$$\left(\Delta P_j A \Delta P_j = \lambda_j \Delta P_j,\ A_I = \frac{A - A^*}{2i}\right),$$

yielding a triangular representation of the operator A. If A is invertible, then it also admits the triangular representation

$$A = \sum_{j=1}^{n} \lambda_j \Delta P_j \left(E + \sum_{j=1}^{n} P_{j-1} (E + P_{j-1} H P_{j-1})^{-1} H \Delta P_j\right)^{-1} \tag{2}$$

$$(H = E - A^* A),$$

whose derivation is rather complicated.

The first step in the theory of triangular representations of nonselfadjoint operators operating in infinite-dimensional spaces was taken by M. S. Livšic [1] in 1954. Using the theory of characteristic functions created by him, he constructed a triangular functional model of a bounded linear operator with nuclear imaginary component. Later on, thanks to the investigations of L. A. Sahnovič [1,2], A. V. Kužel' [1,2], V. T. Poljackiĭ [1] and others, triangular functional models of operators belonging to other classes were found. Simultaneously, in the work of the present author [1-4], I. C. Gohberg and M. G. Kreĭn, [1-6], Ju. I. Ljubič and V. I. Macaev [1,2,3], V. I. Macaev [1,2], V. M. Brodskiĭ [1], and V. M. Brodskiĭ and the present author [1], the theory of abstract triangular representations was formulated. It was proved in particular that every completely continuous operator, and also every bounded operator with a completely continuous imaginary component, whose eigenvalues tend to zero sufficiently rapidly, is representable in an integral form which is the natural analog of the right side of formula (1). Analogously, invertible operators, close in a certain sense to unitary operators, turned out to be connected with formula (2).

We now turn to the consideration of an operator operating in a finite-dimensional space. It is easy to see that its matrix, reduced to normal (Jordan) form, consists of only one Jordan cell if and only if, given any two invariant subspaces of the operator, one is contained in the other. A linear operator, finite or infinite dimensional, will be called unicellular if its invariant subspaces satisfy this last condition.

The proof of the existence of infinite-dimensional unicellular operators was first obtained in 1957 (M. S. Brodskiĭ, [5], W. F. Donoghue [1], G. K. Kalisch [1], and L. A. Sahnovič [3]). Certain classes of such operators were investigated also in the papers of N. K. Nikol'skiĭ [1,2] and L. A. Sahnovič [4]. For Volterra operators with positive nuclear imaginary component the present author and G. È. Kisilevskiĭ [1] found criteria for unicellularity, expressed in terms of the growth of the resolvent.

We will say that the space $\mathfrak{H}$ is the approximate sum of subspaces $\mathfrak{H}_\gamma$ $(\gamma \in \Gamma)$ if 1) the closure of the linear envelope of the subspaces $\mathfrak{H}_\gamma$ $(\gamma \in \Gamma)$ coincides with $\mathfrak{H}$, and 2) for any decomposition of Γ into nonintersecting parts Γ_1 and Γ_2 the equation $\mathfrak{H}^{(\Gamma_1)} \cap \mathfrak{H}^{(\Gamma_2)} = 0$ is satisfied, where $\mathfrak{H}^{(\Gamma_j)}$ is the closure of the linear envelope of the subspaces $\mathfrak{H}_\gamma$ $(\gamma \in \Gamma_j)$. Using the criterion mentioned above for unicellularity and theorems of the theory of characteristic functions, Kisilevskiĭ proved in [1] that a space in which there is given a Volterra operator A with positive nuclear imaginary component is representable in the form of an approximate sum of a finite or countable number of subspaces invariant relative to A, in each of which there is induced a unicellular operator.

In the first chapter of this book we study operator nodes, i.e. collections of the form

$$\Theta = \begin{pmatrix} A & K & J \\ \mathfrak{H} & & \mathfrak{G} \end{pmatrix} \left(J = J^*,\ J^2 = E,\ KJK^* = \frac{A - A^*}{2i} \right),$$

where $\mathfrak{H}$ and $\mathfrak{G}$ are separable Hilbert spaces, and A, K and J are bounded linear operators operating respectively in $\mathfrak{H}$, from $\mathfrak{G}$ into $\mathfrak{H}$, and in $\mathfrak{G}$. To each operator node Θ we assign its characteristic function

$$W_\Theta(\lambda) = E - 2iK^*(A - \lambda E)^{-1} KJ$$

and establish conditions which are necessary and sufficient for a given analytic operator-function to be characteristic for an operator node of some class. Also, we investigate the connection, important in the applications, between the so-called regular divisors of the function $W_\theta(\lambda)$ and the invariant subspaces of the operator A. The results obtained in this chapter are applied in the last two sections of the second chapter and in almost all the sections of the third chapter.

The second chapter is devoted to the theory of abstract triangular representations of Volterra operators and to multiplicative decompositions of characteristic functions of the corresponding operator nodes. Along the way we consider the applications of the theory of triangular representations to the problem of the multiplicative decomposition of analytic operator-functions, originating with V. P. Potapov [1]. The theorems on triangular representations of operators lying in other classes are applied without detailed proofs in the part of the book called "Notes on the literature, and additional remarks". The theory of abstract triangular representations of Volterra operators is presented in the second chapter only to the degree to which it is necessary for the understanding of the third chapter. An exhaustive exposition of the present state of this theory and its numerous applications can be found in the monograph of I. C. Gohberg and M. G. Kreĭn [5].

The third chapter is based on the material of the first two. It contains tests for the unicellularity of Volterra operators and theorems on the decomposition of Volterra operators with positive nuclear imaginary component into unicellular operators.

We assume that the reader is acquainted with the theory of linear operators to the extent of the first six chapters of the course of N. I. Ahiezer and I. M. Glazman [1].

The author is grateful to M. G. Kreĭn, M. S. Livšic, Ju. P. Ginzburg and G. È. Kisilevskiĭ, who read the book in manuscript and made valuable comments. The last four sections of the book were written in collaboration with G. È. Kisilevskiĭ.

M. Brodskiĭ

TABLE OF CONTENTS

CHAPTER I

OPERATOR AND MATRIX NODES. CHARACTERISTIC FUNCTIONS OF NODES

§1. Operator nodes

1. **Definition of operator node.** Suppose given separable Hilbert spaces $\mathfrak{H}$ and $\mathfrak{G}$ and bounded linear operators A, K, and J, operating respectively in $\mathfrak{H}$, from $\mathfrak{G}$ into $\mathfrak{H}$ and in $\mathfrak{G}$. If the relations[1]

$$J = J^*, \quad J^2 = E, \tag{1.1}$$

$$KJK^* = \frac{A - A^*}{2i} \tag{1.2}$$

hold, then the collection consisting of the spaces $\mathfrak{H}$ and $\mathfrak{G}$ and the operators A, K and J is said to be an *operator node* or simply a *node* and is denoted by the symbol

$$\Theta = \begin{pmatrix} A & K & J \\ \mathfrak{H} & & \mathfrak{G} \end{pmatrix}.$$

Represent J in the form

$$J = P^+ - P^-, \tag{1.3}$$

where $P^+ = (E + J)/2$, $P^- = (E - J)/2$. Then $P^+ + P^- = E$, and, in view of (1.1), P^+ and P^- are orthoprojectors. Conversely, if P^+ and P^- are orthoprojectors in $\mathfrak{G}$ and $P^+ + P^- = E$, then the operator $J = P^+ - P^-$ satisfies conditions (1.1).

The spaces $\mathfrak{H}$ and $\mathfrak{G}$ are called respectively *interior* and *exterior*, and the operators A, K, J are *basic*, *canal*, and *directing*.

We recall that the operator $A_R = (A + A^*)/2$ is said to be the *real* and the operator $A_I = (A - A^*)/2i$ to be the *imaginary component* of the operator A. Both components are selfadjoint operators, and $A = A_R + iA_I$.

---

1) The letter E denotes the unit operator throughout.

The range of the operators A_I and K will be denoted by $\mathfrak{R}(A_I)$ and $\mathfrak{R}(K)$. (1.2) implies the relation

$$\mathfrak{R}(A_I) \subseteq \mathfrak{R}(K). \tag{1.4}$$

The subspace $\overline{\mathfrak{R}(A_I)}$ is said to be *non-Hermitian*, and the subspace $\overline{\mathfrak{R}(K)}$ *canal*.

Theorem 1.1. *If A is a bounded linear operator operating in a separable Hilbert space $\mathfrak{H}$, and $\mathfrak{R}$ is any subspace containing $\mathfrak{R}(A_I)$, then there exists a node Θ for which the operator A is basic and the operator $\mathfrak{R}$ is a canal operator.*

Proof. The operator A_I maps $\overline{\mathfrak{R}(A_I)}$ into itself and annihilates its orthogonal complement. Consider the spectral decomposition $\int_a^b t\,dE(t)$ of the operator $A_I^{(0)}$ induced by A_I in $\overline{\mathfrak{R}(A_I)}$. Putting

$$K_0 = \int_a^b |t|^{1/2}\, dE(t), \quad J_0 = \int_a^b \operatorname{sign} t\, dE(t),$$

we get

$$J_0 = J_0^*, \quad J_0^2 = E, \quad K_0 J_0 K_0^* = A_I^{(0)}.$$

Construct the orthogonal sum

$$\mathfrak{G} = \overline{\mathfrak{R}(A_I)} \oplus \mathfrak{H}_1 \oplus \mathfrak{H}_2,$$

where $\mathfrak{H}_1$ and $\mathfrak{H}_2$ are Hilbert spaces whose dimensions coincide with the dimension of the subspace $\mathfrak{H}_0 = \mathfrak{R} \ominus \overline{\mathfrak{R}(A_I)}$. Suppose that U_{12} is some isometric mapping of $\mathfrak{H}_2$ onto $\mathfrak{H}_1$ and that K_1 is a bounded linear mapping of $\mathfrak{H}_1$ into $\mathfrak{H}_0$ which has a range dense in $\mathfrak{H}_0$. Then $K_2 = K_1 U_{12}$ is a mapping of $\mathfrak{H}_2$ into $\mathfrak{H}_0$, while $K_2 K_2^* h = K_1 K_1^* h$ $(h \in \mathfrak{H}_0)$.

Introduce a mapping K of $\mathfrak{G}$ into $\mathfrak{H}$ and a mapping J of $\mathfrak{G}$ into itself, defined by the formulas

$$Kg = \begin{cases} K_0 g, & g \in \overline{\mathfrak{R}(A_I)}, \\ K_1 g, & g \in \mathfrak{H}_1, \\ K_2 g, & g \in \mathfrak{H}_2, \end{cases} \qquad Jg = \begin{cases} J_0 g, & g \in \overline{\mathfrak{R}(A_I)}, \\ g, & g \in \mathfrak{H}_1, \\ -g, & g \in \mathfrak{H}_2. \end{cases}$$

It is easy to see that K has a range which is dense in $\mathfrak{R}$. Moreover, $J = J^*$ and $J^2 = E$. Since

$$K^* h = \begin{cases} K_0^* h, & h \in \overline{\mathfrak{R}(A_I)}, \\ (K_1^* + K_2^*)\, h, & h \in \mathfrak{H}_0, \\ 0, & h \perp \mathfrak{R}, \end{cases}$$

we have $KJK^* = A_I$. Thus

$$\Theta = \begin{pmatrix} A & K & J \\ \mathfrak{H} & & \mathfrak{G} \end{pmatrix}$$

is the desired node. The theorem is proved.

A node

$$\Theta = \begin{pmatrix} A & K & J \\ \mathfrak{H} & & \mathfrak{G} \end{pmatrix}$$

constructed relative to a given bounded linear operator A will be called an *imbedding of A into a node.* It is clear that this operation is not unique.

2. **Completely nonselfadjoint operators.** A bounded linear operator A will be said to be *completely nonselfadjoint* if the space $\mathfrak{H}$ in which it operates cannot be represented in the form of an orthogonal sum of two subspaces $\mathfrak{H}_1$ and $\mathfrak{H}_0$ $(\neq 0)$ with the following properties: 1) $\mathfrak{H}_1$ and $\mathfrak{H}_0$ are invariant relative to A; 2) A induces in $\mathfrak{H}_0$ a selfadjoint operator.

Theorem 1.2. *The closure $\mathfrak{H}_1$ of the linear envelope of vectors of the form*

$$A^n A_I h \qquad (n = 0, 1, \ldots; \ h \in \mathfrak{H}) \tag{1.5}$$

and its orthogonal complement $\mathfrak{H}_0 = \mathfrak{H} \ominus \mathfrak{H}_1$ are invariant relative to A. The operator A induces in $\mathfrak{H}_1$ a completely nonselfadjoint operator, and in $\mathfrak{H}_0$ a selfadjoint operator.

Proof. Since $\mathfrak{H}_1$ is obviously invariant relative to A, therefore $\mathfrak{H}_0$ is invariant relative to A^*. Moreover, since as the range of the operator A_I lies in $\mathfrak{H}_1$, we have $A_I \mathfrak{H}_0 = 0$. Thus $Ah = A^*h$ $(h \in \mathfrak{H}_0)$, so that $\mathfrak{H}_0$ is invariant relative to A as well. At the same time we have proved that A induces a selfadjoint operator in $\mathfrak{H}_0$.

Suppose that A_1 is the operator induced by A in $\mathfrak{H}_1$. We suppose that $\mathfrak{H}_1 = \mathfrak{H}_{11} \oplus \mathfrak{H}_{10}$ $(\mathfrak{H}_{10} \neq 0)$, while: 1) $\mathfrak{H}_{11}$ and $\mathfrak{H}_{10}$ are invariant relative to A_1; 2) A_1 induces in $\mathfrak{H}_{10}$ a selfadjoint operator. Then A induces in $\mathfrak{H}_{10} \oplus \mathfrak{H}_0$ a selfadjoint operator, and therefore

$$A^n A_I h \in \mathfrak{H}_{11} \qquad (n = 0, 1, \ldots; \ h \in \mathfrak{H}),$$

which contradicts the definition of the subspace $\mathfrak{H}_1$.

Corollary 1. *For the operator A to be completely nonselfadjoint it is necessary and sufficient that the closure of the linear envelope of vectors of the form* (1.5) *coincide with $\mathfrak{H}$.*

Corollary 2. *The space $\mathfrak{H}$ can be represented in one and only one way in the form of an orthogonal sum of subspaces $\mathfrak{H}_1$ and $\mathfrak{H}_0$ which are invariant relative to A and in which A induces respectively a completely nonselfadjoint and a selfadjoint operator.*

3. Simple nodes. Consider the operator node

$$\Theta = \begin{pmatrix} A & K & J \\ \mathfrak{H} & & \mathfrak{G} \end{pmatrix}$$

and denote by $\mathfrak{E}_\Theta$ the closure of the linear envelope of vectors of the form

$$A^n K g \qquad (n = 0,\ 1,\ \ldots;\ g \in \mathfrak{G}).$$

The subspaces $\mathfrak{E}_\Theta$ and $\mathfrak{E}_\Theta^{(0)} = \mathfrak{H} \ominus \mathfrak{E}_\Theta$ are called respectively *principal* and *excess*. It easily follows from Theorem 1.2 and relation (1.4) that each of the subspaces $\mathfrak{E}_\Theta$ and $\mathfrak{E}_\Theta^{(0)}$ is invariant relative to A and A^*, and that $Ah = A^*h$ $(h \in \mathfrak{E}_\Theta^{(0)})$.

The node Θ is said to be *simple* if $\mathfrak{E}_\Theta = \mathfrak{H}$, and *excess* otherwise. *For a node to be simple it is sufficient that its basic operator be completely nonselfadjoint.* The converse assertion is generally speaking false. Indeed, putting $\mathfrak{R} = \mathfrak{H}$ in Theorem 1.1, we find that *every bounded linear operator may be imbedded in a simple node.*

Having a node

$$\Theta = \begin{pmatrix} A & K & J \\ \mathfrak{H} & & \mathfrak{G} \end{pmatrix},$$

we may construct a new node

$$\Theta^* = \begin{pmatrix} A^* & K & -J \\ \mathfrak{H} & & \mathfrak{G} \end{pmatrix},$$

which a called *adjoint* to Θ.

The principal subspaces of the nodes Θ and Θ^* coincide. Indeed, since the subspace $\mathfrak{E}_\Theta$ is invariant relative to A^* and $\mathfrak{R}(K) \subseteq \mathfrak{E}_\Theta$, we have

$$A^{*^n} K g \in \mathfrak{E}_\Theta \qquad (n = 0,\ 1,\ \ldots;\ g \in \mathfrak{G}),$$

i.e. $\mathfrak{E}_{\Theta^*} \subseteq \mathfrak{E}_\Theta$. An analogous argument shows that $\mathfrak{E}_\Theta \subseteq \mathfrak{E}_{\Theta^*}$.

Lemma 1.1. *Suppose that*

$$\Theta = \begin{pmatrix} A & K & J \\ \mathfrak{H} & & \mathfrak{G} \end{pmatrix}$$

is an operator node. If the subspace $\mathfrak{H}_0 \subset \mathfrak{H}$ *is invariant relative to* A *and orthogonal to* $\Re(K)$, *then it belongs to the excess subspace* $\mathfrak{E}_\Theta^{(0)}$.

Proof. For any vector $h \in \mathfrak{H}_0$ the equation

$$(h, A^{*^n}Kg) = (A^n h, Kg) = 0 \qquad (n = 0, 1, \ldots; \ g \in \mathfrak{G})$$

holds, which means that $h \perp \mathfrak{E}_{\Theta^*}$. Since $\mathfrak{E}_\Theta = \mathfrak{E}_{\Theta^*}$, we have $h \perp \mathfrak{E}_\Theta$, i.e. $h \in \mathfrak{E}_\Theta^{(0)}$

Theorem 1.3. *For the node*

$$\Theta = \begin{pmatrix} A & K & J \\ \mathfrak{H} & & \mathfrak{G} \end{pmatrix}$$

to be excess, it is necessary and sufficient that there exist a subspace $\mathfrak{H}_0 \subset \mathfrak{H}$ $(\mathfrak{H}_0 \neq 0)$ *which is invariant relative to* A *and orthogonal to* $\Re(K)$.

Proof. If Θ is excess, then $\mathfrak{E}_\Theta^{(0)}$ is distinct from zero, invariant relative to A and orthogonal to $\Re(K)$. The sufficiency follows from Lemma.

The theorem is proved.

Suppose that

$$\Theta = \begin{pmatrix} A & K & J \\ \mathfrak{H} & & \mathfrak{G} \end{pmatrix}$$

is an excess node. Denoting by A_Θ and $A_\Theta^{(0)}$ the operators induced by the operator A in the subspaces $\mathfrak{E}_\Theta$ and $\mathfrak{E}_\Theta^{(0)}$, we obtain the nodes

$$\Theta_{\mathfrak{E}} = \begin{pmatrix} A_\Theta & K & J \\ \mathfrak{E}_\Theta & & \mathfrak{G} \end{pmatrix} \quad \text{and} \quad \Theta_{\mathfrak{E}^{(0)}} = \begin{pmatrix} A_\Theta^{(0)} & 0 & J \\ \mathfrak{E}_\Theta^{(0)} & & \mathfrak{G} \end{pmatrix},$$

which are called respectively the *principal* and *excess parts of the node* Θ. It is easy to see that $\Theta_{\mathfrak{E}}$ is a simple node.

§2. Products of operator nodes

1. Definitions. We shall say that the nodes

$$\Theta = \begin{pmatrix} A & K & J \\ \mathfrak{H} & & \mathfrak{G} \end{pmatrix} \quad \text{and} \quad \Theta' = \begin{pmatrix} A' & K' & J' \\ \mathfrak{H}' & & \mathfrak{G}' \end{pmatrix}$$

are *equal*, and write $\Theta = \Theta'$, if $\mathfrak{H} = \mathfrak{H}'$, $\mathfrak{G} = \mathfrak{G}'$, $A = A'$, $K = K'$ and $J = J'$.

Consider the nodes

$$\Theta_1 = \begin{pmatrix} A_1 & K_1 & J \\ \mathfrak{H}_1 & & \mathfrak{G} \end{pmatrix} \quad \text{and} \quad \Theta_2 = \begin{pmatrix} A_2 & K_2 & J \\ \mathfrak{H}_2 & & \mathfrak{G} \end{pmatrix},$$

for which the exterior spaces and the directing operator coincide, and denote by P_1 and P_2 the orthoprojectors onto $\mathfrak{H}_1$ and $\mathfrak{H}_2$ operating in the space $\mathfrak{H} = \mathfrak{H}_1 \oplus \mathfrak{H}_2$. Introduce the operators

$$A = A_1P_1 + A_2P_2 + 2iK_1JK_2^*P_2, \qquad K = K_1 + K_2, \tag{2.1}$$

operating respectively in $\mathfrak{H}$ and from $\mathfrak{G}$ into $\mathfrak{H}$. The spaces $\mathfrak{H}$, $\mathfrak{G}$ and the operators A, K, J combine to form a node. Indeed, since

$$K_1JK_1^* = \frac{A_1 - A_1^*}{2i}, \qquad K_2JK_2^* = \frac{A_2 - A_2^*}{2i},$$

$$A^* = A_1^*P_1 + A_2^*P - 2iK_2JK_1^*P_1, \qquad K^* = K_1^*P_1 + K_2^*P_2,$$

we have

$$\frac{A - A^*}{2i} = K_1JK_1^*P_1 + K_2JK_2^*P_2 + K_1JK_2^*P_2 + K_2JK_1^*P_1$$
$$= (K_1 + K_2)\,J\,(K_1^*P_1 + K_2^*P_2) = KJK^*.$$

We agree to call the node

$$\Theta = \begin{pmatrix} A_1P_1 + A_2P_2 + 2iK_1JK_2^*P_2 & K_1 + K_2 & J \\ \mathfrak{H} & & \mathfrak{G} \end{pmatrix}$$

the *product* of the nodes Θ_1 and Θ_2 and to write $\Theta = \Theta_1\Theta_2$.

One shows by a direct verification that the following formulas hold:

$$(\Theta_1\Theta_2)\,\Theta_3 = \Theta_1\,(\Theta_2\Theta_3), \tag{2.2}$$

$$(\Theta_1\Theta_2)^* = \Theta_2^*\Theta_1^*. \tag{2.3}$$

If $\Theta_1\Theta_2 = \Theta_1\Theta_3$ or $\Theta_2\Theta_1 = \Theta_3\Theta_1$, then $\Theta_2 = \Theta_3$.

2. **Projections of a node.** We select in the interior space $\mathfrak{H}$ of the node

$$\Theta = \begin{pmatrix} A & K & J \\ \mathfrak{H} & & \mathfrak{G} \end{pmatrix}$$

any subspace $\mathfrak{H}_0$, and define in it an operator $A_0h = P_0Ah$ $(h \in \mathfrak{H}_0)$, where P_0 is the orthoprojector onto $\mathfrak{H}_0$. Moreover, we construct a mapping $K_0 = P_0K$ of the space $\mathfrak{G}$ into $\mathfrak{H}_0$. Since

$$\left.\begin{aligned} &A_0^*h = P_0A^*h, \qquad K_0^*h = K^*h, \\ &K_0JK_0^*h = P_0KJK^*h = P_0\frac{A - A^*}{2i}\,h = \frac{A_0 - A_0^*}{2i}\,h \end{aligned}\right\} \qquad (h \in \mathfrak{H}_0),$$

the node

$$\Theta_0 = \begin{pmatrix} A_0 & K_0 & J \\ \mathfrak{H}_0 & & \mathfrak{G} \end{pmatrix}$$

is an operator node. We will call the node Θ_0 *the projection of* Θ *onto the subspace* $\mathfrak{H}_0$ *and write* $\Theta_0 = \mathrm{pr}_{\mathfrak{H}_0}\Theta$.

We note the relations

$$\mathrm{pr}_{\mathfrak{H}_0}\Theta^* = (\mathrm{pr}_{\mathfrak{H}_0}\Theta)^*, \tag{2.4}$$

$$\mathrm{pr}_{\mathfrak{H}_1}\Theta = \mathrm{pr}_{\mathfrak{H}_1}(\mathrm{pr}_{\mathfrak{H}_2}\Theta) \qquad (\mathfrak{H}_1 \subseteq \mathfrak{H}_2), \tag{2.5}$$

which follow directly from the definition of projection.

If

$$\Theta_1 = \begin{pmatrix} A_1 & K_1 & J \\ \mathfrak{H}_1 & & \mathfrak{G} \end{pmatrix}, \quad \Theta_2 = \begin{pmatrix} A_2 & K_2 & J \\ \mathfrak{H}_2 & & \mathfrak{G} \end{pmatrix}, \quad \Theta = \Theta_1\Theta_2 = \begin{pmatrix} A & K & J \\ \mathfrak{H} & & \mathfrak{G} \end{pmatrix},$$

then, as is shown by formulas (2.1), Θ_1 and Θ_2 are projections of the node Θ onto $\mathfrak{H}_1$ and $\mathfrak{H}_2$ respectively, and $\mathfrak{H}_1$ is invariant relative to A. Conversely, *every node* Θ *is the product of its projections*

$$\Theta_1 = \begin{pmatrix} A_1 & K_1 & J \\ \mathfrak{H}_1 & & \mathfrak{G} \end{pmatrix} \text{ and } \Theta_2 = \begin{pmatrix} A_2 & K_2 & J \\ \mathfrak{H}_2 & & \mathfrak{G} \end{pmatrix}$$

onto an arbitrary subspace $\mathfrak{H}_1$ *invariant relative to* A *and its orthogonal complement* $\mathfrak{H}_2$. Indeed, denoting by P_j the orthoprojector onto $\mathfrak{H}_j$ ($j = 1, 2$), we get

$$P_2AP_1 = 0, \qquad P_1A^*P_2 = 0,$$

$$P_1AP_2 = 2iP_1\frac{A - A^*}{2i}P_2 = 2iP_1KJK^*P_2 = 2iK_1JK_2^*P_2,$$

$$A = (P_1 + P_2)A(P_1 + P_2) = A_1P_1 + A_2P_2 + 2iK_1JK_2^*P_2.$$

In particular, *each node is equal to the product of its principal and excess parts.*

The node Θ_0 will be called a *left* (*right*) *divisor* of the node Θ if it is a projection of Θ onto a subspace $\mathfrak{H}_0 \subseteq \mathfrak{H}$ which is invariant relative to A (A^*).

Suppose that Θ is an operator node and that the subspaces

$$0 = \mathfrak{H}_0 \subset \mathfrak{H}_1 \subset \mathfrak{H}_2 \subset \ldots \subset \mathfrak{H}_n = \mathfrak{H}$$

are invariant relative to A. Inasmuch as the subspace $\mathfrak{H}_{k-1}$ is invariant relative to the operator A_k induced by A in $\mathfrak{H}_k$ ($k = 2, 3, \cdots, n-1$), from formula (2.5) we obtain

$$\Theta = \mathrm{pr}_{\mathfrak{H}_1}\Theta\, \mathrm{pr}_{\mathfrak{H}_2 \ominus \mathfrak{H}_1}\Theta \ldots \mathrm{pr}_{\mathfrak{H}_n \ominus \mathfrak{H}_{n-1}}\Theta. \tag{2.6}$$

3. Product of simple nodes.

Lemma 2.1. *If*

$$\Theta = \begin{pmatrix} A & K & J \\ \mathfrak{H} & & \mathfrak{G} \end{pmatrix}$$

is the product of the nodes

$$\Theta_1 = \begin{pmatrix} A_1 & K_1 & J \\ \mathfrak{H}_1 & & \mathfrak{G} \end{pmatrix} \quad and \quad \Theta_2 = \begin{pmatrix} A_2 & K_2 & J \\ \mathfrak{H}_2 & & \mathfrak{G} \end{pmatrix},$$

then

$$\mathfrak{E}^{(0)}_{\Theta_j} = \mathfrak{E}^{(0)}_{\Theta} \cap \mathfrak{H}_j \qquad (j = 1,\ 2). \tag{2.7}$$

Proof. The subspace $\mathfrak{E}^{(0)}_{\Theta_1}$ is invariant relative to A and orthogonal to $\mathfrak{R}(K_1)$. It is also orthogonal to $\mathfrak{R}(K)$, since $\mathfrak{R}(K_1)$ is the projection of $\mathfrak{R}(K)$ onto $\mathfrak{H}_1$. By Lemma 1.1, $\mathfrak{E}^{(0)}_{\Theta_1} \subseteq \mathfrak{E}^{(0)}_{\Theta} \cap \mathfrak{H}_1$. On the other hand, the subspace $\mathfrak{E}^{(0)}_{\Theta} \cap \mathfrak{H}_1$ is invariant relative to A_1 and orthogonal to $\mathfrak{R}(K_1)$. Applying Lemma 1.1 to the node Θ_1, we find that $\mathfrak{E}^{(0)}_{\Theta} \cap \mathfrak{H}_1 \subseteq \mathfrak{E}^{(0)}_{\Theta_1}$. Thus equation (2.7) is proved for $j = 1$. It remains to be noted that $\Theta^* = \Theta^*_2\Theta^*_1$, so that, from what has already been proved, $\mathfrak{E}^{(0)}_{\Theta^*_1} = \mathfrak{E}^{(0)}_{\Theta^*} \cap \mathfrak{H}_2$. Inasmuch as $\mathfrak{E}^{(0)}_{\Theta^*_2} = \mathfrak{E}^{(0)}_{\Theta_2}$ and $\mathfrak{E}^{(0)}_{\Theta^*} = \mathfrak{E}^{(0)}_{\Theta}$, we have $\mathfrak{E}^{(0)}_{\Theta_2} = \mathfrak{E}^{(0)}_{\Theta} \cap \mathfrak{H}_2$.

Theorem 2.1. *If* $\Theta = \Theta_1\Theta_2 \cdots \Theta_n$, *then*

$$\mathfrak{E}^{(0)}_{\Theta_j} = \mathfrak{E}^{(0)}_{\Theta} \cap \mathfrak{H}_j \qquad (j = 1,\ 2,\ \ldots,\ n),$$

where $\mathfrak{H}_j$ *is the interior space of the node* Θ_j.

Proof. In view of Lemma 2.1

$$\mathfrak{E}^{(0)}_{\Theta_j} = \mathfrak{E}^{(0)}_{\Theta_1\Theta_2\ldots\Theta_j} \cap \mathfrak{H}_j$$

and

$$\mathfrak{E}^{(0)}_{\Theta_1\Theta_2\ldots\Theta_j} = \mathfrak{E}^{(0)}_{\Theta} \cap (\mathfrak{H}_1 \oplus \mathfrak{H}_2 \oplus \ \ldots \ \oplus \mathfrak{H}_j).$$

Accordingly,

$$\mathfrak{E}^{(0)}_{\Theta_j} = \mathfrak{E}^{(0)}_{\Theta} \cap (\mathfrak{H}_1 \oplus \mathfrak{H}_2 \oplus \ \ldots \ \oplus \mathfrak{H}_j) \cap \mathfrak{H}_j = \mathfrak{E}^{(0)}_{\Theta} \cap \mathfrak{H}_j.$$

As a consequence of Theorem 2.1 we obtain the following assertion.

Theorem 2.2. *If* $\Theta = \Theta_1\Theta_2 \cdots \Theta_n$ *is a simple node, then all the nodes* Θ_j $(j = 1, 2, \cdots, n)$ *are simple.*

Suppose that A is a bounded linear operator operating in $\mathfrak{H}$. A subspace $\mathfrak{H}_0 \subseteq \mathfrak{H}$ will be said to be *semi-invariant* if it is representable in the form $\mathfrak{H}_0 = \mathfrak{H}_2 \ominus \mathfrak{H}_1$, where $\mathfrak{H}_1$ and $\mathfrak{H}_2$ are invariant relative to A. In particular, all subspaces invariant relative to A and their complements are semi-invariant.

Theorem 2.2 and formula (2.6) imply the following.

Theorem 2.2′. *The projection of the simple node*

$$\Theta = \begin{pmatrix} A & K & J \\ \mathfrak{H} & & \mathfrak{G} \end{pmatrix}$$

onto any semi-invariant (relative to A) subspace is again a simple node.

We note that the assertion converse to Theorem 2.2 is false: a product of simple nodes may turn out to be an excess node. Moreover, the following holds.

Theorem 2.3. *Suppose given a selfadjoint bounded operator A_0 in a separable Hilbert space $\mathfrak{H}_0$. There exist simple nodes Θ_1 and Θ_2 such that A_0 is a basic operator of the excess portion of the product $\Theta_1\Theta_2$.*

Proof. Consider the orhtogonal sum $\tilde{\mathfrak{H}} = \mathfrak{H}_0 \oplus \mathfrak{H}$, where $\mathfrak{H}$ is some Hilbert space whose dimension is equal to the dimension of the space $\mathfrak{H}_0$. Suppose that we are given any isometric mapping U of $\mathfrak{H}_0$ onto $\mathfrak{H}$ and that the operator $A = UA_0U^{-1}$ is imbedded in the simple node

$$\Theta = \begin{pmatrix} A & K & J \\ \mathfrak{H} & & \mathfrak{G} \end{pmatrix}.$$

It is not hard to see that the node

$$\tilde{\Theta} = \Theta\Theta_0 = \begin{pmatrix} \tilde{A} & K & J \\ \tilde{\mathfrak{H}} & & \mathfrak{G} \end{pmatrix}, \quad \text{where} \quad \Theta_0 = \begin{pmatrix} A_0 & 0 & J \\ \mathfrak{H}_0 & & \mathfrak{G} \end{pmatrix},$$

has principal part Θ and excess part Θ_0.

The subspace $\mathfrak{H}_1$ of the space $\tilde{\mathfrak{H}}$ consisting of vectors of the form $f + Uf$ $(f \in \mathfrak{H}_0)$ is invariant relative to $\tilde{A}$, since

$$\tilde{A}(f + Uf) = A_0 f + AUf = A_0 f + UA_0 f \qquad (f \in \mathfrak{H}_0).$$

Accordingly,

$$\tilde{\Theta} = \Theta_1\Theta_2 \quad (\Theta_1 = \mathrm{pr}_{\mathfrak{H}_1}\tilde{\Theta},\ \Theta_2 = \mathrm{pr}_{\mathfrak{H}_2}\tilde{\Theta},\quad \mathfrak{H}_2 = \tilde{\mathfrak{H}} \ominus \mathfrak{H}_1).$$

Inasmuch as $\mathfrak{H}_0 = \mathfrak{G}^{(0)}_{\tilde{\Theta}}$ and $\mathfrak{H}_j \cap \mathfrak{H}_0 = 0$ $(j = 1, 2)$, by Lemma 2.1 Θ_1 and Θ_2 are simple nodes.

4. Spectrum of the product of nodes.

Theorem 2.4. *Suppose that*

$$\Theta = \begin{pmatrix} A & K & J \\ \mathfrak{H} & & \mathfrak{G} \end{pmatrix}$$

is the product of the nodes

$$\Theta_1 = \begin{pmatrix} A_1 & K_1 & J \\ \mathfrak{H}_1 & & \mathfrak{G} \end{pmatrix} \text{ and } \Theta_2 = \begin{pmatrix} A_2 & K_2 & J \\ \mathfrak{H}_2 & & \mathfrak{G} \end{pmatrix}.$$

If the point λ in the complex plane is regular for the operators A_1 and A_2, then it is regular for the operator A as well, while

$$(A-\lambda E)^{-1} = (A_1-\lambda E)^{-1}P_1 + (A_2-\lambda E)^{-1}P_2 - 2i(A_1-\lambda E)^{-1}K_1JK_2^*(A_2-\lambda E)^{-1}P_2, \tag{2.8}$$

where P_1 and P_2 are orthoprojectors onto $\mathfrak{H}_1$ and $\mathfrak{H}_2$.

Proof. From the first of equations (2.1) we have

$$A-\lambda E = (A_1-\lambda E)P_1 + (A_2-\lambda E)P_2 + 2iK_1JK_2^*P_2. \tag{2.9}$$

Now it is easy to verify that the right side of (2.8) is an operator which is both left and right invertible for the operator (2.9). The theorem is proved.

We agree that by *the spectrum of a node* we will mean the spectrum of its basic operator. Theorem 2.4 means that *the spectrum of the product of two nodes is contained in the union of the spectra of the factors.*

Lemma 2.2. *Suppose that $\mathfrak{H}_0$ is an invariant subspace of the bounded linear operator A operating in the space $\mathfrak{H}$, and O is a bounded open connected piece of the complex plane all of whose points are regular for A. If at least one point $\lambda_0 \in O$ is regular for the operator A_0 induced by A in $\mathfrak{H}_0$, then all the points of O have the same property.*

Proof. We note in preparation that the point $\lambda \in O$ is regular for the operator A_0 if and only if $(A-\lambda E)^{-1}\mathfrak{H}_0 \subseteq \mathfrak{H}_0$.

For any point $\lambda_1 \in O$ we construct in O a polygon L joining λ_0 to λ_1. We choose $\epsilon > 0$ and points

$$\lambda_0 = \mu_0,\ \mu_1,\ \mu_2,\ \ldots,\ \mu_n = \lambda_1$$

on L so that the disks $k_j(|\lambda - \mu_j| < \epsilon,\ j = 0, 1, \cdots, n-1)$ lie in O and so that the inequalities $|\mu_{j+1} - \mu_j| < \epsilon$ are satisfied for $j = 0, 1, \cdots, n-1$. Since

$(A-\mu_0 E)^{-1}\mathfrak{H}_0 \subseteq \mathfrak{H}_0$ and the following expansion, which converges in norm,

$$R_\lambda = R_{\mu_0} + (\lambda-\mu_0)R^2_{\mu_0} + (\lambda-\mu_0)^2 R^3_{\mu_0} + \ldots$$
$$(\lambda \in k_0,\ R_\lambda = (A-\lambda E)^{-1})$$

holds, we have $(A-\lambda E)^{-1}\mathfrak{H}_0 \subseteq \mathfrak{H}_0$ $(\lambda \in k_0)$, and, in particular, $(A-\mu_1 E)^{-1}\mathfrak{H}_0 \subseteq \mathfrak{H}_0$. Analogously, in view of the decomposition

$$R_\lambda = R_{\mu_1} + (\lambda-\mu_1)R^2_{\mu_1} + (\lambda-\mu_1)^2 R^3_{\mu_1} + \ldots \qquad (\lambda \in k_1),$$

we find that $(A-\mu_2 E)^{-1}\mathfrak{H}_0 \subseteq \mathfrak{H}_0$. Continuing the process, we arrive at the relation $(A-\mu_n E)^{-1}\mathfrak{H}_0 \subseteq \mathfrak{H}_0$.

The lemma is proved.

The following is an easy consequence of Theorem 2.4, Lemma 2.2, and equation (2.3).

Theorem 2.5. *Suppose that* Θ *is the product of the nodes* Θ_1 *and* Θ_2. *If the set of regular points of the operator* A *is connected, then the spectrum of* Θ *is equal to the union of the spectra of* Θ_1 *and* Θ_2.

5. **Unitarily equivalent nodes.** An operator A_1, operating in a space $\mathfrak{H}_1$, is said to be *unitarily equivalent* to the operator A_2 operating in $\mathfrak{H}_2$ if there exists an isometric mapping U of $\mathfrak{H}_1$ onto $\mathfrak{H}_2$ such that $UA_1 = A_2 U$.

We will say that *the node*

$$\Theta_1 = \begin{pmatrix} A_1 & K_1 & J \\ \mathfrak{H}_1 & & \mathfrak{G} \end{pmatrix}$$

is unitarily equivalent to the node

$$\Theta_2 = \begin{pmatrix} A_2 & K_2 & J \\ \mathfrak{H}_2 & & \mathfrak{G} \end{pmatrix},$$

if there exists an isometric mapping U of the space $\mathfrak{H}_1$ onto $\mathfrak{H}_2$ such that

$$UA_1 = A_2 U, \qquad UK_1 = K_2. \tag{2.10}$$

Obviously, the relation of unitary equivalence is reflexive, symmetric and transitive. It is also easy to see that if one of two unitarily equivalent nodes is simple then so is the other.

If Θ is a simple node, and if for some unitary operator U the equations

$$UA = AU, \qquad UK = K$$

are satisfied, then

$$UA^nKg = A^nKg \quad (n=0,\ 1,\ \ldots;\ g\in\mathfrak{G}),$$

which means that $U = E$. Using this remark, we arrive at the following conclusion.

If

$$\Theta_1 = \begin{pmatrix} A_1 & K_1 & J \\ \mathfrak{H}_1 & & \mathfrak{G} \end{pmatrix} \quad and \quad \Theta_2 = \begin{pmatrix} A_2 & K_2 & J \\ \mathfrak{H}_2 & & \mathfrak{G} \end{pmatrix}$$

are unitarily equivalent simple nodes, then the isometric mapping satisfying conditions (2.10) *is defined uniquely.*

Theorem 2.6. *Suppose that*

$$\Theta_{12} = \Theta_1\Theta_2,\quad \Theta_{34} = \Theta_3\Theta_4,\quad \Theta_j = \begin{pmatrix} A_j & K_j & J \\ \mathfrak{H}_j & & \mathfrak{G} \end{pmatrix}.$$

If the nodes Θ_1 *and* Θ_2 *are unitarily equivalent respectively to* Θ_3 *and* Θ_4, *then* Θ_{12} *is unitarily equivalent to* Θ_{34}.

Proof. Denote by P_j an orthoprojector onto $\mathfrak{H}_j$, operating in $\mathfrak{H}_{12} = \mathfrak{H}_1 \oplus \mathfrak{H}_2$ for $j = 1, 2$ and in $\mathfrak{H}_{34} = \mathfrak{H}_3 \oplus \mathfrak{H}_4$ for $j = 3, 4$. Since

$$\Theta_{12} = \begin{pmatrix} A_{12} & K_{12} & J \\ \mathfrak{H}_{12} & & \mathfrak{G} \end{pmatrix},\quad \Theta_{34} = \begin{pmatrix} A_{34} & K_{34} & J \\ \mathfrak{H}_{34} & & \mathfrak{G} \end{pmatrix},$$

$$A_{12} = A_1P_1 + A_2P_2 + 2iK_1JK_2^*P_2,$$

$$A_{34} = A_3P_3 + A_4P_4 + 2iK_3JK_4^*P_4$$

and, in view of the hypothesis of the theorem, there exist isometric operators U_{31} and U_{42} mapping $\mathfrak{H}_1$ onto $\mathfrak{H}_3$ and $\mathfrak{H}_2$ onto $\mathfrak{H}_4$ respectively, and such that

$$U_{31}A_1 = A_3U_{31},\quad U_{42}A_2 = A_4U_{42},\quad U_{31}K_1 = K_3,\quad U_{42}K_2 = K_4,$$

it follows that the operator $U = U_{31}P_1 + U_{42}P_2$, isometrically mapping $\mathfrak{H}_{12}$ onto $\mathfrak{H}_{34}$, satisfies the relation

$$\begin{aligned} UA_{12} &= U_{31}A_1P_1 + U_{42}A_2P_2 + 2iU_{31}K_1JK_2^*P_2 \\ &= A_3U_{31}P_1 + A_4U_{42}P_2 + 2iK_3JK_4^*U_{42}P_2 = A_{34}U. \end{aligned}$$

Moreover,

$$UK_{12} = (U_{31}P_1 + U_{42}P_2)(K_1 + K_2) = U_{31}K_1 + U_{42}K_2 = K_3 + K_4 = K_{34}.$$

Theorem 2.7. *If the node* $\Theta^{(1)}$ *is unitarily equivalent to the node* $\Theta^{(2)}$ *and* $\Theta^{(1)} = \Theta_1\Theta_2$, *then* $\Theta^{(2)} = \Theta_3\Theta_4$, *where* Θ_3 *and* Θ_4 *are unitarily equivalent respectively to* Θ_1 *and* Θ_2.

Proof. Let

$$\Theta^{(j)} = \begin{pmatrix} A^{(j)} & K^{(j)} & J \\ \mathfrak{H}^{(j)} & & \mathfrak{G} \end{pmatrix} \quad (j = 1,\ 2), \quad \Theta_j = \begin{pmatrix} A_j & K_j & J \\ \mathfrak{H}_j & & \mathfrak{G} \end{pmatrix} \quad (j = 1,\ 2)$$

and let U be an isometric mapping of $\mathfrak{H}^{(1)}$ onto $\mathfrak{H}^{(2)}$ for which $UA^{(1)} = A^{(2)}U$, $UK^{(1)} = K^{(2)}$. Then $\mathfrak{H}^{(2)} = \mathfrak{H}_3 \oplus \mathfrak{H}_4$, where $\mathfrak{H}_3 = U\mathfrak{H}_1$ and $\mathfrak{H}_4 = U\mathfrak{H}_2$, and $\mathfrak{H}_3$ is invariant relative to $A^{(2)}$. It is easy to see that the nodes $\Theta_3 = \mathrm{pr}_{\mathfrak{H}_3}\Theta^{(2)}$ and $\Theta_4 = \mathrm{pr}_{\mathfrak{H}_4}\Theta^{(2)}$ satisfy the requirements of the theorem.

§ 3. Characteristic operator-functions of operator nodes

1. **Multiplication theorem.** Given an operator node

$$\Theta = \begin{pmatrix} A & K & J \\ \mathfrak{H} & & \mathfrak{G} \end{pmatrix}$$

we construct the function

$$W_\Theta(\lambda) = E - 2iK^*(A - \lambda E)^{-1}KJ \tag{3.1}$$

of the complex variable λ. This function is said to be the *characteristic operator-function* (COF) of the node Θ. It is obviously defined and holomorphic on the set G_A of regular points of the operator A, and its values are bounded linear operators operating in the exterior space $\mathfrak{G}$.

Theorem 3.1. *Suppose that*

$$\Theta = \begin{pmatrix} A & K & J \\ \mathfrak{H} & & \mathfrak{G} \end{pmatrix}$$

is the product of the nodes

$$\Theta_1 = \begin{pmatrix} A_1 & K_1 & J \\ \mathfrak{H}_1 & & \mathfrak{G} \end{pmatrix} \quad \text{and} \quad \Theta_2 = \begin{pmatrix} A_2 & K_2 & J \\ \mathfrak{H}_2 & & \mathfrak{G} \end{pmatrix}.$$

If the point λ is regular for the operators A_1 and A_2, then

$$W_{\Theta_1\Theta_2}(\lambda) = W_{\Theta_1}(\lambda)\, W_{\Theta_2}(\lambda). \tag{3.2}$$

Proof. Denote by P_j $(j = 1,\ 2)$ the orthoprojector onto $\mathfrak{H}_j$ operating in the space $\mathfrak{H} = \mathfrak{H}_1 \oplus \mathfrak{H}_2$. Applying Theorem 2.4, we obtain

$$\begin{aligned}W_{\Theta_1\Theta_2}(\lambda) &= E - 2iK^*(A-\lambda E)^{-1}KJ\\ &= E - 2iK^*P_1(A_1-\lambda E)^{-1}P_1KJ - 2iK^*P_2(A_2-\lambda E)^{-1}P_2KJ\\ &\quad + (2i)^2K^*P_1(A_1-\lambda E)^{-1}K_1JK_2^*(A_2-\lambda E)^{-1}P_2KJ\\ &= E - 2iK_1^*(A_1-\lambda E)^{-1}K_1J - 2iK_2^*(A_2-\lambda E)^{-1}K_2J\\ &\quad + (2i)^2K_1^*(A_1-\lambda E)^{-1}K_1JK_2^*(A_2-\lambda E)^{-1}K_2J\\ &= \left[E - 2iK_1^*(A_1-\lambda E)^{-1}K_1J\right]\left[E - 2iK_2^*(A_2-\lambda E)^{-1}K_2J\right] = W_{\Theta_1}(\lambda)\,W_{\Theta_2}(\lambda).\end{aligned}$$

Corollary. *If the basic operator node*

$$\Theta = \begin{pmatrix} A & K & J \\ \mathfrak{H} & & \mathfrak{G}\end{pmatrix}$$

has invariant subspaces $0 = \mathfrak{H}_0 \subset \mathfrak{H}_1 \subset \cdots \subset \mathfrak{H}_n = \mathfrak{H}$ *and the basic operators of the nodes* $\Theta_j = \mathrm{pr}_{\mathfrak{H}_j \ominus \mathfrak{H}_{j-1}}\Theta$ $(j = 1, 2, \cdots, n)$ *are regular at the point* λ, *then*

$$W_\Theta(\lambda) = W_{\Theta_1}(\lambda)\,W_{\Theta_2}(\lambda)\,\ldots\,W_{\Theta_n}(\lambda). \tag{3.3}$$

The proof follows from formula (2.6).

2. **Criteria for unitary equivalence of nodes.** If the nodes

$$\Theta_1 = \begin{pmatrix} A_1 & K_1 & J \\ \mathfrak{H}_1 & & \mathfrak{G}\end{pmatrix} \text{ and } \Theta_2 = \begin{pmatrix} A_2 & K_2 & J \\ \mathfrak{H}_2 & & \mathfrak{G}\end{pmatrix}$$

are unitarily equivalent, then the set G_{A_1} of regular points of A_1 coincides with the set G_{A_2} of regular points of A_2, and $W_{\Theta_1}(\lambda) = W_{\Theta_2}(\lambda)$ $(\lambda \in G_{A_1})$. Indeed, in view of (2.10),

$$\begin{aligned}W_{\Theta_2}(\lambda) &= E - 2iK_2^*(A_2-\lambda E)^{-1}K_2J\\ &= E - 2iK_1^*U^{-1}\left[U(A_1-\lambda E)^{-1}U^{-1}\right]UK_1J\\ &= E - 2iK_1^*(A_1-\lambda E)^{-1}K_1J = W_{\Theta_1}(\lambda).\end{aligned}$$

Theorem 3.2. *Suppose that*

$$\Theta_1 = \begin{pmatrix} A_1 & K_1 & J \\ \mathfrak{H}_1 & & \mathfrak{G}\end{pmatrix} \textit{ and } \Theta_2 = \begin{pmatrix} A_2 & K_2 & J \\ \mathfrak{H}_2 & & \mathfrak{G}\end{pmatrix}$$

are simple nodes. If in some neighborhood G *of the point at infinity* $W_{\Theta_1}(\lambda) \equiv W_{\Theta_2}(\lambda)$, *then* Θ_1 *and* Θ_2 *are unitarily equivalent.*

Proof. In the theorem it is given that

$$K_1^*(A_1-\lambda E)^{-1}K_1 = K_2^*(A_2-\lambda E)^{-1}K_2 \qquad (\lambda \in G).$$

Inasmuch as

$$\begin{aligned}&(A_j-\lambda E)^{-1}-(A_j^*-\bar{\mu}E)^{-1}\\&=(A_j^*-\bar{\mu}E)^{-1}\left[(A_j^*-\bar{\mu}E)-(A_j-\lambda E)\right](A_j-\lambda E)^{-1}\\&=(\lambda-\bar{\mu})(A_j^*-\bar{\mu}E)^{-1}(A_j-\lambda E)^{-1}\\&\qquad-2i(A_j^*-\bar{\mu}E)^{-1}K_jJK_j^*(A_j-\lambda E)^{-1}\quad (j=1,\ 2)\end{aligned}$$

which means that

$$\begin{aligned}&(\lambda-\bar{\mu})K_1^*(A_1^*-\bar{\mu}E)^{-1}(A_1-\lambda E)^{-1}K_1\\&\quad=K_1^*(A_1-\lambda E)^{-1}K_1-K_1^*(A_1^*-\bar{\mu}E)^{-1}K_1\\&\quad+2iK_1^*(A_1^*-\bar{\mu}E)^{-1}K_1JK_1^*(A_1-\lambda E)^{-1}K_1\\&\quad=K_2^*(A_2-\lambda E)^{-1}K_2-K_2^*(A_2^*-\bar{\mu}E)^{-1}K_2\\&\quad+2iK_2^*(A_2^*-\bar{\mu}E)^{-1}K_2JK_2^*(A_2-\lambda E)^{-1}K_2\\&\qquad\qquad=(\lambda-\bar{\mu})K_2^*(A_2^*-\bar{\mu}E)^{-1}(A_2-\lambda E)^{-1}K_2,\end{aligned}$$

ir follows that

$$\begin{aligned}&K_1^*(A_1^*-\bar{\mu}E)^{-1}(A_1-\lambda E)^{-1}K_1\\&\qquad=K_2^*(A_2^*-\bar{\mu}E)^{-1}(A_2-\lambda E)^{-1}K_2\quad(\lambda,\ \mu\in G).\end{aligned}$$

Using the expansion

$$(A_j-\lambda E)^{-1}=-\frac{E}{\lambda}-\frac{A_j}{\lambda^2}-\frac{A_j^2}{\lambda^3}-\ \ldots\qquad(|\lambda|>\|A_j\|),$$

we arrive at the equations

$$\begin{gathered}(A_1^mK_1g,\ A_1^nK_1g')=(A_2^mK_2g,\ A_2^nK_2g')\\(m,\ n=0,\ 1,\ \ldots;\ g,\ g'\in\mathfrak{G}).\end{gathered}\tag{3.4}$$

Denote by $\mathfrak{H}_j^{(0)}$ $(j=1,\ 2)$ the linear envelope of vectors of the form $A_j^mK_jg$ $(m=0,\ 1,\cdots;\ g\in\mathfrak{G})$ and consider the mapping U of the set $\mathfrak{H}_1^{(0)}$ onto $\mathfrak{H}_2^{(0)}$ which assigns to each vector of the form $\sum_{n=0}^{l}A_1^nK_1g_n$ the vector $\sum_{n=0}^{l}A_2^nK_2g_n$. In view of (3.4) the mapping U is isometric. Since $\mathfrak{H}_1^{(0)}$ and $\mathfrak{H}_2^{(0)}$ are dense in $\mathfrak{H}_1$ and $\mathfrak{H}_2$ respectively, U can be extended by continuity to an isometric mapping of the entire space $\mathfrak{H}_1$ onto all of $\mathfrak{H}_2$. Denoting the extended mapping again by U, we obtain the equation

$$UA_1^nK_1=A_2^nK_2\qquad(n=0,\ 1,\ \ldots).$$

Thus $UK_1=K_2$, and, moreover,

$$UA_1A_1^nK_1=A_2A_2^nK_2=A_2UA_1^nK_1\qquad(n=0,\ 1,\ \ldots),$$

i.e. $UA_1=A_2U$.

Corollary. *Suppose that*

$$\Theta_1=\begin{pmatrix} A_1 & K_1 & J \\ \mathfrak{H}_1 & & \mathfrak{G} \end{pmatrix} \quad \textit{and} \quad \Theta_2=\begin{pmatrix} A_2 & K_2 & J \\ \mathfrak{H}_2 & & \mathfrak{G} \end{pmatrix}$$

are simple nodes. If in some neighborhood of the point at infinity $W_{\Theta_1}(\lambda) = W_{\Theta_2}(\lambda)$, *then* $G_{A_1} = G_{A_2}$ *and* $W_{\Theta_1}(\lambda) \equiv W_{\Theta_2}(\lambda)$ $(\lambda \in G_{A_1})$.

Note that in Theorem 3.2 we cannot get along without the requirement of simplicity of the nodes. For example, let us consider an excess node Θ with principal part $\Theta_{\mathfrak{G}}$ and excess part $\Theta_{\mathfrak{G}(0)}$. Obviously Θ and $\Theta_{\mathfrak{G}}$ cannot be unitarily equivalent. At the same time $W_{\Theta_{\mathfrak{G}(0)}}(\lambda) \equiv E$, and, by Theorem 3.1, there exists a neighborhood of the point at infinity in which

$$W_\Theta(\lambda) = W_{\Theta_{\mathfrak{G}}}(\lambda)\, W_{\Theta_{\mathfrak{G}(0)}}(\lambda) = W_{\Theta_{\mathfrak{G}}}(\lambda).$$

Theorem 3.3. *Suppose that*

$$\Theta_1=\begin{pmatrix} A_1 & K_1 & J \\ \mathfrak{H}_1 & & \mathfrak{G} \end{pmatrix} \quad \textit{and} \quad \Theta_2=\begin{pmatrix} A_2 & K_2 & J \\ \mathfrak{H}_2 & & \mathfrak{G} \end{pmatrix}$$

are left divisors of the simple node

$$\Theta=\begin{pmatrix} A & K & J \\ \mathfrak{H} & & \mathfrak{G} \end{pmatrix}.$$

If $W_{\Theta_1} \equiv W_{\Theta_2}(\lambda)$, *then* $\Theta_1 = \Theta_2$.

Proof. In view of the equations $\Theta_1 = \mathrm{pr}_{\mathfrak{H}_1}\Theta$ and $\Theta_2 = \mathrm{pr}_{\mathfrak{H}_2}\Theta$, it suffices to show that $\mathfrak{H}_1 = \mathfrak{H}_2$.

Consider the relations

$$\Theta = \Theta_1\Theta_3 = \Theta_2\Theta_4$$
$$(\Theta_3 = \mathrm{pr}_{\mathfrak{H}_3}\Theta,\ \Theta_4 = \mathrm{pr}_{\mathfrak{H}_4}\Theta,\ \mathfrak{H}_3 = \mathfrak{H} \ominus \mathfrak{H}_1,\ \mathfrak{H}_4 = \mathfrak{H} \ominus \mathfrak{H}_2).$$

For all λ lying in some neighborhood G of the point at infinity,

$$W_\Theta(\lambda) = W_{\Theta_1}(\lambda)\, W_{\Theta_3}(\lambda) = W_{\Theta_2}(\lambda)\, W_{\Theta_4}(\lambda),$$

and, since $\lim_{\lambda\to\infty} \| W_{\Theta_j}(\lambda) - E \| = 0$, the operator $W^{-1}_{\Theta_1}(\lambda) = W^{-1}_{\Theta_2}(\lambda)$ exists. Accordingly, $W_{\Theta_3}(\lambda) = W_{\Theta_4}(\lambda)$ $(\lambda \in G)$, and, from Theorem 3.2, the nodes Θ_1 and Θ_3 are unitarily equivalent to Θ_2 and Θ_4 respectively. By Theorem 2.6 there exists a unitary operator U for which

$$UA = AU, \qquad UK = K, \qquad U\mathfrak{H}_1 = \mathfrak{H}_2.$$

Since Θ is a simple node, we have $U = E$, so that $\mathfrak{H}_1 = \mathfrak{H}_2$.

3. Analytic properties of a COF. Suppose that Θ is some node and that G_A is the set of regular points of the operator A.

The following formula holds:

$$W_\Theta(\lambda)JW_\Theta^*(\mu)-J$$
$$=2i(\bar{\mu}-\lambda)K^*(A-\lambda E)^{-1}(A^*-\bar{\mu}E)^{-1}K \quad (\lambda,\ \mu\in G_A). \tag{3.5}$$

Indeed, since

$$(A-\lambda E)^{-1}-(A^*-\bar{\mu}E)^{-1}$$
$$=(A-\lambda E)^{-1}[(A^*-\bar{\mu}E)-(A-\lambda E)](A^*-\bar{\mu}E)^{-1}$$
$$=(\lambda-\bar{\mu})(A-\lambda E)^{-1}(A^*-\bar{\mu}E)^{-1}-2i(A-\lambda E)^{-1}KJK^*(A^*-\bar{\mu}E)^{-1},$$

we have

$$W_\Theta(\lambda)JW_\Theta^*(\mu)-J$$
$$=\left[E-2iK^*(A-\lambda E)^{-1}KJ\right]J\left[E+2iJK^*(A^*-\bar{\mu}E)^{-1}K\right]-J$$
$$=-2iK^*\left[(A-\lambda E)^{-1}-(A^*-\bar{\mu}E)^{-1}\right.$$
$$\left.+2i(A-\lambda E)^{-1}KJK^*(A^*-\bar{\mu}E)^{-1}\right]K$$
$$=2i(\bar{\mu}-\lambda)K^*(A-\lambda E)^{-1}(A^*-\bar{\mu}E)^{-1}K.$$

In particular, if λ and $\bar{\lambda}$ lie in G_A, then

$$W_\Theta(\lambda)JW_\Theta^*(\bar{\lambda})-J=0. \tag{3.6}$$

Moreover, at each point $\lambda\in G_A$

$$W_\Theta(\lambda)JW_\Theta(\lambda)-J=4\operatorname{Im}\lambda K^*(A-\lambda E)^{-1}(A^*-\bar{\lambda}E)^{-1}K, \tag{3.7}$$

which means that

$$W_\Theta(\lambda)JW_\Theta^*(\lambda)-J\geqslant 0 \quad (\operatorname{Im}\lambda>0,\ \lambda\in G_A), \tag{3.8}$$
$$W_\Theta(\lambda)JW_\Theta^*(\lambda)-J\leqslant 0 \quad (\operatorname{Im}\lambda<0,\ \lambda\in G_A). \tag{3.9}$$

Analogously, using the equation

$$(A-\lambda E)^{-1}-(A^*-\bar{\mu}E)^{-1}=(\lambda-\bar{\mu})(A^*-\bar{\mu}E)^{-1}(A-\lambda E)^{-1}$$
$$-2i(A^*-\bar{\mu}E)^{-1}KJK^*(A-\lambda E)^{-1}, \tag{3.10}$$

we obtain the relation

$$W_\Theta^*(\mu)JW_\Theta(\lambda)-J=2i(\bar{\mu}-\lambda)JK^*(A^*-\bar{\mu}E)^{-1}(A-\lambda E)^{-1}KJ, \tag{3.11}$$

which shows that

$$W_\Theta^*(\bar{\lambda})JW_\Theta(\lambda)-J=0 \quad (\lambda,\ \bar{\lambda}\in G_A), \tag{3.12}$$

$$W^*_\Theta(\lambda)JW_\Theta(\lambda)-J\geqslant 0 \quad (\operatorname{Im}\lambda>0,\ \lambda\in G_A), \tag{3.13}$$
$$W^*_\Theta(\lambda)JW_\Theta(\lambda)-J\leqslant 0 \quad (\operatorname{Im}\lambda<0,\ \lambda\in G_A). \tag{3.14}$$

If λ and $\bar\lambda$ lie in G_A, then by (3.6) and (3.12) the operator $W_\Theta(\lambda)$ has a bounded inverse:

$$W_\Theta^{-1}(\lambda)=JW^*_\Theta(\bar\lambda)J. \tag{3.15}$$

Since there exists a neighborhood of the point at infinity in which the resolvent of A decomposes into a series

$$(A-\lambda E)^{-1}=-\frac{E}{\lambda}-\frac{A}{\lambda^2}-\dots, \tag{3.16}$$

which converges in norm, it follows that in this same neighborhood

$$W_\Theta(\lambda)=E+\frac{2i}{\lambda}K^*KJ+\dots \tag{3.17}$$

4. **Bilinear transformation of a COF.** We assign to the node Θ the operator-function

$$V_\Theta(\lambda)=K^*(A_R-\lambda E)^{-1}K \quad \left(A_R=\frac{A+A^*}{2}\right). \tag{3.18}$$

The function $V_\Theta(\lambda)$ is holomorphic on the set G_{A_R} of regular points of the operator A_R, and its values, as are those of the function $W_\Theta(\lambda)$, are operators operating in $\mathfrak{G}$. We note that G_{A_R} contains all nonreal points.

It follows from the equation

$$V_\Theta(\lambda)-V^*_\Theta(\lambda)=2i\operatorname{Im}\lambda K^*(A_R-\bar\lambda E)^{-1}(A_R-\lambda E)^{-1}K \tag{3.19}$$

that

$$\frac{V_\Theta(\lambda)-V^*_\Theta(\lambda)}{2i}\geqslant 0\ (\operatorname{Im}\lambda>0),\quad \frac{V_\Theta(\lambda)-V^*_\Theta(\lambda)}{2i}\leqslant 0\ (\operatorname{Im}\lambda<0) \tag{3.20}$$

and

$$V_\Theta(\lambda)=V^*_\Theta(\lambda) \quad (\operatorname{Im}\lambda=0,\ \lambda\in G_{A_R}). \tag{3.21}$$

At each point of the set $G_A^{(0)}=G_A\cap G_{A_R}$ *there exist the operators* $(W_\Theta(\lambda)+E)^{-1}$ *and* $(E+iV_\Theta(\lambda)J)^{-1}$, *while*

$$\begin{aligned} V_\Theta(\lambda)&=i(W_\Theta(\lambda)+E)^{-1}(W_\Theta(\lambda)-E)J\\ &=i(W_\Theta(\lambda)-E)(W_\Theta(\lambda)+E)^{-1}J, \end{aligned} \tag{3.22}$$

$$W_\Theta(\lambda) = (E + iV_\Theta(\lambda) J)^{-1} (E - iV_\Theta(\lambda) J)$$
$$= (E - iV_\Theta(\lambda) J)(E + iV_\Theta(\lambda) J)^{-1}. \tag{3.23}$$

Indeed, since

$$(A_R - \lambda E)^{-1} - (A - \lambda E)^{-1} = i (A - \lambda E)^{-1} A_I (A_R - \lambda E)^{-1} \tag{3.24}$$

and $A_I = KJK^*$, we have

$$K^* (A_R - \lambda E)^{-1} K - K^* (A - \lambda E)^{-1} K$$
$$= iK^* (A - \lambda E)^{-1} KJK^* (A_R - \lambda E)^{-1} K.$$

Thus

$$V_\Theta(\lambda) + \frac{i}{2} (E - W_\Theta(\lambda)) J = \frac{1}{2} (E - W_\Theta(\lambda)) V_\Theta(\lambda),$$

so that

$$(W_\Theta(\lambda) + E)(E + iV_\Theta(\lambda) J) = 2E. \tag{3.25}$$

Analogously, starting from the relations

$$(A_R - \lambda E)^{-1} - (A - \lambda E)^{-1} = i (A_R - \lambda E)^{-1} A_I (A - \lambda E)^{-1}, \tag{3.26}$$

we get

$$(E + iV_\Theta(\lambda) J)(W_\Theta(\lambda) + E) = 2E. \tag{3.27}$$

In view of (3.25) and (3.27) each of the operators $W_\Theta(\lambda) + E$ and $E + iV_\Theta(\lambda) J$ has a bounded inverse for $\lambda \in G_A^{(0)}$. Formulas (3.22) and (3.23) follow easily from (3.25).

§4. Integral representations of certain analytic operator-functions

1. **Stieltjes operator integral.** In this section we shall use Stieltjes integrals of the type $\int_a^b \phi(t)\, dF(t)$ $(-\infty \le a < b \le +\infty)$, where $\phi(t)$ $(-\infty < t < +\infty)$ is a scalar complex-valued function and $F(t)$ $(-\infty < t < +\infty)$ is a function with values in the set of bounded linear operators, operating in some Hilbert space $\mathfrak{G}$.

We shall say that the integral $\int_a^b \phi(t)\, dF(t)$ $(-\infty < a < b < +\infty)$ exists, and write $\mathcal{J} = \int_a^b \phi(t)\, dF(t)$, if $\mathcal{J}$ is a bounded linear operator for which

$$\left\| \mathcal{J} - \sum_{j=1}^{n} \varphi(\xi_j) (F(t_j) - F(t_{j-1})) \right\| \to 0$$
$$(a = t_0 \leqslant \xi_1 \leqslant t_1 \leqslant \ldots \leqslant t_{n-1} \leqslant \xi_n \leqslant t_n = b)$$

as $\max(t_j - t_{j-1}) \to 0$.

If we write $\mathcal{J} = \int_{-\infty}^{\infty} \phi(t)\, dF(t)$, this means that the integral $\int_a^b \phi(t)\, dF(t)$

exists for any finite a and b and that

$$\left\| \mathcal{J} - \int_a^b \varphi(t)\, dF(t) \right\| \to 0$$

as $a \to -\infty$ and $b \to +\infty$.

Lemma 4.1. *Suppose that α_j and β_j $(j = 1, 2, \cdots, n)$ are complex numbers, and H_j are positive operators. If $|\alpha_j| \leq |\beta_j|$, then*

$$\left\| \sum_{j=1}^{n} \alpha_j H_j \right\| \leqslant \left\| \sum_{j=1}^{n} |\beta_j| H_j \right\|.$$

Proof. The assertion of the lemma follows immediately from the inequality

$$\left| \left(\sum_{j=1}^{n} \alpha_j H_j f,\ g \right) \right| \leqslant \sum_{j=1}^{n} |\alpha_j| \, |(H_j f,\ g)|$$

$$\leqslant \sum_{i=1}^{n} \left(|\beta_j|^{1/2} \| H_j^{1/2} f \| \right) \left(|\beta_j|^{1/2} \| H_j^{1/2} g \| \right)$$

$$\leqslant \left(\sum_{j=1}^{n} |\beta_j| H_j f,\ f \right)^{1/2} \left(\sum_{j=1}^{n} |\beta_j| H_j g,\ g \right)^{1/2}.$$

Theorem 4.1. *If $\phi(t)$ $(-\infty < a \leq t \leq b < +\infty)$ is continuous and $F(t)$ $(a \leq t \leq b)$ is a nondecreasing function, then the integral $\int_a^b \phi(t)\, dF(t)$ exists.*

Proof. The set of bounded linear operators operating in the given space is dense in the uniform topology. Hence it is sufficient to show that for each $\epsilon > 0$ there exists a $\delta > 0$ satisfying the following condition: if σ_1 is a decomposition of the segment $[a, b]$ such that the lengths of all the partial segments are less than δ, and σ_2 is a continuation of the subdivision σ_1, then the norm of the difference of arbitrary integral sums corresponding to the subdivisions σ_1 and σ_2 is less than ϵ.

For the given $\epsilon > 0$ we choose $\delta > 0$ so that if $|t' - t''| < \delta$ the inequality

$$|\varphi(t') - \varphi(t'')| < \frac{\varepsilon}{\| F(b) - F(a) \|}$$

holds. If

$$a = t_0 < t_1 < \ldots < t_n = b \qquad (t_j - t_{j-1} < \delta,\ j = 1, 2, \ldots, n)$$

and

$$a = t_1^{(0)} < t_1^{(1)} < \ldots < t_1^{(k_1)} = t_1 < \ldots < t_{n-1} = t_n^{(0)} < t_n^{(1)} < \ldots < t_n^{(k_n)} = b,$$

then, from Lemma 4.1,

$$\left\| \sum_{j=1}^{n} \varphi(\xi_j)\,\Delta F_j - \sum_{j=1}^{n} \sum_{k=1}^{k_j} \varphi(\xi_j^{(k)})\,\Delta F_j^{(k)} \right\|$$

$$= \left\| \sum_{j=1}^{n} \sum_{k=1}^{k_j} \varphi(\xi_j)\,\Delta F_j^{(k)} - \sum_{j=1}^{n} \sum_{k=1}^{k_j} \varphi(\xi_j^{(k)})\,\Delta F_j^{(k)} \right\|$$

$$\leqslant \frac{\varepsilon}{\| F(b) - F(a) \|} \left\| \sum_{j=1}^{n} \sum_{k=1}^{k_j} \Delta F_j^{(k)} \right\| = \varepsilon$$

$$(t_{j-1} \leqslant \xi_j \leqslant t_j,\ t_j^{(k-1)} \leqslant \xi_j^{(k)} \leqslant t_j^{(k)},\ \Delta F_j = F(t_j) - F(t_{j-1}),$$
$$\Delta F_j^{(k)} = F(t_j^{(k)}) - F(t_j^{(k-1)})).$$

Lemma 4.2. *Suppose that the scalar functions $\phi(t)$ and $\psi(t)$ are continuous on the finite segment $[a, b]$. If $|\phi(t)| \leq |\psi(t)|$ $(a \leq t \leq b)$ and $F(t)$ is a nondecreasing operator-function, then*

$$\left\| \int_a^b \varphi(t)\,dF(t) \right\| \leqslant \left\| \int_a^b |\psi(t)|\,dF(t) \right\|. \tag{4.1}$$

Proof. By Lemma 4.1

$$\left\| \sum_{j=1}^{n} \varphi(\xi_j)\,\Delta F_j \right\| \leqslant \left\| \sum_{j=1}^{n} |\psi(\xi_j)|\,\Delta F_j \right\| \tag{4.2}$$

$$(a = t_0 \leqslant \xi_1 \leqslant t_1 \leqslant \ldots \leqslant t_{n-1} \leqslant \xi_n \leqslant t_n = b;$$
$$\Delta F_j = F(t_j) - F(t_{j-1})).$$

Passing to the limit in (4.2) as $\max(t_j - t_{j-1}) \to 0$, we get (4.1).

Theorem 4.2. *Suppose that $\phi(t)$ $(-\infty < t < +\infty)$ is continuous and that $F(t)$ $(-\infty < t < +\infty)$ is a nondecreasing function. Suppose that at least one of the following conditions holds:*

1) $\lim_{t \to -\infty} \phi(t) = \lim_{t \to +\infty} \phi(t) = 0$ *and the function $\|F(t)\|$ $(-\infty < t < +\infty)$ is bounded.*

2) *The function $\phi(t)$ $(-\infty < t < +\infty)$ is bounded, and $F(t)$ has uniform limits as t tends to $-\infty$ or $+\infty$.*

Then the integral $\int_{-\infty}^{+\infty} \phi(t)\,dF(t)$ exists.

The proof follows easily from Theorem 4.1, Lemma 4.2, and the definitions of the integral $\int_{-\infty}^{+\infty} \phi(t)\,dF(t)$.

2. Generalized Helly theorem. We need the following Helly theorem:

a) *Suppose that on a finite interval* $[a, b]$ *there is given a sequence of scalar functions* $f_n(t)$. *If all the functions* $f_n(t)$ *and all their total variations are uniformly bounded, then there exists a subsequence* $f_{n_j}(t)$ *which converges at each point* $t \in [a, b]$.

b) *Suppose that on a finite interval* $[a, b]$ *there is given a continuous scalar function* $\phi(t)$ *and a sequence of scalar functions* $f_n(t)$. *If the total variations of all the functions* $f_n(t)$ *are uniformly bounded and if* $f_n(t)$ *converges to a finite function* $f(t)$ *at each point* $t \in [a, b]$, *then* $f(t)$ *is of bounded variation and*

$$\lim_{n\to\infty} \int_a^b \varphi(t)\, df_n(t) = \int_a^b \varphi(t)\, df(t).$$

Theorem 4.3. *Suppose that* $F^{(n)}(t)$ $(-\infty < a \leq t \leq b < +\infty)$ *is a sequence of functions whose values are bounded linear operators operating in a separable Hilbert space* $\mathfrak{G}$. *If* $F^{(n)}(t)$ *are nondecreasing functions and if there exists a bounded linear operator* F_0 *such that* $0 \leq F^{(n)}(t) \leq F_0$ $(a \leq t \leq b,\ n = 1, 2, \cdots)$, *then it is possible to extract a subsequence from the sequence of* $F^{(n)}$ *which weakly converges for each value of* t.

Proof. Denote by $\mathfrak{G}_0$ some countable set which is dense in $\mathfrak{G}$. Fix vectors $g, h \in \mathfrak{G}_0$ and consider the sequence of scalar functions $(F^{(n)}(t)g, h)$. Using the inequalities

$$|(F^{(n)}(t)\, g,\ h)| \leqslant \| F_0 \| \| g \| \| h \|,$$

$$\sum_{j=1}^{m} |(\Delta F_j^{(n)} g,\ h)| \leqslant \sum_{j=1}^{m} (\Delta F_j^{(n)} g,\ g)^{1/2} (\Delta F_j^{(n)} h,\ h)^{1/2}$$

$$\leqslant \left(\sum_{j=1}^{m} (\Delta F_j^{(n)} g,\ g)\right)^{1/2} \left(\sum_{j=1}^{m} (\Delta F_i^{(n)} h,\ h)\right)^{1/2} \leqslant 2 \| F_0 \| \| g \| \| h \| \tag{4.3}$$

$$(a = t_0 < t_1 < \ldots < t_m = b,\ \ \Delta F_j^{(n)} = F^{(n)}(t_j) - F^{(n)}(t_{j-1}))$$

and Theorem a), we select a subsequence of it which converges for each fixed t. Using the diagonal process we may now form a subsequence $F^{(n_j)}(t)$ such that the sequence $(F^{(n_j)}(t)g, h)$ will converge at each point t for any $g, h \in \mathfrak{G}_0$. But then, as is easily verified, it will converge for any $g, h \in \mathfrak{G}$.

Theorem 4.4. *Suppose that the sequence of continuous scalar functions* $\phi_n(t)$ $(-\infty < a \leq t \leq b < +\infty)$ *converges uniformly to a function* $\phi(t)$, *and that the sequence of nondecreasing operator-functions* $F_n(t)$ $(a \leq t \leq b)$ *satisfies the condition* $0 \leq F_n(t) \leq F_0$ $(n = 1, 2, \cdots)$, *where* F_0 *is some bounded operator, and* *converges weakly for each* $t \in [a, b]$ *to* $F(t)$. *Then the sequence* $\int_a^b \phi_n(t)dF_n(t)$ *converges weakly to* $\int_a^b \phi(t)\,dF(t)$.

Proof. Since the sequence $(F_n(t)g, h)$ converges to $(F(t)g, h)$ and, from estimate (4.3), the total variations of the functions $(F_n(t)g, h)$ are uniformly bounded, by Theorem b) we obtain

$$\int_a^b \varphi(t)\,d(F_n(t)g, h) \to \int_a^b \varphi(t)\,d(F(t)g, h).$$

In other words, the sequence $\int_a^b \phi(t)\,dF_n(t)$ converges weakly to $\int_a^b \phi(t)\,dF(t)$. It remains to be noted that

$$\left\| \int_a^b \varphi_n(t)\,dF_n(t) - \int_a^b \varphi(t)\,dF_n(t) \right\| \leqslant 2 \max_{a \leqslant t \leqslant b} |\varphi_n(t) - \varphi(t)| \, \| F_0 \| \to 0.$$

3. Integral representations.

Theorem 4.5. *For the function* $T(z)$ *of the complex variable* z $(|z| < 1)$, *with values which are bounded linear operators in a separable Hilbert space, to admit the representation*

$$T(z) = iT_0 + \int_0^{2\pi} \frac{e^{it} + z}{e^{it} - z}\,dF(t) \qquad (|z| < 1), \tag{4.4}$$

where T_0 *is a selfadjoint operator and* $F(t)$ $(0 \le t \le 2\pi)$ *is a nondecreasing nonnegative operator-function, it is necessary and sufficient that* $T(z)$ *should be holomorphic in the disk* $|z| < 1$ *and that it should have a nonnegative real component* $T_R(z) = (T(z) + T^*(z))/2$.

Proof. The necessity of the condition of the theorem is verified directly. To verify the sufficiency, we note that by the Schwarz formula

$$T(z) = iT_I(0) + \frac{1}{2\pi}\int_0^{2\pi} \frac{re^{it} + z}{re^{it} - z}\,T_R(re^{it})\,dt \qquad (|z| < r < 1).$$

Putting

$$F_r(t) = \frac{1}{2\pi}\int_0^t T_R(re^{is})\,ds,$$

we get

$$T(z) = iT_I(0) + \int_0^{2\pi} \frac{re^{it} + z}{re^{it} - z}\,dF_r(t) \qquad (|z| < r < 1).$$

Fix an increasing sequence of positive numbers r_j converging to unity. Each of the functions $F_j(t) = F_{r_j}(t)$ is nondecreasing and satisfies the relations $F_j(0) = 0$, $F_j(t) \leq F_j(2\pi) = T_R(0)$. By Theorem 4.3 there exists a subsequence $F_{j_k}(t)$, weakly convergent to some nondecreasing function $F(t)$. In view of Theorem 4.4

$$T(z) = iT_I(0) + \int_0^{2\pi} \frac{e^{it}+z}{e^{it}-z}\, dF(t) \qquad (|z|<1). \tag{4.5}$$

Theorem 4.6. *Suppose that the operator-function $T(z)$ is holomorphic in the disk $|z| < 1$ and that it has a nonnegative real component. If the function $\|T_R(x)\|$ is bounded on the interval* (0, 1), *then formula* (4.5) *may be rewritten in the form*

$$T(z) = iT_I(0) + \lim_{\substack{\varepsilon_1\to 0\\ \varepsilon_2\to 0}} \int_{\varepsilon_1}^{2\pi-\varepsilon_2} \frac{e^{it}+z}{e^{it}-z}\, dF(t) \qquad (|z|<1), \tag{4.6}$$

where the passage to the limit is realized in the sense of convergence in norm.

Proof. By formula (4.4)

$$T_R(x) = \int_0^{2\pi} \frac{1-x^2}{|e^{it}-x|^2}\, dF(t) \qquad (-1<x<1),$$

and therefore

$$T_R(x) \geqslant \int_0^{\varepsilon} \frac{1-x^2}{|e^{it}-x|^2}\, dF(t) \geqslant \frac{1-x^2}{|e^{i\varepsilon}-x|^2}(F(\varepsilon)-F(0))$$
$$(0<x<1,\ \varepsilon\leqslant\pi).$$

There exists a constant C such that $\|T_R(x)\| \leq C$ $(0 < x < 1)$. Therefore

$$\|F(\varepsilon)-F(0)\| \leqslant C\,\frac{|e^{i\varepsilon}-x|^2}{1-x^2}. \tag{4.7}$$

Letting first ϵ tend to zero in (4.7), and then x to unity, we get

$$\lim_{\varepsilon\to 0} \|F(\varepsilon)-F(0)\| = 0.$$

Analogous arguments show that

$$\lim_{\varepsilon\to 0} \|F(2\pi)-F(2\pi-\varepsilon)\| = 0.$$

The assertion of the theorem now follows from the equation

$$T(z) = iT_I(0) + \int_0^{\varepsilon_1} \frac{e^{it}+z}{e^{it}-z}\,dF(t)$$

$$+ \int_{\varepsilon_1}^{2\pi-\varepsilon_2} \frac{e^{it}+z}{e^{it}-z}\,dF(t) + \int_{2\pi-\varepsilon_2}^{2\pi} \frac{e^{it}+z}{e^{it}-z}\,dF(t),$$

inasmuch as the norms of the first and third integrals tend to zero as ϵ_1 and ϵ_2 tend to zero respectively.

Theorem 4.7. *Suppose that the function $V(\lambda)$, whose values are bounded linear operators operating in a separable Hilbert space, is holomorphic in the upper halfplane and has a nonnegative imaginary component $V_I(\lambda) = (V(\lambda) - V^*(\lambda))/2i$. If $\|V_I(i\tau)\|$ is bounded on the interval $1 \le \tau < \infty$, then there exists a nondecreasing nonnegative operator-function $G(s)$ $(-\infty < S < +\infty)$, having uniform limits as s tends to $+\infty$ or $-\infty$, such that*

$$V(\lambda) = V_R(i) + \int_{-\infty}^{\infty} \frac{1+s\lambda}{s-\lambda}\,dG(s) \qquad (\operatorname{Im}\lambda > 0). \tag{4.8}$$

Proof. Put $T(z) = -iV(\lambda)$, where $z = (\lambda - i)/(\lambda + i)$. Then:

1) $T(z)$ is holomorphic in the disk $|z| < 1$;

2) $(T(z) + T^*(z))/2 = V(\lambda) - V^*(\lambda))/2i \ge 0$;

3) $\|T_R(x)\|$ is bounded on the interval $(0,1)$.

By the preceding theorem

$$V(\lambda) = -T_I(0) + i \lim_{\substack{\varepsilon_1\to 0\\ \varepsilon_2\to 0}} \int_{\varepsilon_1}^{2\pi-\varepsilon_2} \frac{e^{it}+z}{e^{it}-z}\,dF(t)$$

$$= V_R(i) + \lim_{\substack{\varepsilon_1\to 0\\ \varepsilon_2\to 0}} \int_{\varepsilon_1}^{2\pi-\varepsilon_2} \frac{\lambda \operatorname{ctg}\frac{t}{2} - 1}{\lambda + \operatorname{ctg}\frac{t}{2}}\,dF(t).$$

Putting $s = -\operatorname{ctg}(t/2)$, $G(s) = F(-2 \operatorname{arc\,ctg} s)$, we get formula (4.8).

In the proof of Theorem 4.6 it was shown that $F(t)$ has uniform limits as t tends to 0 or 2π. Accordingly, there exist uniform limits for the function $G(s)$ as well, when s tends to $-\infty$ or $+\infty$.

Theorem 4.8. *Suppose that the values of the function $V(\lambda)$ $(\operatorname{Im}\lambda > 0)$ are bounded linear operators operating in a separable Hilbert space. For $V(\lambda)$ to admit the representation*

$$V(\lambda) = \int_{-\infty}^{+\infty} \frac{dF(s)}{s-\lambda} \qquad (\operatorname{Im}\lambda > 0), \tag{4.9}$$

where $F(s)$ $(-\infty < S < +\infty)$ *is a nondecreasing function with a bounded norm, it is necessary and sufficient that it should be holomorphic in the entire upper halfplane and have a nonnegative imaginary component, and that the function* $\|\tau V(i\tau)\|$ *should be bounded on the interval* $1 < \tau < \infty$.

Proof. The necessity of the conditions of the theorem is obvious. We proceed to the proof of sufficiency. Inasmuch as $V(\lambda)$ satisfies the hypotheses of Theorem 4.7, formula (4.8) holds, so that

$$\tau V(i\tau) = \tau V_R(i) + \tau(1-\tau^2)\int_{-\infty}^{+\infty} \frac{s}{s^2+\tau^2}\,dG(s) + i\tau^2\int_{-\infty}^{+\infty} \frac{1+s^2}{s^2+\tau^2}\,dG(s).$$

There exists a constant M such that

$$\left\|\tau V_R(i) + \tau(1-\tau^2)\int_{-\infty}^{+\infty} \frac{s}{s^2+\tau^2}\,dG(s)\right\| \leqslant M, \tag{4.10}$$

$$\left\|\tau^2\int_{-\infty}^{+\infty} \frac{1+s^2}{s^2+\tau^2}\,dG(s)\right\| \leqslant M \quad (1<\tau<\infty). \tag{4.11}$$

In view of (4.10)

$$\lim_{\tau\to\infty}\left\|V_R(i) - \int_{-\infty}^{+\infty} \frac{(\tau^2-1)s}{s^2+\tau^2}\,dG(s)\right\| = 0,$$

so that, replacing $V_R(i)$ by

$$V(\lambda) - \int_{-\infty}^{+\infty} \frac{1+s\lambda}{s-\lambda}\,dG(s),$$

we get

$$\lim_{\tau\to\infty}\left\|V(\lambda) - \int_{-\infty}^{+\infty} \frac{(\tau^2+s\lambda)(1+s^2)}{(s^2+\tau^2)(s-\lambda)}\,dG(s)\right\| = 0. \tag{4.12}$$

On the other hand, from (4.11),

$$\tau^2\int_a^b \frac{1+s^2}{s^2+\tau^2}\,dG(s) \leqslant ME \quad (-\infty<a<b<+\infty,\ 1<\tau<\infty).$$

Letting τ tend to ∞, we arrive at the estimate

$$\int_a^b (1+s^2)\,dG(s) \leqslant ME \qquad (-\infty<a<b<+\infty),$$

which makes it possible to introduce the nondecreasing function

$$F(s)=\lim_{a\to-\infty}\int_a^s(1+t^2)\,dG(t),$$

the passage to the limit being realized in the sense of strong convergence. Obviously, $\|F(s)\|\le M$ $(-\infty<s<+\infty)$. Formula (4.12) may now be rewritten in the form

$$\lim_{\tau\to\infty}\left\|V(\lambda)-\int_{-\infty}^{+\infty}\frac{\tau^2+s\lambda}{(s^2+\tau^2)(s-\lambda)}\,dF(s)\right\|=0.$$

Since

$$\lim_{\tau\to\infty}\left\|\int_{-\infty}^{+\infty}\frac{1}{s-\lambda}\,dF(s)-\int_{-\infty}^{+\infty}\frac{\tau^2+s\lambda}{(s^2+\tau^2)(s-\lambda)}\,dF(s)\right\|=0,$$

we have

$$V(\lambda)=\int_{-\infty}^{+\infty}\frac{dF(s)}{s-\lambda}.$$

Theorem 4.9. *For the function $V(\lambda)$, whose values are bounded linear operators operating in a separable Hilbert space, to admit outside the finite interval $[a,b]$ of the real axis the representation*

$$V(\lambda)=\int_a^b\frac{dF(s)}{s-\lambda},\tag{4.13}$$

where $F(s)$ $(a\le s\le b)$ is a nonnegative nondecreasing operator-function, it is necessary and sufficient that $V(\lambda)$ satisfy the following conditions: 1) *it is holomorphic outside $[a,b]$,* 2) $V(\infty)=0$, 3) *in the upper halfplane it has a nonnegative imaginary component,* 4) *it has selfadjoint values on the intervals $(-\infty,a)$ and $(b,+\infty)$ of the real axis.*

Proof. As in the preceding theorems, we restrict ourselves to the proof of sufficiency. In view of conditions 1), 2) and 3), the preceding theorem is applicable to the function $V(\lambda)$. Accordingly there exists a nonnegative nondecreasing function $F(s)$ $(-\infty<s<+\infty)$ *such that*

$$V(\lambda)=\int_{-\infty}^{+\infty}\frac{dF(s)}{s-\lambda}\qquad(\operatorname{Im}\lambda>0).$$

We may suppose that $F(s)$ is continuous to the left at a and to the right at b. Therefore the theorem will be proved if we verify that $F(s)$ has a positive value on each of the intervals $(-\infty,a)$ and $(b,+\infty)$.

Suppose that $[c, d]$ is a segment of the real axis not having points in common with $[a, b]$. Since

$$V_I(\sigma + i\tau) = \int_{-\infty}^{\infty} \frac{\tau}{(s-\sigma)^2 + \tau^2}\, dF(s) \geqslant \int_{c}^{d} \frac{\tau}{(s-\sigma)^2 + \tau^2}\, dF(s) \qquad (\tau > 0),$$

we have

$$\int_c^d V_I(\sigma + i\tau)\, d\sigma \geqslant \int_c^d \int_c^d \frac{\tau}{(s-\sigma)^2 + \tau^2}\, d\sigma\, dF(s)$$
$$= \int_c^d \left(\operatorname{arctg} \frac{d-s}{\tau} + \operatorname{arctg} \frac{s-c}{\tau}\right) dF(s)$$
$$\geqslant \operatorname{arctg} \frac{d-c}{\tau} (F(d) - F(c)).$$

Passing to the limit as $\tau \to 0$ and taking account of condition 4), we obtain the equation $F(c) = F(d)$.

§5. Divisors of a characteristic operator-function

1. **The class Ω_J.** Suppose that the linear operator J, operating in Hilbert space $\mathfrak{G}$, satisfies the conditions $J = J^*$ and $J^2 = E$. We will say that the function of a complex variable $W(\lambda)$ whose values are bounded linear operators in $\mathfrak{G}$ *lies in the class* Ω_J if it has the following properties:

(I) $W(\lambda)$ is holomorphic in some neighborhood G_W of the point at infinity.

(II) $\lim_{\lambda \to \infty} \|W(\lambda) - E\| = 0$.

(III) For all $\lambda \in G_2$ the operator $W(\lambda) + E$ has a bounded inverse, while the operator-function

$$V(\lambda) = i(W(\lambda) + E)^{-1}(W(\lambda) - E)J = i(W(\lambda) - E)(W(\lambda) + E)^{-1} J \qquad (5.1)$$

is analytically continuable into a region G_V constituting the entire extended complex plane with the exception of some bounded set of real points.

$$\frac{V(\lambda) - V^*(\lambda)}{2i} \geqslant 0 \qquad (\operatorname{Im} \lambda > 0). \qquad (5.2)$$

$$V(\lambda) = V^*(\lambda) \qquad (\operatorname{Im} \lambda = 0,\ \lambda \in G_V). \qquad (5.3)$$

In view of (5.1)

$$(W(\lambda) + E)(E + iV(\lambda)J)$$
$$= (E + iV(\lambda)J)(W(\lambda) + E) = 2E \qquad (\lambda \in G_W). \qquad (5.4)$$

Thus at each point $\lambda \in G_W$ the operator

$$E + iV(\lambda)J$$

has a bounded inverse, and

$$W(\lambda) = (E + iV(\lambda)J)^{-1}(E - iV(\lambda)J)$$
$$= (E - iV(\lambda)J)(E + iV(\lambda)J)^{-1}. \qquad (5.5)$$

It follows from the formulas of §3 that the COF of any node Θ lies in the class Ω_J.

Theorem 5.1. *If the operator-function $W(\lambda)$ lies in the class Ω_J, then there exists a node Θ with directing operator J such that $W_\Theta(\lambda) \equiv W(\lambda)$ in some neighborhood of the point at infinity.*[1)]

Proof. From conditions (I)–(V) it follows that the function $V(\lambda)$ satisfies the requirements of Theorem 4.9. So there exists a nonnegative nondecreasing function $F(t)$ $(-\infty < a \leq t \leq b < +\infty)$ such that

$$V(\lambda) = \int_a^b \frac{dF(t)}{t - \lambda} \qquad (\lambda \notin [a, b]).$$

Clearly there is no loss of generality in setting $F(a) = 0$.

By the generalized Naĭmark theorem (see Appendix I) we have

$$F(t) = K^* E(t) K \qquad (a \leqslant t \leqslant b),$$

where $E(t)$ is an orthogonal resolution of unity in some Hilbert space $\mathfrak{H}$, and K is a bounded linear operator operating from $\mathfrak{G}$ into $\mathfrak{H}$.

Suppose we are given in $\mathfrak{H}$ the operator

$$A = \int_a^b t\, dE(t) + iKJK^*$$

Consider the node

$$\Theta = \begin{pmatrix} A & K & J \\ \mathfrak{H} & & \mathfrak{G} \end{pmatrix}.$$

Since

$$V(\lambda) = K^* \int_a^b \frac{dE(t)}{t - \lambda} K = K^* \left(\frac{A + A^*}{2} - \lambda E \right)^{-1} K = V_\Theta(\lambda),$$

1) We suppose throughout, unless expressly stated otherwise, that independently of the possibility of analytic continuation a characteristic operator-function $W_\Theta(\lambda)$ of the node Θ is defined only at the regular points of the operator A.

by formulas (5.5) and (3.23) there exists a neighborhood of the point at infinity in which

$$W(\lambda) = (E + iV(\lambda)J)^{-1}(E - iV(\lambda)J)$$
$$= (E + iV_\Theta(\lambda)J)^{-1}(E - iV_\Theta(\lambda)J) = W_\Theta(\lambda).$$

Corollary 1. *If the function $W(\lambda)$ lies in the class Ω_J, then in some neighborhood of the point at infinity it satisfies relations* (3.6), (3.8), (3.9) *and* (3.12)–(3.15).

Corollary 2. *There exists a neighborhood of the point at infinity in which the function $W(\lambda)$ decomposes into a series of the form*

$$W(\lambda) = E + \frac{2i}{\lambda} HJ + \ldots, \tag{5.6}$$

where $H \geq 0$. If $H = 0$, then $W(\lambda) \equiv E$.

Corollary 3. *Along with each two operator-functions the class Ω_J contains their product. If $W_1(\lambda) \in \Omega_J$, $W_2(\lambda) \in \Omega_J$ and $W_1(\lambda)W_2(\lambda) = E$ $(\lambda \in G_{W_1} \cap G_{W_2})$, then $W_1(\lambda) \equiv W_2(\lambda) \equiv E$.*

The proof of the first assertion follows from Theorem 3.1, and that of the second from Corollary 2.

We have already mentioned that the COF of an excess node and its principal part coincide in some neighborhood of the point at infinity. This leads to the following assertion.

Theorem 5.2. *If $W(\lambda) \in \Omega_J$, then there exists a simple node Θ with a directing operator J such that in some neighborhood of the point at infinity $W_\theta(\lambda) \equiv W(\lambda)$.*

2. Divisors and regular divisors of functions of the class Ω_J. We will say that the function $W_1(\lambda) \in \Omega_J$ is a *left (right) divisor* of the function $W_2(\lambda) \in \Omega_J$ if there exists a function $W_{12}(\lambda) \in \Omega_J$ $(W_{21}(\lambda) \in \Omega_J)$ such that in some neighborhood of the point at infinity

$$W_2(\lambda) = W_1(\lambda) W_{12}(\lambda) \quad (W_2(\lambda) = W_{21}(\lambda) W_1(\lambda)).$$

The notation $W_1(\lambda) \succ W_2(\lambda)$ will mean that $W_1(\lambda)$ is a left divisor of the function $W_2(\lambda)$.

The relation $\prec$ converts Ω_J into a partially ordered set. In fact, if $W_1(\lambda) \prec W_2(\lambda)$ and $W_2(\lambda) \prec W_1(\lambda)$, then

$$W_2(\lambda) = W_1(\lambda) W_{12}(\lambda) \quad (W_{12}(\lambda) \in \Omega_J),$$
$$W_1(\lambda) = W_2(\lambda) W_{21}(\lambda) \quad (W_{21}(\lambda) \in \Omega_J),$$

so that $W_1(\lambda) = W_1(\lambda) W_{12}(\lambda) W_{21}(\lambda)$. Thus $W_{12}(\lambda) W_{21}(\lambda) = E$, so that in view of Corollary 3 of Theorem 5.1 we have $W_{12}(\lambda) = E$ and $W_1(\lambda) = W_2(\lambda)$. Moreover, if $W_1(\lambda) \prec W_2(\lambda)$ and $W_2(\lambda) \prec W_3(\lambda)$, then

$$\left.\begin{aligned} W_2(\lambda) &= W_1(\lambda) W_{12}(\lambda) \quad (W_{12}(\lambda) \in \Omega_J), \\ W_3(\lambda) &= W_2(\lambda) W_{23}(\lambda) \quad (W_{23}(\lambda) \in \Omega_J), \\ W_3(\lambda) &= W_1(\lambda) W_{13}(\lambda) \quad (W_{13}(\lambda) = W_{12}(\lambda) W_{23}(\lambda) \in \Omega_J) \end{aligned}\right\} \tag{5.7}$$

so that $W_1(\lambda) \prec W_3(\lambda)$.

Suppose that $W_1(\lambda)$ is a left divisor of $W_2(\lambda)$:

$$W_2(\lambda) = W_1(\lambda) W_{12}(\lambda) \quad (W_{12}(\lambda) \in \Omega_J).$$

The function $W_1(\lambda)$ is said to be a *regular left divisor* of $W_2(\lambda)$ if the product $\Theta_1\Theta_{12}$ of simple nodes

$$\Theta_1 = \begin{pmatrix} A_1 & K_1 & J \\ \mathfrak{H}_1 & & \mathfrak{G} \end{pmatrix} \quad \text{and} \quad \Theta_{12} = \begin{pmatrix} A_{12} & K_{12} & J \\ \mathfrak{H}_{12} & & \mathfrak{G} \end{pmatrix},$$

for which $W_{\Theta_1}(\lambda) = W_1(\lambda)$ and $W_{\Theta_{12}}(\lambda) = W_{12}(\lambda)$ is a simple node.

We note that this definition is not connected with the choice of the nodes Θ_1 and Θ_{12}. If they are replaced by simple nodes Θ'_1 and Θ'_{12}, then by Theorems 3.2 and 2.6 the nodes $\Theta_1\Theta_{12}$ and $\Theta'_1\Theta'_{12}$ will be unitarily equivalent, and therefore the simplicity of one of them implies the simplicity of the other.

Analogously one defines a *regular right divisor.*

We will write $W_1(\lambda) \ll W_2(\lambda)$ if $W_1(\lambda)$ is a regular left divisor of $W_2(\lambda)$.

The relation $\ll$ *also partially orders the set* Ω_J. In fact, if $W_1(\lambda) \ll W_2(\lambda)$ and $W_2(\lambda) \ll W_3(\lambda)$, then relation (5.7) holds. Suppose that Θ_1, Θ_{12} and Θ_{23} are simple nodes, with directing operator J, for which

$$W_{\Theta_1}(\lambda) = W_1(\lambda), \quad W_{\Theta_{12}}(\lambda) = W_{12}(\lambda), \quad W_{\Theta_{23}}(\lambda) = W_{23}(\lambda). \tag{5.8}$$

By the definition of regular left divisor, $\Theta_1\Theta_{12}$ is a simple node. In addition, in view of Theorem 3.1,

$$W_{\Theta_1\Theta_{12}}(\lambda) = W_{\Theta_1}(\lambda) W_{\Theta_{12}}(\lambda) = W_1(\lambda) W_{12}(\lambda) = W_2(\lambda),$$

so that the node $\Theta_1\Theta_{12}\Theta_{23}$ is simple as well. Theorem 2.2 now implies the simplicity of the node $\Theta_{12}\Theta_{23}$. Inasmuch as $W_{\Theta_{12}\Theta_{23}}(\lambda) = W_{13}(\lambda)$, we have $W_1(\lambda) \ll W_3(\lambda)$. The set of left divisors of a given function $W(\lambda) \in \Omega_J$, generally speaking, is wider than the set of its regular left divisors. We give an example. Suppose that $\mathfrak{H}$ is a one-dimensional and $\mathfrak{G}$ a two-dimensional set, $h \in \mathfrak{H}$ ($\|h\| = 1$), and that $\{g_1, g_2\}$ is an orthonormal basis in $\mathfrak{G}$. We introduce operators K and J, putting

$$Kg_1 = Kg_2 = h, \quad Jg_1 = g_1, \quad Jg_2 = -g_2.$$

Then $K^*h = g_1 + g_2$, $KJK^* = 0$, so that

$$\Theta = \begin{pmatrix} 0 & K & J \\ \mathfrak{H} & & \mathfrak{G} \end{pmatrix}$$

is a simple node. It is not hard to see that all the functions

$$W^{(\sigma)}(\lambda) = E + \frac{2i\sigma}{\lambda} K^*KJ \qquad (0 \leqslant \sigma \leqslant 1)$$

are left divisors of the function

$$W_\Theta(\lambda) = E + \frac{2i}{\lambda} K^*KJ.$$

On the other hand, $W_\Theta(\lambda)$ has only two regular left divisors: $W^{(0)}(\lambda) = E$ and $W^{(1)}(\lambda) = W_\Theta(\lambda)$.

Theorem 5.3. *If* $W_j(\lambda) \in \Omega_J$ ($j = 1, 2, 3$) *and*

$$W_1(\lambda) \prec W_2(\lambda) \prec W_3(\lambda), \qquad W_1(\lambda) \lll W_3(\lambda),$$

then $W_1(\lambda) \lll W_2(\lambda)$.

Proof. Formulas (5.7) hold, and there exist simple nodes Θ_1, Θ_{12} and Θ_{23}, with directing operator J, for which relations (5.8) are satisfied. Consider the node $\Theta_3 = \Theta_1\Theta_{12}\Theta_{23}$ and denote by Θ'_{13} the principal part of the node $\Theta_{13} = \Theta_{12}\Theta_{23}$. Since

$$W_{\Theta'_{13}}(\lambda) = W_{\Theta_{13}}(\lambda) = W_{\Theta_{12}}(\lambda)\, W_{\Theta_{23}}(\lambda) = W_{13}(\lambda)$$

and, by the hypothesis of the theorem, $\Theta_1\Theta'_{13}$ is the principal part of the node Θ_3, it follows that the excess subspace $\mathfrak{G}^{(0)}_{\Theta_3}$ of the node Θ_3 is orthogonal to the interior space $\mathfrak{H}_1$ of the node Θ_1, i.e. it lies in the orthogonal sum $\mathfrak{H}_{12} \oplus \mathfrak{H}_{23}$ of the interior spaces of the nodes Θ_{12} and Θ_{23}. If now the subspace $\mathfrak{G}^{(0)}_{\Theta_3}$ had a nonzero intersection with $\mathfrak{H}_1 \oplus \mathfrak{H}_{12}$, then this intersection would lie in $\mathfrak{H}_{12}$, and according to Theorem 2.1 the node Θ_{12} would not be simple. Thus $\mathfrak{G}^{(0)}_{\Theta_3} \cap (\mathfrak{H}_1 \oplus \mathfrak{H}_{12}) = 0$, which means that $\Theta_1\Theta_{12}$ is a simple node.

Corollary. *In the set of all regular left divisors of the function* $W(\lambda) \in \Omega_J$ *the relations* $\prec$ *and* $\preccurlyeq$ *are equivalent.*

Theorem 5.4. *Suppose that* $W(\lambda) \in \Omega_J$ *and that* Θ *is a simple node with directing operator* J *such that* $W_\Theta(\lambda) = W(\lambda)$. *For the function* $W_1(\lambda)$ *to be a regular left divisor of* $W(\lambda)$, *it is necessary and sufficient that it be a COF of some left divisor of* Θ.

Proof. If $W_1(\lambda)$ is a regular left divisor of $W(\lambda)$ and Θ'_1 and Θ'_2 are simple nodes for which

$$W_{\Theta'_1}(\lambda) = W_1(\lambda), \qquad W_{\Theta'_2}(\lambda) = W_2(\lambda) = W_1^{-1}(\lambda)\, W(\lambda),$$

then the node $\Theta' = \Theta'_1\Theta'_2$ is also simple and $W_{\Theta'}(\lambda) = W_\Theta(\lambda)$. In view of Theorem 3.2 the nodes Θ and Θ' are unitarily equivalent, and by Theorem 2.7 $\Theta = \Theta_1\Theta_2$, where Θ_1 and Θ_2 are unitarily equivalent respectively to Θ'_1 and Θ'_2. Accordingly, $W_{\Theta_1}(\lambda) = W_{\Theta'_1}(\lambda) = W_1(\lambda)$.

The sufficiency of the condition follows from Theorems 2.2 and 3.1.

Theorem 5.5. *Suppose that* Θ_1 *and* Θ_2 *are left divisors of the simple node* Θ. *For the node* Θ_1 *to be a left divisor of the node* Θ_2, *it is necessary and sufficient that the function* $W_{\Theta_1}(\lambda)$ *be a left divisor of the function* $W_{\Theta_2}(\lambda)$.

Proof. The necessity follows from Theorem 3.1. If $W_{\Theta_1}(\lambda) \prec W_{\Theta_2}(\lambda)$, then in view of Theorem 5.4 and as a consequence of Theorem 5.3 we have $W_{\Theta_1}(\lambda) \preccurlyeq W_{\Theta_2}(\lambda)$. Applying Theorem 5.4 again, we find that $W_\Theta(\lambda) = W_{\Theta_1}(\lambda)$, where Θ'_1 is some left divisor of the node Θ_2. Since Θ_1 and Θ'_1 are left divisors of Θ, from Theorem 3.3, $\Theta_1 = \Theta'_1$.

3. **Invariant subspaces of an operator and regular divisor of the characteristic operator-function.** We imbed a given operator A into the simple node

$$\Theta = \begin{pmatrix} A & K & J \\ \mathfrak{H} & & \mathfrak{G} \end{pmatrix}$$

and consider a mapping Φ which assigns to each subspace $\mathfrak{H}_0 \subseteq \mathfrak{H}$ invariant relative to A the operator-function $W_{\Theta_0}(\lambda)$, where $\Theta_0 = \mathrm{pr}_{\mathfrak{H}_0}\Theta$. The function $W_\Theta(\lambda)$ lies in the class Ω_J, and $W_{\Theta_0}(\lambda)$ is a regular left divisor of it.

Theorem 5.6. *The mapping* Φ *has the following properties*:

1. *Each regular left divisor of* $W_\Theta(\lambda)$ *lies in the range of the mapping* Φ.
2. *The mapping* Φ *is one-to-one.*

3. *If* $\mathfrak{H}_1$ *and* $\mathfrak{H}_2$ *are invariant subspaces of the operator* A *and* $W_1(\lambda)$ *and* $W_2(\lambda)$ *are the regular left divisors of* $W_\Theta(\lambda)$ *corresponding to them under the mapping* Φ, *then* $W_1(\lambda)$ *is a left divisor of* $W_2(\lambda)$ *if and only if* $\mathfrak{H}_1 \subseteq \mathfrak{H}_2$.

Proof. Assertions 1, 2 and 3 follow from Theorems 5.4, 3.3 and 5.5 respectively.

A portion Ω'_J of the set Ω_J is said to be *ordered* if for any two functions $W_1(\lambda) \in \Omega'_J$ and $W_2(\lambda) \in \Omega'_J$ one of the relations $W_1(\lambda) \prec W_2(\lambda)$ or $W_2(\lambda) \prec W_1(\lambda)$ holds.

The bounded linear operator A is said to be *unicellular* if one of any two invariant subspaces of A is included in the other.

The following is an easy consequence of Theorem 5.6.

Theorem 5.7. *Suppose that* $W(\lambda)$ *is the COF of the simple node* Θ. *For the operator* A *to be unicellular, it is necessary and sufficient that the set of regular left divisors of* $W(\lambda)$ *be ordered.*

Theorem 5.8. *Suppose that* $W_j(\lambda)$ $(j = 1, 2)$ *are regular left divisors of the function* $W(\lambda) \in \Omega_J$, *that* Θ *is a simple node with the COF* $W(\lambda)$, *and that the* $\mathfrak{H}_j$ $(j = 1, 2)$ *are subspaces invariant relative to* A *for which*

$$W_j(\lambda) = W_{\Theta_j}(\lambda) \qquad \left(\Theta_j = \mathrm{pr}_{\mathfrak{H}_j}\Theta = \begin{pmatrix} A_j & K_j & J \\ \mathfrak{H}_j & & \mathfrak{G} \end{pmatrix}\right).$$

For $\mathfrak{H}_1$ *and* $\mathfrak{H}_2$ *to be mutually orthogonal and for them to satisfy the condition* $\mathfrak{H}_1 \oplus \mathfrak{H}_2 = \mathfrak{H}$, *it is necessary and sufficient that in some neighborhood of the point at infinity*

$$W(\lambda) = W_1(\lambda)\, W_2(\lambda) = W_1(\lambda) + W_2(\lambda) - E. \tag{5.9}$$

Proof. If $\mathfrak{H} = \mathfrak{H}_1 \oplus \mathfrak{H}_2$, then $\Theta = \Theta_1\Theta_2$, so that, from Theorem 3.1, $W(\lambda) = W_1(\lambda) W_2(\lambda)$. Denoting by P_j the orthoprojector onto $\mathfrak{H}_j$, we get

$$K_1 J K_2^* = P_1 K J K^* P_2 = P_1 A_I P_2 = 0.$$

Accordingly,

$$(W_1(\lambda) - E)(W_2(\lambda) - E) = -4K_1^*(A_1 - \lambda E)^{-1} K_1 J K_2^* (A_2 - \lambda E)^{-1} K_2 J = 0,$$

i.e. $W_1(\lambda) W_2(\lambda) = W_1(\lambda) + W_2(\lambda) - E$.

Conversely, suppose that equation (5.9) holds. Then

$$K_1^*K_1JK_2^*K_2 = \frac{1}{4}\lim_{\lambda\to\infty}[\lambda^2(W_1(\lambda)-E)(W_2(\lambda)-E)J] = 0,$$

$$[(K_1JK_2^*)(K_1JK_2^*)^*]^2 = K_1JK_2^*K_2J(K_1^*K_1JK_2^*K_2)JK_1^* = 0,$$

so that $P_1A_IP_2 = K_1JK_2^* = 0$. Thus, $P_1AP_2 = P_1A^*P_2$, so that in view of the relations $A^nP_j = P_jA^nP_j$ $(j = 1, 2;\ n = 1, 2, \cdots)$ we will have

$$P_1A^nP_2 = P_1A^{*n}P_2 \qquad (n = 1, 2, \ldots).$$

On the other hand,

$$\begin{aligned} E - 2iK^*(A-\lambda E)^{-1}KJ \\ = E - 2iK_1^*(A_1-\lambda E)^{-1}K_1J - 2iK_2^*(A_2-\lambda E)^{-1}K_2J \\ = E - 2iK^*(A-\lambda E)^{-1}P_1KJ - 2iK^*(A-\lambda E)^{-1}P_2KJ, \end{aligned}$$

so that

$$K^*A^n(E-P_1-P_2)K = 0 \qquad (n = 0, 1, \ldots).$$

Since Θ is a simple node,

$$(E-P_1-P_2)K = 0 \tag{5.10}$$

and finally

$$P_1A^nP_2K = P_1A^{*n}(E-P_1)K = 0 \qquad (n = 0, 1, \ldots),$$

which proves the mutual orthogonality of the subspaces $\mathfrak{H}_1$ and $\mathfrak{H}_2$. It also follows from (5.10) that $\mathfrak{H}_1 \oplus \mathfrak{H}_2 = \mathfrak{H}$, inasmuch as $A^nKg \in \mathfrak{H}_1 \oplus \mathfrak{H}_2$ $(n = 0, 1, \cdots;\ g \in \mathfrak{G})$.

§6. The classes Ω_J^+ and Ω_J^-

Suppose again that J is a bounded linear operator operating in the Hilbert space $\mathfrak{G}$ and satisfying the conditions $J = J^*$ and $J^2 = E$. We will say that the function $W(\lambda)$, whose values are bounded linear operators in $\mathfrak{G}$, lies in the class Ω_J^+ if:

(I$^+$) $W(\lambda)$ is holomorphic in a region G_W gotten by removing from the extended complex plane some bounded set which lies in the closed upper halfplane;

(II$^+$) $\lim_{\lambda\to\infty}\|W(\lambda) - E\| = 0$;

(III$^+$) $W^*(\lambda)JW(\lambda) - J \le 0$ $(\operatorname{Im}\lambda < 0)$;

(IV$^+$) $W^*(\lambda)JW(\lambda) - J = 0$ $(\operatorname{Im}\lambda = 0,\ \lambda \in G_W)$,

and to the class Ω_J^- if

(I^-) $W(\lambda)$ is holomorphic in a region G_W gotten by removing from the extended complex plane some bounded set which lies in the closed lower halfplane;

(II^-) $\lim_{\lambda\to\infty}\|W(\lambda)-E\| = 0$;

(III^-) $W^*(\lambda)JW(\lambda)-J \geq 0$ $(\operatorname{Im}\lambda > 0)$;

(IV^-) $W^*(\lambda)JW(\lambda)-J = 0$ $(\operatorname{Im}\lambda = 0,\ \lambda\in G_W)$.

If the functions $W_1(\lambda)$ *and* $W_2(\lambda)$ *lie in the class* Ω_J^+ (Ω_J^-), *then the function* $W(\lambda)=W_1(\lambda)W_2(\lambda)$ *lies in the same class.*

The proof follows from the equation

$$W^*(\lambda)JW(\lambda)-J$$
$$= W_2^*(\lambda)\left(W_1^*(\lambda)JW_1(\lambda)-J\right)W_2(\lambda)+W_2^*(\lambda)JW(\lambda)-J. \tag{6.1}$$

We will show below that Ω_J^+ and Ω_J^- are in the class Ω_J.

Lemma 6.1. *Suppose that a holomorphic function* $T(\lambda)$ *is given in the upper halfplane and that its values are bounded linear operators operating in some Hilbert space. Suppose that* $T(\lambda)$ *has a bounded inverse for all* $\lambda=\sigma+i\tau$ *which lie in the region* $a<\sigma<b$, $\tau>0$. *If the operator* $T^*(\lambda)T(\lambda)$ *has a bounded inverse for any* λ *of the upper halfplane, then* $T(\lambda)$ *has the same property.*[1)]

Proof. Put $S(\lambda)=T(\lambda+i\tau_0)$ $(\tau_0>0)$. The functions $S(\lambda)$ and $S^*(\bar\lambda)$ are holomorphic in the strip $|\tau|<\tau_0$, and the operator $S^*(\bar\lambda)S(\lambda)$ has a bounded inverse for all real λ. Accordingly, the function $(S^*(\bar\lambda)S(\lambda))^{-1}$ is defined and holomorphic in some region G which lies in the strip $|\tau|<\tau_0$ and contains the entire real axis. Since each of the operators $S(\lambda)$ and $S^*(\bar\lambda)$ has a bounded inverse in the rectangle $(a<\sigma<b,\ -\tau_0<\tau<\tau_0)$, we have

$$(S^*(\bar\lambda)S(\lambda))^{-1}S^*(\bar\lambda)S(\lambda)=E,\quad S(\lambda)(S^*(\bar\lambda)S(\lambda))^{-1}S^*(\bar\lambda)=E$$
$$(a<\sigma<b,\ \tau=0). \tag{6.2}$$

By the uniqueness theorem, identities (6.2) are valid at all points of the region G. This means in particular that for any real λ the operator $(S^*(\bar\lambda)S(\lambda))^{-1}S^*(\bar\lambda)$ is an inverse for the operator $T(\lambda+i\tau_0)=S(\lambda)$.

Lemma 6.2. *Suppose that the function* $W(\lambda)$, *whose values are bounded linear operators in the Hilbert space* $\mathfrak{G}$, *is holomorphic in the upper halfplane. If*

---

1) The assertion of the lemma remains in force if the requirement that the function $T(\lambda)$ be invertible in the region $a<\sigma<b$, $\tau>0$ is replaced by the requirement that it be invertible at only one point of the upper halfplane (see I. C. Gohberg and M. G. Kreĭn [7], Theorems 2.4 and 7.1).

$$\lim_{\lambda\to\infty} \| W(\lambda) - E \| = 0 \tag{6.3}$$

and

$$W^*(\lambda) J W(\lambda) - J \geqslant 0 \quad (\operatorname{Im}\lambda > 0), \tag{6.4}$$

where J is some operator satisfying the requirements $J = J^*$ and $J^2 = E$, then

$$W(\lambda) J W^*(\lambda) - J \geqslant 0 \quad (\operatorname{Im}\lambda > 0) \tag{6.5}$$

and

$$\left| \frac{((W(\lambda) - W(\mu)) f, g)}{\lambda - \mu} \right|^2 \leqslant$$
$$\leqslant \frac{((W^*(\lambda) J W(\lambda) - J) f, f)}{2 \operatorname{Im}\lambda} \, \frac{((W(\mu) J W^*(\mu) - J) g, g)}{2 \operatorname{Im}\mu} \tag{6.6}$$
$$(\operatorname{Im}\lambda > 0, \quad \operatorname{Im}\mu > 0; \quad f, g \in \mathfrak{G}).$$

Proof. By formula (1.3) there exist orthoprojectors P^+ and P^- such that

$$J = P^+ - P^-, \quad P^+ + P^- = E.$$

We introduce the function

$$Y_\rho(\lambda) = W(\lambda) + \rho P^+ + \frac{1}{\rho} P^- \quad (0 < \rho < 1)$$

and show that for any fixed λ and ρ

$$\inf_{\|h\|=1} \left(Y_\rho^*(\lambda) Y_\rho(\lambda) h, h\right) > 0. \tag{6.7}$$

Supposing the contrary, we find λ_0, ρ_0 and a sequence h_n ($\|h_n\| = 1$), for which $\|Y_{\rho_0}(\lambda_0) h_n\| \to 0$. But then

$$((W^*(\lambda_0) J W(\lambda_0) - J) h_n, h_n)$$
$$= (\rho_0^2 - 1)(P^+ h_n, h_n) + \left(1 - \frac{1}{\rho_0^2}\right)(P^- h_n, h_n) + \varepsilon_n$$
$$\leqslant (\rho_0^2 - 1)(P^+ h_n, h_n) + (\rho_0^2 - 1)(P^- h_n, h_n) + \varepsilon_n = \rho_0^2 - 1 + \varepsilon_n \quad (\varepsilon_n \to 0)$$

and hence for sufficiently large n

$$((W^*(\lambda_0) J W(\lambda_0) - J) h_n, h_n) < 0,$$

which contradicts (6.4). Inequality (6.7) means that for any λ and ρ the operator $Y_\rho^*(\lambda) Y_\rho(\lambda)$ has a bounded inverse. Using condition (6.3) and Lemma 6.1, we find that the operator $Y_\rho(\lambda)$ also is a bounded invertible operator for all possible λ and ρ.

Put

$$W_\rho(\lambda) = \left(\frac{1}{\rho} P^+ + \rho P^-\right) W(\lambda).$$

Since

$$W_\rho(\lambda) + E = \left(\frac{1}{\rho} P^+ + \rho P^-\right) Y_\rho(\lambda),$$

there exists a bounded operator $(W_\rho(\lambda) + E)^{-1}$ ($\operatorname{Im}\lambda > 0$, $0 < \rho < 1$). The function

$$V_\rho(\lambda) = i(W_\rho(\lambda) + E)^{-1}(W_\rho(\lambda) - E)J$$

satisfies the relations

$$\frac{V_\rho(\lambda) - V_\rho(\mu)}{2i} = (W_\rho(\mu) + E)^{-1}(W_\rho(\lambda) - W_\rho(\mu))(W_\rho(\lambda) + E)^{-1}J, \tag{6.8}$$

$$\frac{V_\rho(\lambda) - V_\rho^*(\lambda)}{2i} = J\left(W_\rho^*(\lambda) + E\right)^{-1}\left(W_\rho^*(\lambda)JW_\rho(\lambda) - J\right)(W_\rho(\lambda) + E)^{-1}J, \tag{6.9}$$

$$\frac{V_\rho(\lambda) - V_\rho^*(\lambda)}{2i} = (W_\rho(\lambda) + E)^{-1}\left(W_\rho(\lambda)JW_\rho^*(\lambda) - J\right)\left(W_\rho^*(\lambda) + E\right)^{-1}. \tag{6.10}$$

From (6.4), the inequality

$$W_\rho^*(\lambda)JW_\rho(\lambda) - J = W^*(\lambda)JW(\lambda) - J + W^*(\lambda)\left[\left(\frac{1}{\rho^2} - 1\right)P^+ + (1-\rho^2)P^-\right]W(\lambda) \geqslant 0$$

and (6.9) it follows that

$$\frac{V_\rho(\lambda) - V_\rho^*(\lambda)}{2i} \geqslant 0 \quad (\operatorname{Im}\lambda > 0). \tag{6.11}$$

From (6.11) and (6.10) we have

$$W_\rho(\lambda)JW_\rho^*(\lambda) - J \geqslant 0.$$

Passing to the limit as $\rho \to 1$, we obtain (6.5).

It is easy to see that the function $V_\rho(\lambda)$ satisfies the hypotheses of Theorem 4.7. Accordingly there exists a nonnegative nondecreasing operator-function $G_\rho(s)$ ($-\infty < s < +\infty$), having uniform limits as s tends to $-\infty$ and $+\infty$, and such that

$$V_\rho(\lambda) = \frac{V_\rho(i) + V_\rho^*(i)}{2} + \int_{-\infty}^{+\infty} \frac{1 + s\lambda}{s - \lambda}\, dG_\rho(s) \quad (\operatorname{Im}\lambda > 0).$$

Therefore

$$V_\rho(\lambda) - V_\rho(\mu) = (\lambda - \mu) \int_{-\infty}^{+\infty} \frac{1+s^2}{(s-\lambda)(s-\mu)}\, dG_\rho(s),$$

$$V_\rho(\lambda) - V_\rho^*(\lambda) = (\lambda - \bar\lambda) \int_{-\infty}^{+\infty} \frac{1+s^2}{|s-\lambda|^2}\, dG_\rho(s),$$

$$\left| \frac{((V_\rho(\lambda) - V_\rho(\mu))f,\, g)}{\lambda - \mu} \right|^2 = \left| \int_{-\infty}^{+\infty} \frac{1+s^2}{(s-\lambda)(s-\mu)}\, d(G_\rho(s) f,\, g) \right|^2$$

$$\leqslant \int_{-\infty}^{+\infty} \frac{1+s^2}{|s-\lambda|^2}\, d(G_\rho(s) f,\, f) \int_{-\infty}^{+\infty} \frac{1+s^2}{|s-\mu|^2}\, d(G_\rho(s) g,\, g),$$

so that

$$\left| \frac{((V_\rho(\lambda) - V_\rho(\mu))f,\, g)}{\lambda - \mu} \right|^2 \leqslant \frac{((V_\rho(\lambda) - V_\rho^*(\lambda))f,\, f)}{\lambda - \bar\lambda} \frac{((V_\rho(\mu) - V_\rho^*(\mu))g,\, g)}{\mu - \bar\mu}.$$

Replacing f and g in the last inequality respectively by $J(W_\rho(\lambda) + E)f$ and $(W_\rho^*(\mu) + E)g$ and using relations (6.8)–(6.10), we get

$$\left| \frac{((W_\rho(\lambda) - W_\rho(\mu))f,\, g)}{\lambda - \mu} \right|^2 \leqslant \frac{((W_\rho^*(\lambda) J W_\rho(\lambda) - J)f,\, f)}{2\,\mathrm{Im}\,\lambda} \frac{((W_\rho(\mu) J W_\rho^*(\mu) - J)g,\, g)}{2\,\mathrm{Im}\,\mu}. \tag{6.12}$$

Inequality (6.6) is obtained from (6.12) as $\rho \to 1$.

Theorem 6.1. *If the function $W(\lambda)$ lies in one of the classes Ω_J^+, Ω_J^-, then there exists a simple node Θ with directing operator J such that in some neighborhood of the point at infinity $W_\Theta(\lambda) \equiv W(\lambda)$.*

Proof. By Theorem 5.2 it is sufficient to show that $W(\lambda) \in \Omega_J$. Consider first the case when $W(\lambda) \in \Omega_J^-$. It follows from condition (II^-) that the following limit exists in the sense of the uniform norm:

$$\lim_{\lambda \to \infty} [\lambda (W(\lambda) - E)] = T.$$

Using the equation

$$W^*(\lambda) J W(\lambda) - J = (W^*(\lambda) - E) J (W(\lambda) - E) + J(W(\lambda) - E) + (W^*(\lambda) - E) J,$$

and also conditions (III^-) and (IV^-), we find the relations

$$i(T^*J - JT) = \lim_{\tau \to +\infty} [\tau(W^*(i\tau) JW(i\tau) - J)] \geqslant 0, \tag{6.13}$$

$$T^*J + JT = \lim_{\sigma \to \infty} [\sigma(W^*(\sigma) JW(\sigma) - J)] = 0, \tag{6.14}$$

$$(\lambda = \sigma + i\tau).$$

If we put $H = (1/2i)\, TJ$, it follows from (6.13) and (6.14) that $H \geq 0$ and

$$\lim_{\tau \to +\infty} [\tau(W^*(i\tau) JW(i\tau) - J)] = 4JHJ. \tag{6.15}$$

Using analogous arguments we arrive at the formula

$$\lim_{\tau \to +\infty} [\tau(W(i\tau) JW^*(i\tau) - J)] = 4H. \tag{6.16}$$

The function $W(\lambda)$ satisfies the hypotheses of Lemma 6.2. From (6.6)

$$|((W(i\tau) - W(\mu))f,\ g)|^2$$

$$\leqslant \frac{|i\tau - \mu|^2}{4\tau \operatorname{Im} \mu} ((W^*(i\tau) JW(i\tau) - J)f,\ f)((W(\mu) JW^*(\mu) - J)g, g)$$

$$(\tau > 0, \quad \operatorname{Im} \mu > 0),$$

so that in view of (6.15)

$$|((E - W(\mu))f,\ g)|^2 = \lim_{\tau \to +\infty} |((W(i\tau) - W(\mu))f,\ g)|^2$$

$$\leqslant (JHJf,\ f) \frac{((W(\mu) JW^*(\mu) - J)g,\ g)}{\operatorname{Im} \mu},$$

$$\|(W^*(\mu) - E)g\|^2 \leqslant \|H\| \frac{((W(\mu) JW^*(\mu) - J)g,\ g)}{\operatorname{Im} \mu} \qquad (\operatorname{Im} \mu > 0) \tag{6.17}$$

Changing the places of λ and μ in our arguments, we get

$$\|(W(\lambda) - E)f\|^2 \leqslant \|H\| \frac{((W^*(\lambda) JW(\lambda) - J)f,\ f)}{\operatorname{Im} \lambda} \qquad (\operatorname{Im} \lambda > 0). \tag{6.18}$$

We will show that

$$\inf_{\|f\|=1} \|(W(\lambda) + E)f\| > 0 \qquad (\operatorname{Im} \lambda > 0) \tag{6.19}$$

and

$$\inf_{\|f\|=1} \|(W^*(\lambda) + E)f\| > 0 \qquad (\operatorname{Im} \lambda > 0). \tag{6.20}$$

We suppose that there exists a point λ_0 $(\operatorname{Im} \lambda_0 > 0)$ and a sequence f_n $(\|f_n\| = 1)$, such that $\|(W(\lambda_0) + E)f_n\| \to 0$. Since

$$W^*(\lambda) JW(\lambda) - J = (W^*(\lambda) + E) J (W(\lambda) + E)$$

$$- J(W(\lambda) + E) - (W^*(\lambda) + E) J,$$

we see that

$$((W^*(\lambda_0) JW(\lambda_0) - J)f_n,\ f_n) \to 0$$

and, from (6.18), $\|(W(\lambda_0) - E)f_n\| \to 0$. But then

$$\|f_n\| = \frac{1}{2}\|(W(\lambda_0)+E)f_n - (W(\lambda_0)-E)f_n\| \to 0,$$

which is impossible. Thus inequality (6.19) is proved. Inequality (6.20) is proved analogously using (6.17).

Taking account of (6.19), (6.20) and (II$^-$), we arrive at the conclusion that the function $(W(\lambda)+E)^{-1}$ is defined and holomorphic in a region G_W containing the entire upper halfplane and some neighborhood of the point at infinity. The function

$$V(\lambda) = i(W(\lambda)+E)^{-1}(W(\lambda)-E)J \qquad (\lambda \in G_W)$$

satisfies the relation

$$\frac{V(\lambda)-V^*(\lambda)}{2i} = J(W^*(\lambda)+E)^{-1}(W^*(\lambda)JW(\lambda)-J)(W(\lambda)+E)^{-1}J,$$

so that

$$\frac{V(\lambda)-V^*(\lambda)}{2i} \geqslant 0 \ (\operatorname{Im}\lambda > 0), \ V(\lambda) = V^*(\lambda) \ (\operatorname{Im}\lambda = 0, \ \lambda \in G_W).$$

In view of the last equation and the principle of symmetry the function $V(\lambda)$ may be continued analytically into the entire lower halfplane, putting $V(\lambda) = V^*(\bar{\lambda})$ $(\operatorname{Im}\lambda < 0)$.

Now suppose that $W(\lambda) \in \Omega_J^+$. Then the function $W_1(\lambda) = W(-\lambda)$ belongs to the class Ω_{-J}^-, and by what was proved above it coincides in the neighborhood of the point at infinity with the COF of some simple node

$$\Theta_1 = \begin{pmatrix} A & K & -J \\ \mathfrak{H} & & \mathfrak{G} \end{pmatrix}.$$

It remains to note that

$$W_{\Theta_1}(\lambda) = W_{\Theta}(\lambda), \text{ where } \Theta = \begin{pmatrix} -A & K & J \\ \mathfrak{H} & & \mathfrak{G} \end{pmatrix}.$$

The theorem is proved.

If

$$\Theta = \begin{pmatrix} A & K & J \\ \mathfrak{H} & & \mathfrak{G} \end{pmatrix}$$

is an operator node and if the spectrum of the operator A lies in the upper closed halfplane, then obviously $W_\Theta(\lambda) \in \Omega_J^+$. Conversely, if $W(\lambda) \in \Omega_J^+$, then, as was proved in the last theorem, $W(\lambda)$ coincides in the neighborhood of the point at infinity with the COF of some node Θ. Under these circumstances, can the spectrum of the operator A lie partially or completely in the lower halfplane? A negative answer will be obtained in the following sections for certain special cases.

§7. Dissipative nodes

1. **Dissipative operators.** A bounded linear operator A is said to be *dissipative* if $A_I \geq 0$.

Suppose that the dissipative operator A induces in some subspace $\mathfrak{H}_0$ invariant relative to A a selfadjoint operator A_0. Then

$$\left\| \left(\frac{A-A^*}{2i} \right)^{1/2} h \right\|^2 = \left(\frac{A-A^*}{2i} h,\ h \right) = \operatorname{Im}(A_0 h,\ h) \qquad (h \in \mathfrak{H}_0),$$

so that $Ah = A^*h$ $(h \in \mathfrak{H}_0)$. Therefore it follows that *a dissipative operator A is completely nonselfadjoint if and only if there does not exist a subspace invariant relative to A in which A induces a selfadjoint operator.*

Theorem 7.1. *The spectrum of a dissipative operator A lies in the upper closed halfplane.*

Proof. Suppose that $\lambda = \sigma + i\tau$ $(\tau < 0)$. Then

$$\left.\begin{array}{l} \|(A-\lambda E)h\|^2 = \|(A-\sigma E)h\|^2 - 2\tau(A_I h,\ h) + \tau^2 \geqslant \tau^2, \\ \|(A^*-\bar{\lambda} E)h\|^2 = \|(A^*-\sigma E)h\|^2 - 2\tau(A_I h,\ h) + \tau^2 \geqslant \tau^2 \end{array}\right\} \quad (\|h\| = 1),$$

and therefore

$$\inf_{\|h\|=1} \|(A-\lambda E)h\| > 0, \qquad \inf_{\|h\|=1} \|(A^*-\bar{\lambda}E)h\| > 0.$$

It follows from the last two inequalities that λ is a regular point of the operator A.

2. **Dissipative nodes.** The node

$$\Theta = \begin{pmatrix} A & K & J \\ \mathfrak{H} & & \mathfrak{G} \end{pmatrix}$$

is said to be dissipative if $J = E$. The basic operator of a dissipative node is dissipative, since

$$(A_I h,\ h) = (KK^*h,\ h) = \|K^*h\|^2 \geqslant 0.$$

The converse assertion is generally speaking false. Using the method indicated in Theorem 1.1, it is easy to construct a nondissipative node with a dissipative basic operator. However, putting $\mathfrak{R} = \overline{\mathfrak{R}(A_I)}$ in the proof of that theorem, we arrive at the following conclusion.

Every dissipative operator may be imbedded in a dissipative node. In particular, every completely nonselfadjoint dissipative operator may be imbedded in a simple dissipative node.

Theorem 7.2. *The canal subspace of a dissipative node*

$$\Theta = \begin{pmatrix} A & K & E \\ \mathfrak{H} & & \mathfrak{G} \end{pmatrix}$$

coincides with the non-Hermitian subspace.

Proof. If $h \perp \mathfrak{R}(A_I)$, then

$$\| K^* h \|^2 = (KK^* h, h) = (A_I h, h) = 0$$

so that $h \perp \mathfrak{R}(K)$. Thus $\overline{\mathfrak{R}(K)} \subseteq \overline{\mathfrak{R}(A_I)}$, which along with (1.4) leads to the equation $\overline{\mathfrak{R}(K)} = \overline{\mathfrak{R}(A_I)}$.

Corollary 1. *For a dissipative node to be simple, it is necessary and sufficient that its basic operator be completely nonselfadjoint.*

Corollary 2. *If P_0 is an orthoprojector onto a semi-invariant subspace $\mathfrak{H}_0$ of a completely nonselfadjoint dissipative operator A, then the operator $A_0 = P_0 A$, considered in $\mathfrak{H}_0$, is also completely nonselfadjoint.*

Below we will carry out the proof of a sufficiency test for the simplicity of the product of two simple dissipative nodes.

Lemma 7.1. *If A is a bounded selfadjoint operator and $\lim_{n\to\infty} \|A^n h\|^{1/n} = 0$, then $Ah = 0$.*

Proof. No generality will be lost if we suppose $\|h\| = 1$. In this case

$$\| Ah \| = (A^2 h, h)^{1/2} \leqslant \| A^2 h \|^{1/2} = (A^4 h, h)^{1/4}$$
$$\leqslant \| A^4 h \|^{1/4} \leqslant \ldots \leqslant \| A^{2k} h \|^{1/2k} \to 0.$$

Lemma 7.2. *If for the node*

$$\Theta = \begin{pmatrix} A & K & J \\ \mathfrak{H} & & \mathfrak{G} \end{pmatrix}$$

there exists an entire operator-function coinciding in the neighborhood of zero with the function $W_\Theta(1/\mu)$, then

$$\lim_{m\to\infty} \| A^m K \|^{1/m} = 0.$$

Proof. Since in the neighborhood of zero

$$W_\Theta\left(\frac{1}{\mu}\right) = E + 2i\sum_{n=1}^{\infty} \mu^n K^* A^{n-1} KJ \tag{7.1}$$

and the series (7.1) converges for all μ, we have $\|K^* A^n K\|^{1/n} \to 0$. From the estimates

$$\begin{aligned} \| K^*A^{*m} A^n K - K^* A^{*m+1} A^{n-1} K \| &= \| K^* A^{*m} (A - A^*) A^{n-1} K \| \\ &= 2 \| K^* A^{*m} KJK^* A^{n-1} K \| \leqslant 2 \| K^* A^{*m} K \| \| K^* A^{n-1} K \| \\ &\qquad (m = 0, 1, \ldots; \ n = 1, 2, \ldots), \end{aligned}$$

$$\begin{aligned} \| K^* A^{*m} A^m K - K^* A^{*2m} K \| &\leqslant \\ &\leqslant \sum_{p=0}^{m-1} \| K^* A^{*m+p} A^{m-p} K - K^* A^{*m+p+1} A^{m-p-1} K \| \\ &\leqslant 2 \sum_{p=0}^{m-1} \| K^* A^{*m+p} K \| \| K^* A^{m-p-1} K \| \end{aligned}$$

and the boundedness of the sequence

$$\| K^* A^n K \| \quad (n = 1, 2, \ldots)$$

it follows that there exists a constant M for which

$$\| K^* A^{*m} A^m K \| \leqslant M \sum_{p=m}^{2m} \| K^* A^p K \|.$$

Accordingly

$$\| A^m K \|^{1/m} = \| K^* A^{*m} A^m K \|^{1/2m} \leqslant M^{1/2m} \left(\sum_{p=m}^{2m} \| K^* A^p K \| \right)^{1/2m} \to 0.$$

Theorem 7.3. *Suppose that*

$$\Theta_1 = \begin{pmatrix} A_1 & K_1 & E \\ \mathfrak{H}_1 & & \mathfrak{G} \end{pmatrix} \text{ and } \Theta_2 = \begin{pmatrix} A_2 & K_2 & E \\ \mathfrak{H}_2 & & \mathfrak{G} \end{pmatrix}$$

are simple dissipative nodes. If there exists an entire function coinciding in the neighborhood of zero with one of the functions $W_{\Theta_1}(1/\mu)$, $W_{\Theta_2}(1/\mu)$, then $\Theta = \Theta_1\Theta_2$ is a simple node.

Proof. Suppose that there exists an entire function coinciding in the neighborhood of zero with $W_{\Theta_1}(1/\mu)$. If $\mathfrak{H}_1^{(0)}$ is the linear envelope of vectors of the form $A_1^n K_1 g$ $(n = 0, 1, \cdots; g \in \mathfrak{G})$ and $h \in \mathfrak{H}_1^{(0)}$, then $h = \sum_{p=0}^{r} A_1^p K_1 g_p$, and, from Lemma 7.2,

$$\|A_1^n h\|^{1/n} \leqslant \sum_{p=0}^{r} \|A_1^{n+p} K_1 g_p\|^{1/n} \to 0.$$

Denote by P_0 the orthoprojector operating in $\mathfrak{H} = \mathfrak{H}_1 \oplus \mathfrak{H}_2$ onto the excess subspace $\mathfrak{E}_\Theta^{(0)}$ of the node Θ, and by A_0 the selfadjoint operator induced by the basic operator of Θ in $\mathfrak{E}_\Theta^{(0)}$. Inasmuch as

$$P_0 A_1 h = A_0 P_0 h \quad (h \in \mathfrak{H}_1),$$

so that

$$\|A_0^n P_0 h\|^{1/n} = \|P_0 A_1^n h\|^{1/n} \leqslant \|A_1^n h\|^{1/n} \to 0 \qquad (h \in \mathfrak{H}_1^{(0)}),$$

in view of Lemma 7.1 we have

$$P_0 A_1 h = A_0 P_0 h = 0 \qquad (h \in \mathfrak{H}_1^{(0)}).$$

Since $\mathfrak{H}_1^{(0)}$ is dense in $\mathfrak{H}_1$, we get $P_0 A_1 \mathfrak{H}_1 = 0$. The range of the operator A_1 is also dense in $\mathfrak{H}_1$, since in the contrary case the operator A_1 would not be completely nonselfadjoint. Thus $P_0 \mathfrak{H}_1 = 0$, so that $\mathfrak{E}_\Theta^{(0)} \subset \mathfrak{H}_2$. In view of Lemma 2.1, $\mathfrak{E}_\Theta^{(0)} = 0$.

The second assertion of the theorem may be obtained analogously, replacing the nodes Θ_1 and Θ_2 respectively by the nodes

$$\begin{pmatrix} -A_1^* & K_1 & E \\ \mathfrak{H}_1 & & \mathfrak{G} \end{pmatrix} \text{ and } \begin{pmatrix} -A_2^* & K_2 & E \\ \mathfrak{H}_2 & & \mathfrak{G} \end{pmatrix}.$$

3. The COF's of dissipative nodes. The COF $W(\lambda) = W_\Theta(\lambda)$ of the dissipative node Θ lies, obviously, in the class Ω_E^+. Accordingly,

($\mathrm{I}_{\mathfrak{G}}^+$) $W(\lambda)$ is holomorphic in a region G_W obtained by deleting from the extended complex plane some bounded set lying in the closed upper halfplane;

($\mathrm{II}_{\mathfrak{G}}^+$)
$$\lim_{\lambda \to \infty} \|W(\lambda) - E\| = 0;$$

($\mathrm{III}_{\mathfrak{G}}^+$)
$$W^*(\lambda) W(\lambda) - E \leqslant 0 \qquad (\operatorname{Im} \lambda < 0);$$

($\mathrm{IV}_{\mathfrak{G}}^+$)
$$W^*(\lambda) W(\lambda) - E = 0 \qquad (\operatorname{Im} \lambda = 0,\ \lambda \in G_W).$$

The set of operator functions satisfying conditions ($\mathrm{I}_{\mathfrak{G}}^+$)–($\mathrm{IV}_{\mathfrak{G}}^+$) will be denoted by $\Omega_{\mathfrak{G}}^+$.

Theorem 7.4. *If the function $W(\lambda)$ lies in the class $\Omega_{\mathfrak{G}}^+$, then there exists a simple dissipative node Θ such that in some region containing the entire lower* *halfplane and the neighborhood of the point at infinity the equation* $W_\Theta(\lambda) = W(\lambda)$ *is satisfied.*

The proof follows easily from Theorems 6.1 and 7.1.

Theorem 7.5. *If for the dissipative node*

$$\Theta = \begin{pmatrix} A & K & E \\ \mathfrak{H} & & \mathfrak{G} \end{pmatrix}$$

there exists an orthoprojector P *such that* $K^*K \leq \sigma P$ $(\sigma > 0)$, *then*

$$W_\Theta(\lambda)\, Q = Q W_\Theta(\lambda) = Q \qquad (Q = E - P,\ \lambda \in G_A).$$

Proof. Since $QK^*KQ = 0$, we have $KQ = QK^* = 0$. Accordingly,

$$W_\Theta(\lambda)\, Q = \left(E - 2iK^*(A - \lambda E)^{-1} K\right) Q = Q$$

and analogously $QW_\Theta(\lambda) = Q$.

The theorem is proved.

Consider the dissipative node Θ and denote by $\mathfrak{G}_0$ the subspace consisting of all vectors $g \in \mathfrak{G}$ for which $Kg = 0$. Suppose that K_1 is a mapping of the subspace $\mathfrak{G}_1 = \mathfrak{G} \ominus \mathfrak{G}_0$ into $\mathfrak{H}$, defined by the equation $K_1 g = Kg$ $(g \in \mathfrak{G}_1)$. Since, obviously, $K_1^* h = K^* h$ $(h \in \mathfrak{H})$, it follows that

$$\Theta_1 = \begin{pmatrix} A & K_1 & E \\ \mathfrak{H} & & \mathfrak{G} \end{pmatrix}$$

is also a node. It is easy to see that

$$W_\Theta(\lambda)\, g = \begin{cases} W_{\Theta_1}(\lambda)\, g & (g \in \mathfrak{G}_1), \\ g & (g \in \mathfrak{G}_0). \end{cases} \tag{7.2}$$

The node Θ is said to be *minimal* if $\mathfrak{G}_0 = 0$.

The characteristic operator-functions of the minimal nodes

$$\Theta_1 = \begin{pmatrix} A & K_1 & E \\ \mathfrak{H} & & \mathfrak{G}_1 \end{pmatrix} \text{ and } \Theta_2 = \begin{pmatrix} A & K_2 & E \\ \mathfrak{H} & & \mathfrak{G}_2 \end{pmatrix}$$

are unitarily equivalent.

Indeed, denote by U_0 the mapping assigning to the vector $K_1^* h$ $(h \in \mathfrak{H})$ the vector $K_2^* h$. From the equation $K_1 K_1^* = K_2 K_2^*$ and the minimality of the nodes Θ_1 and Θ_2 it follows that U_0 can be extended by continuity to an operator U defined in the entire space $\mathfrak{G}_1$ and carrying $\mathfrak{G}_1$ isometrically onto the entire space $\mathfrak{G}_2$. Also

$$UW_{\Theta_1}(\lambda) = U(E - 2iK_1^*(A - \lambda E)^{-1} K_1)$$
$$= (E - 2iK_2^*(A - \lambda E)^{-1} K_2)\, U = W_{\Theta_2}(\lambda)\, U.$$

§8. Dissipative exponential nodes

1. Operator-functions of the class $\Omega_{\mathfrak{G}}^{(\exp)}$. We recall that an entire function $\Phi(\lambda)$ whose values are bounded linear operators is said to be a function *of exponential type* if there exist positive constants α and β such that for all λ

$$\|\Phi(\lambda)\| \leqslant \beta e^{\alpha|\lambda|}. \tag{8.1}$$

We assign the number α to a set $\mathfrak{A}$ if there exists a β such that (8.1) holds. The greatest lower bound of $\mathfrak{A}$ is called the *type* of the function $\Phi(\lambda)$.

We will say that the function $W(\lambda)$ of a complex variable whose values are bounded linear operators operating in the space $\mathfrak{G}$ *lies in the class* $\Omega_{\mathfrak{G}}^{(\exp)}$ if:

($\mathrm{I}_{\mathfrak{G}}^{(\exp)}$) $W(\lambda)$ is an entire function of exponential type of $\mu = 1/\lambda$;

($\mathrm{II}_{\mathfrak{G}}^{(\exp)}$) $$\lim_{\lambda\to\infty} \| W(\lambda) - E \| = 0;$$

($\mathrm{III}_{\mathfrak{G}}^{(\exp)}$) $$W^*(\lambda)\, W(\lambda) - E \geqslant 0 \quad (\operatorname{Im}\lambda > 0);$$

($\mathrm{IV}_{\mathfrak{G}}^{(\exp)}$) $$W^*(\lambda)\, W(\lambda) - E = 0 \quad (\operatorname{Im}\lambda = 0,\ \lambda \neq 0).$$

The type of the function $W(1/\mu)$ will be denoted by the symbol $\sigma[W]$.

Formula (6.1) shows that along with any two functions $W_1(\lambda)$ and $W_2(\lambda)$ the class $\Omega_{\mathfrak{G}}^{(\exp)}$ contains the function $W(\lambda) = W_1(\lambda)W_2(\lambda)$. Obviously,

$$\sigma[W_1W_2[\leqslant \sigma[W_1] + \sigma[W_2]. \tag{8.2}$$

Theorem 8.1. *If $W(\lambda) \in \Omega_{\mathfrak{G}}^{(\exp)}$, then for all $\lambda \neq 0$*

$$\|W(\lambda)\| \leqslant e^{\sigma\left|\operatorname{Im}\frac{1}{\lambda}\right|} \quad (\sigma = \sigma[W]). \tag{8.3}$$

Proof. We employ the following theorem of S. N. Bernšteĭn [1] (see also B. Ja. Levin [1]).

Suppose that $f(z)$ is an entire scalar function of exponential type σ. If $f(z)$ is bounded on the real axis:

$$|f(x)| \leqslant M \quad (-\infty < x < +\infty),$$

then $|(d/dx)f(x)| \leq \sigma M$, and accordingly

$$\left|\frac{d^n}{dx^n} f(x)\right| \leqslant \sigma^n M \qquad (-\infty < x < +\infty,\ n = 1, 2, \ldots).$$

For any $g, g' \in \mathfrak{G}$ the type of the scalar function $(W(1/\mu)g, g')$ is not greater than σ. Moreover, according to condition $(\mathrm{IV}_{\mathfrak{G}}^{(\exp)})$, it satisfies the inequality

$$\left|\left(W\left(\frac{1}{x}\right)g,\ g'\right)\right| \leqslant \|g\|\|g'\| \qquad (-\infty < x < +\infty)$$

on the real axis. Therefore

$$\left|\left(W\left(\frac{1}{\mu}\right)g,\ g'\right)\right| = \left|\left(W\left(\frac{1}{x}\right)g,\ g'\right) + iy\frac{d}{dx}\left(W\left(\frac{1}{x}\right)g,\ g'\right)\right.$$
$$\left. + \frac{(iy)^2}{2!}\frac{d^2}{dx^2}\left(W\left(\frac{1}{x}\right)g,\ g'\right) + \ldots\right| \leqslant e^{\sigma|y|}\|g\|\|g'\|$$
$$(\mu = x + iy),$$

i.e. $\|W(1/\mu)\| \le e^{\sigma|\mathrm{Im}\,\mu|}$.

Corollary. *If* $W(\lambda) \in \Omega_{\mathfrak{G}}^{(\exp)}$ *and* $\sigma[W] = 0$, *then* $W(\lambda) \equiv E$.

Indeed, from (8.3), for all μ the inequality

$$\left|\left(W\left(\frac{1}{\mu}\right)g,\ g'\right)\right| \leqslant \|g\|\|g'\|$$

holds. Applying the Liouville theorem, we find that $W(\lambda)$ does not depend on λ. In view of $(\mathrm{II}_{\mathfrak{G}}^{(\exp)})$ $W(\lambda) \equiv E$.

Lemma 8.1. *If* $W(\lambda) \in \Omega_{\mathfrak{G}}^{(\exp)}$, *then for each* $\lambda \neq 0$ *the operator* $W(\lambda)$ *has a bounded inverse, and*

$$W^{-1}(\lambda) = W^*(\bar{\lambda}).$$

Proof. From condition $(\mathrm{IV}_{\mathfrak{G}}^{(\exp)})$ we have

$$W^*(\bar{\lambda})\, W(\lambda) = E \qquad (\mathrm{Im}\,\lambda = 0,\ \lambda \neq 0).$$

Inasmuch as the left side of this equation is holomorphic in the entire extended complex plane with the deletion of the point zero, by the uniqueness theorem

$$W^*(\bar{\lambda})\, W(\lambda) = E \qquad (\lambda \neq 0).$$

In view of condition $(\mathrm{II}_{\mathfrak{G}}^{(\exp)})$ there exists a neighborhood of the point at infinity in which the operator $W^*(\bar{\lambda})$ is not only a left but also a right inverse for the operator $W(\lambda)$. Accordingly, again by the uniqueness theorem,

$$W(\lambda)\, W^*(\bar{\lambda}) = E \qquad (\lambda \neq 0).$$

Theorem 8.2. *Suppose that $W(\lambda) \in \Omega_{\mathfrak{G}}^{(\exp)}$. If the type of the function $W(1/\mu)$ is equal to σ, then $W(\lambda)$ is a regular left divisor of the function $e^{i\sigma/\lambda}E$.*

Proof. It is easy to see that the function

$$W_1(\lambda) = e^{\frac{i\sigma}{\lambda}} W^{-1}(\lambda)$$

satisfies conditions (I^-), (II^-) and (IV^-) (for $J = E$), introduced in §6. In view of (8.3)

$$W^*(\lambda)\, W(\lambda) \leqslant e^{2\sigma \left| \operatorname{Im} \frac{1}{\lambda} \right|} E,$$

so that

$$W^{-1*}(\lambda)\, W^{-1}(\lambda) \geqslant e^{-2\sigma \left| \operatorname{Im} \frac{1}{\lambda} \right|} E,$$

$$W_1^*(\lambda)\, W_1(\lambda) \geqslant e^{-2\sigma \operatorname{Im} \frac{1}{\lambda}} e^{-2\sigma \left| \operatorname{Im} \frac{1}{\lambda} \right|} E = E \qquad (\operatorname{Im} \lambda > 0).$$

Thus $W_1(\lambda)$ satisfies condition (III^-) as well. By Theorem 6.1 there exist simple nodes

$$\Theta = \begin{pmatrix} A & K & E \\ \mathfrak{H} & & \mathfrak{G} \end{pmatrix} \quad \text{and} \quad \Theta_1 = \begin{pmatrix} A_1 & K_1 & E \\ \mathfrak{H}_1 & & \mathfrak{G} \end{pmatrix}$$

such that in some neighborhood of the point at infinity $W_\Theta(\lambda) = W(\lambda)$ and $W_{\Theta_1}(\lambda) = W_1(\lambda)$. By Theorem 7.3 the product $\Theta\Theta_1$ is a simple node.

2. Universal operator and universal node. Consider in the space $L_2(0, l)$ the operator

$$\mathcal{J}_l f(x) = 2i \int_x^l f(t)\, dt. \tag{8.4}$$

The relations

$$\left.\begin{aligned} &\mathcal{J}_l^* f(x) = -2i \int_0^x f(t)\, dt, \qquad \frac{\mathcal{J}_l - \mathcal{J}_l^*}{2i} f(x) \equiv \int_0^l f(t)\, dt, \\ &\left(\frac{\mathcal{J}_l - \mathcal{J}_l^*}{2i} f(x),\ f(x) \right) = \left| \int_0^l f(t)\, dt \right|^2 \geqslant 0 \end{aligned}\right\} \tag{8.5}$$

show that it is dissipative and that it has a one-dimensional imaginary component, mapping $L_2(0, l)$ into the collection of functions which are constant on $(0, l)$.

Since the series

$$\sum_{n=0}^{\infty} \frac{1}{\lambda^{n+1}} \mathcal{J}_l^n f(x)$$

$$= \frac{1}{\lambda} f(x) + \frac{2i}{\lambda^2} \left[\int_x^l f(t)\,dt + \frac{2i}{\lambda} \int_x^l (t-x) f(t)\,dt + \left(\frac{2i}{\lambda}\right)^2 \int_x^l \frac{(t-x)^2}{2!} f(t)\,dt + \ldots \right]$$

converges uniformly for any fixed $\lambda \neq 0$, the operator $\mathcal{J}_l$ has no spectrum points other than zero, and its resolvent is defined by the formula

$$(\mathcal{J}_l - \lambda E)^{-1} f(x) = -\frac{1}{\lambda} f(x) - \frac{2i}{\lambda^2} \int_x^l e^{\frac{2i}{\lambda}(t-x)} f(t)\,dt. \tag{8.6}$$

From the inequality

$$\left| \int_x^l e^{\frac{2i}{\lambda}(t-x)} f(t)\,dt \right|^2 \leqslant \int_x^l \left| e^{\frac{2i}{\lambda}(t-x)} \right|^2 dt \int_x^l |f(t)|^2 dt \leqslant l e^{\frac{4l}{|\lambda|}} \| f(t) \|^2$$

there follows the estimate

$$\|(\mathcal{J}_l - \lambda E)^{-1}\| \leqslant \frac{1}{|\lambda|} + \frac{2l}{|\lambda|^2} e^{\frac{2l}{|\lambda|}}, \tag{8.7}$$

from which it is clear that $(\mathcal{J}_l - (1/\mu) E)^{-1}$ is a function of exponential type (its type is not higher than $2l$). On the other hand,

$$\left(\left(\mathcal{J}_l - \frac{1}{\mu} E \right)^{-1} h(x),\ h(x) \right) = \frac{1}{2il} \left(1 - e^{2il\mu}\right) \qquad (h(x) \equiv l^{-1/2}), \tag{8.8}$$

so that the function $(\mathcal{J}_l - (1/\mu) E)^{-1}$ is equal to $2l$.

Denote by $L_2^{(r)}(0, l)$ $(r \leq \infty)$ the Hilbert space consisting of the one-rowed matrices

$$f(x) = \| f^{(1)}(x) \quad f^{(2)}(x) \ldots f^{(r)}(x) \|, \tag{8.9}$$

where $f^{(j)}(x)$ is a function measurable on $(0, l)$ and satisfying the condition

$$\int_0^l f(x) f^*(x)\,dx = \sum_{j=1}^{r} \int_0^l |f^{(j)}(x)|^2\,dx < \infty. \tag{8.10}$$

The scalar product in $L_2^{(r)}(0, l)$ is defined by the formula

$$(f, g) = \int_0^l f(x) g^*(x)\,dx = \sum_{j=1}^{r} \int_0^l f^{(j)}(x) \overline{g^{(j)}(x)}\,dx. \tag{8.11}$$

The space $L_2^{(r)}(0, l)$ for $r = \infty$ will be denoted by $\widetilde{L}_2(0, l)$.

Suppose that $\widetilde{\mathcal{J}}_l$ is an operator in $\widetilde{L}_2(0, l)$, assigning to the matrix

$$f(x) = \| f^{(1)}(x) \quad f^{(2)}(x) \ldots \|$$

the matrix

$$\widetilde{\mathcal{J}}_l f(x) = \| \mathcal{J}_l f^{(1)}(x) \quad \mathcal{J}_l f^{(2)}(x) \ldots \|.$$

The space $\widetilde{L}_2(0, l)$ is an orthogonal sum of a countable set of subspaces invariant relative to $\widetilde{\mathcal{J}}_l$, in each of which there is induced an operator $\mathcal{J}_l$. Therefore it easily follows that $\widetilde{\mathcal{J}}_l$ is a dissipative operator having no spectrum points other than zero, and that $(\widetilde{\mathcal{J}}_l - (1/\mu) E)^{-1}$ is a function of exponential type $2l$.

We imbed the operator $\widetilde{\mathcal{J}}_l$ in a dissipative node. To this end we construct in $\widetilde{L}_2(0, l)$ an orthonormalized sequence

$$h_j = \| \underbrace{0 \ldots 0\, l^{-1/2}}_{j} 0 \ldots \| \quad (j = 1, 2, \ldots)$$

and consider in the space l_2 the orthonormalized basis

$$g_j = (\underbrace{0, \ldots, 0, 1}_{j}, 0, \ldots) \quad (j = 1, 2, \ldots).$$

The condition $\widetilde{K}_l g_j = l^{1/2} h_j$ uniquely defines a bounded linear operator $\widetilde{K}_l$ operating from l_2 into $\widetilde{L}_2(0, l)$. Since obviously $(1/2i)(\widetilde{\mathcal{J}}_l - \widetilde{\mathcal{J}}_l^*) = \widetilde{K}_l \widetilde{K}_l^*$, we see that

$$\widetilde{\Theta}_l = \begin{pmatrix} \widetilde{\mathcal{J}}_l & \widetilde{K}_l & E \\ \widetilde{L}_2(0, l) & & l_2 \end{pmatrix}$$

is an operator node.

The linear envelope of functions $\mathcal{J}_l^n h(x)$ $(n = 0, 1, \cdots; h(x) \equiv \text{const})$ contains all polynomials, and accordingly is dense in $L_2(0, l)$. Therefore $\widetilde{\Theta}_l$ is a simple node.

According to formula (8.8)

$$((\widetilde{\mathcal{J}}_l - \lambda E)^{-1} h_j, h_k) = \begin{cases} \dfrac{1}{2il}\left(1 - e^{\frac{2il}{\lambda}}\right) & (j = k), \\ 0 & (j \neq k). \end{cases} \tag{8.12}$$

Hence

$$\left(W_{\tilde{\Theta}_l}(\lambda) g_j, g_k\right) = (g_j, g_k) - 2il\left((\tilde{\mathfrak{J}}_l - \lambda E)^{-1} h_j, h_k\right) = \begin{cases} e^{\frac{2il}{\lambda}} & (j = k), \\ 0 & (j \neq k). \end{cases}$$

Thus

$$W_{\tilde{\Theta}_l}(\lambda) = e^{\frac{2il}{\lambda} E} = e^{\frac{2il}{\lambda}} E. \tag{8.13}$$

The operator $\tilde{\mathfrak{J}}_l$ and the node $\tilde{\Theta}_l$ will be said to be *universal.*[1)]

3. **Operators of the class** $\Lambda^{(\exp)}$. An operator A will be assigned to the class $\Lambda^{(\exp)}$ if it satisfies the following conditions: 1) $A_I \geq 0$, 2) A has no spectrum point other than zero, 3) $(A - (1/\mu)E)^{-1}$ is a function of exponential type. The type of the function $(A - (1/\mu)E)^{-1}$ will be denoted by $\sigma[A]$.

Lemma 8.2. *If the COF of the simple node*

$$\Theta = \begin{pmatrix} A & K & E \\ \mathfrak{H} & & l_2 \end{pmatrix}$$

coincides in the neighborhood of the point at infinity with some function $W(\lambda)$ *of class* $\Omega_{l_2}^{(\exp)}$, *then the node* Θ *is unitarily equivalent to the projection of the universal node*

$$\tilde{\Theta}_l = \begin{pmatrix} \tilde{\mathfrak{J}}_l & \tilde{K}_l & E \\ L_2(0, l) & & l_2 \end{pmatrix} \quad \left(l = \frac{1}{2}\sigma[W]\right)$$

onto one of the invariant subspaces of the operator $\tilde{\mathfrak{J}}_l$.

Proof. According to Theorem 8.2 $W(\lambda) \ll e^{i\sigma/\lambda} E$ $(\sigma = \sigma[W])$. On the other hand, as was shown in the preceding subsection, the function $e^{i\sigma/\lambda} E$ is characteristic for the node $\tilde{\Theta}_l$. On the basis of Theorem 5.4 $W(\lambda)$ coincides with the COF of the projection of the node $\tilde{\Theta}_l$ onto some subspace invariant relative to $\tilde{\mathfrak{J}}_l$. To complete the proof it remains to refer to Theorem 3.2.

Theorem 8.3. *If the basic operator of the node*

$$\Theta = \begin{pmatrix} A & K & E \\ \mathfrak{H} & & \mathfrak{G} \end{pmatrix}$$

lies in the class $\Lambda^{(\exp)}$, *then* $W_\Theta(\lambda) \in \Omega_{\mathfrak{G}}^{(\exp)}$. *Conversely, each function* $W(\lambda) \in \Omega_{\mathfrak{G}}^{(\exp)}$ *is characteristic for some simple node*

---

1) The introduction of the term "universal" is justified by a study of the operators generated in the invariant subspaces of the operator $\tilde{\mathfrak{J}}_l$ (see Theorem 8.4).

$$\Theta = \begin{pmatrix} A & K & E \\ \mathfrak{H} & & \mathfrak{G} \end{pmatrix}, \text{ where } A \in \Lambda^{(\exp)},\ \sigma[A] = \sigma[W].$$

Proof. The first assertion is obvious. If $W(\lambda) \in \Omega_{\mathfrak{G}}^{(\exp)}$, then by Theorem 6.1 there exists a simple node Θ such that in some neighborhood of the point at infinity $W_\Theta(\lambda) = W(\lambda)$. Suppose that U is an isometric mapping of the space $\mathfrak{G}$ into the portion $l_2^{(0)}$ of the space l_2. Putting

$$K_1 g = \begin{cases} KU^{-1}g & (g \in l_2^{(0)}), \\ 0 & (g \perp l_2^{(0)}), \end{cases}$$

we obtain a simple node

$$\Theta_1 = \begin{pmatrix} A & K_1 & E \\ \mathfrak{H} & & l_2 \end{pmatrix}$$

with COF

$$W_\Theta(\lambda) g = \begin{cases} UW_\Theta(\lambda)U^{-1}g & (g \in l_2^{(0)}), \\ g & (g \perp l_2^{(0)}). \end{cases}$$

Construct the function

$$W_1(\lambda) g = \begin{cases} UW(\lambda)U^{-1}g & (g \in l_2^{(0)}), \\ g & (g \perp l_2^{(0)}). \end{cases}$$

Since $W_1(\lambda)$ lies in the class $\Omega_{l_2}^{(\exp)}$ and coincides in the neighborhood of the point at infinity with $W_{\Theta_1}(\lambda)$, by Lemma 8.2 there exists an invariant subspace of the universal operator

$$\tilde{\mathcal{J}}_l \left(l = \frac{1}{2}\sigma[W_1] = \frac{1}{2}\sigma[W]\right),$$

such that the operator induced in it is unitarily equivalent to A. Hence $A \in \Lambda^{(\exp)}$.

By the definition of the COF it follows that $\sigma[W] \leq \sigma[A]$. At the same time

$$\left\| \left(A - \frac{1}{\mu}E\right)^{-1} \right\| \leqslant \left\| \left(\tilde{\mathcal{J}}_l - \frac{1}{\mu}E\right)^{-1} \right\|,$$

and therefore $\sigma[A] \leq \sigma[W]$.

Corollary. *Suppose that A is a completely nonselfadjoint operator of the class $\Lambda^{(\exp)}$ operating in the space $\mathfrak{H}$. If $\sigma[A] = 0$, then $\mathfrak{H} = 0$.*

Theorem 8.4. *If A is a completely nonselfadjoint operator of the class $\Lambda^{(\exp)}$ then there exists a subspace invariant relative to the universal operator $\mathcal{J}_l$ $(l = \sigma[A]/2)$, in which an operator unitarily equivalent to A is induced.*

Proof. We imbed A in a dissipative node

$$\Theta=\begin{pmatrix} A & K & E \\ \mathfrak{H} & & l_2 \end{pmatrix}.$$

By Theorem 8.3 $W_\Theta(l) \in \Omega_{l_2}^{(\exp)}$ and the type of the function $W_\Theta(1/\mu)$ is equal to $\sigma[a]$. The theorem now follows directly from Lemma 8.2.

§9. Quasi-Hermitian nodes

1. Quasi-Hermitian operators. A bounded linear operator is said to be quasi-Hermitian if its imaginary component is completely continuous. An operator is quasi-Hermitian if and only if it is representable in the form of the sum of a self-adjoint operator and a completely continuous operator.

Theorem 9.1. *A limit point of the spectrum of a quasi-Hermitian operator A lies in the spectrum of its real component. Each nonreal point of the spectrum of A is an eigenvalue of finite multiplicity.*

Proof. Suppose that $\rho > \|A\|$. Since $\|A_R\| \le \|A\|$, the spectrum of A_R lies in the disk $|\lambda| < \rho$. Excluding from this disk the spectrum of A_R, we obtain a region G in which the function $(A_R - \lambda E)^{-1}$ is holomorphic. The obvious equation

$$A-\lambda E=(A_R-\lambda E)\left[E+i\,(A_R-\lambda E)^{-1} A_I\right]$$

and Theorem 1 of Appendix I, applied to the function

$$K(\lambda)=i\,(A_R-\lambda E)^{-1} A_I,$$

show that the spectrum of A has no limit points in G. This proves the first assertion of the theorem.

Suppose that λ_0 lies in the nonreal spectrum of A. By the same Theorem 1 of Appendix I there exists a neighborhood of the point λ_0 in which the decomposition

$$(A-\lambda E)^{-1}=\frac{A_{-n}}{(\lambda-\lambda_0)^n}+\ \dots\ +\frac{A_{-1}}{\lambda-\lambda_0}+A_0+(\lambda-\lambda_0)\,A_1+\ \dots$$

holds, where $A_{-n},\cdots,A_{-1}$ are finite-dimensional operators. By a theorem of F. Riesz (see Riesz and Sz.-Nagy [1] the space $\mathfrak{H}$ in which the operator A is given is uniquely representable in the form of a direct sum of subspaces $\mathfrak{H}_1$ and $\mathfrak{H}_2$ invariant relative to A and such that in $\mathfrak{H}_1$ an operator is induced with λ_0 the only point of its spectrum, and in $\mathfrak{H}_2$ an operator whose spectrum does not contain λ_0. Here the operator P_1 of projection onto $\mathfrak{H}_1$ parallel to $\mathfrak{H}_2$ is calculated according to the formula

$$P_1 = -\frac{1}{2\pi i}\int_\gamma (A-\lambda E)^{-1}\,d\lambda,$$

where γ is a sufficiently small positive oriented circumference with center at λ_0. In the case at hand

$$P_1 = -\frac{1}{2\pi i}\int_\gamma \frac{A_{-1}}{\lambda-\lambda_0}\,d\lambda = -A_{-1},$$

so that $\mathfrak{H}_1$ is finite-dimensional. Accordingly there exists a vector $h \in \mathfrak{H}$ $(h \neq 0)$ such that $Ah = \lambda_0 h$. Since each eigenvector corresponding to the number λ_0 lies in $\mathfrak{H}_1$, the multiplicity of the eigenvalue λ_0 is finite.

2. **Quasi-Hermitian nodes.** The node

$$\Theta = \begin{pmatrix} A & K & J \\ \mathfrak{H} & & \mathfrak{G} \end{pmatrix}$$

is said to be *quasi-Hermitian* if the operator K is completely continuous. Since $A_I = KJK^*$, the basic operator of a quasi-Hermitian node is quasi-Hermitian. Generally speaking the converse assertion is false. However, *every quasi-Hermitian operator may be imbedded in a simple quasi-Hermitian node.* To verify this, put $\mathfrak{R} = \mathfrak{H}$ in the proof of Theorem 1.1 and choose the operator K so that it is completely continuous.

The product of quasi-Hermitian nodes, and also projections of quasi-Hermitian nodes onto arbitrary subspaces, are again quasi-Hermitian nodes.

Theorem 9.2. *The spectrum of the product of two quasi-Hermitian nodes is the sum of the spectra of the factors.*

Proof. By Theorem 9.1 the set of regular points of a quasi-Hermitian operator is connected. The assertion accordingly follows from Theorem 2.5.

3. **Characteristic operator-functions of quasi-Hermitian nodes.**

Lemma 9.1. *If*

$$\Theta = \begin{pmatrix} A & K & J \\ \mathfrak{H} & & \mathfrak{G} \end{pmatrix}$$

is a simple node and λ_0 *is a point of the spectrum of the operator* A_R, *then there exists no disk* $|\lambda - \lambda_0| < \rho$ *in which it is possible to give a holomorphic function coinciding at nonreal points with the function* $V_\Theta(\lambda) = K^*(A_R - \lambda E)^{-1}K$.

Proof. We suppose the contrary. Using the formula

$$V_\Theta(\lambda) = \int_{-\infty}^{+\infty} \frac{d\,(K^* E(s) K)}{s-\lambda},$$

where $E(s)$ is the orthogonal resolution of unity corresponding to the operator A_R, and repeating the arguments applied in the proof of Theorem 4.9, we get

$$K^* \Delta E K = 0 \quad (\Delta E = E(s'') - E(s'),\ \ \lambda_0 - \rho < s' < s'' < \lambda_0 + \rho).$$

Thus $\Delta E K = 0$, and inasmuch as

$$\Delta E A_R = A_R \Delta E, \quad \Delta E A_I = \Delta E K J K^* = 0,$$

we have

$$\Delta E A^n K = \Delta E (A_R + i A_I)^n K = 0 \qquad (n = 0,\ 1,\ \ldots). \tag{9.1}$$

In view of the simplicity of the node Θ and equations (9.1), the function $E(s)$ has a constant value on the interval $(\lambda_0 - \rho, \lambda_0 + \rho)$. Accordingly, λ_0 is a regular point of the operator A_R.

Theorem 9.3. *Suppose that Θ is a simple quasi-Hermitian node and that G_A is the set of regular points of the operator A. If the point λ_0 lies in the spectrum of A, then there does not exist a disk $|\lambda - \lambda_0| < \rho$ in which it is possible to give a holomorphic function coinciding with $W_\Theta(\lambda)$ at the points $\lambda \in G_A$.*

Proof. Suppose that in the disk $|\lambda - \lambda_0| < \rho$ there is defined a holomorphic function $\widetilde{W}(\lambda)$ such that $\widetilde{W}(\lambda) = W_\Theta(\lambda)$ at the points λ of that disk which lie in G_A. By Theorem 9.1 the set G_A is dense in the complex plane. Since the operators $W_\Theta(\lambda) - E$ $(\lambda \in G_A)$ are completely continuous, the operators $\widetilde{W}(\lambda) - E$ are also all completely continuous.

We shall show that in the disk $|\lambda - \lambda_0| < \rho$ there exists a contour γ enclosing λ_0, all of whose points are regular for the operator A. This assertion follows immediately from Theorem 9.1 if λ_0 is a nonreal point. In the case of real λ_0, on applying Theorem 1 of Appendix II to the function

$$\widetilde{W}(\lambda) + E = 2(E + \widetilde{K}(\lambda)) \quad \left(\widetilde{K}(\lambda) = \tfrac{1}{2}(\widetilde{W}(\lambda) - E)\right),$$

we find a disk

$$|\lambda - \lambda_1| < \rho_1, \quad |\lambda - \lambda_2| < \rho_2,$$
$$(\lambda_0 - \rho < \lambda_1 - \rho_1 < \lambda_1 + \rho_1 < \lambda_0 < \lambda_2 - \rho_2 < \lambda_2 + \rho_2 < \lambda_0 + \rho),$$

inside which the function

$$\widetilde{V}(\lambda) = i(\widetilde{W}(\lambda) + E)^{-1}(\widetilde{W}(\lambda) - E) J$$

is holomorphic. By formula (3.22) and Lemma 9.1 the intervals $\Delta_1 = (\lambda_1 - \rho_1, \lambda_1 + \rho)$ and $\Delta_2 = (\lambda_2 - \rho_2, \lambda_2 + \rho_2)$ do not contain points of the spectrum of the operator A_R, and in view of Theorem 9.1 there exist intervals $\Delta'_1 \subset \Delta_1$ and $\Delta'_2 \subset \Delta_2$ in which there are no points of the spectrum of A. The required contour γ may now be constructed by passing it through any two points chosen in the intervals Δ'_1 and Δ'_2.

Denote by $\mathfrak{H}'$ and $\mathfrak{H}''$ the linear envelopes of the sets of vectors of the form

$$A^n Kg \ (n = 0, 1, \ldots; \ g \in \mathfrak{G}), \quad A^{*n} Kg \ (n = 0, 1, \ldots; \ g \in \mathfrak{G}).$$

(The reader will note in passing that $\mathfrak{H}' = \mathfrak{H}''$.) The equations

$$K^* (A - \lambda E)^{-1} K = \frac{i}{2} (W_\Theta(\lambda) - E) J$$

and

$$\begin{aligned} &A^m (A - \lambda E)^{-1} A^n \\ &\quad = A^{m+n-1} + \lambda A^{m+n-2} + \ \ldots \ + \lambda^{m+n-1} E + \lambda^{m+n} (A - \lambda E)^{-1} \\ &\qquad (m + n > 0) \end{aligned}$$

show that the function $((A - \lambda E)^{-1} h', h'')$, for any $h' \in \mathfrak{H}'$ and $h'' \in \mathfrak{H}''$, has no singularities in the disk $|\lambda - \lambda_0| < \rho$. Hence

$$\int_\gamma ((A - \lambda E)^{-1} h', h'') \, d\lambda = 0.$$

On the other hand, in view of the continuity of the resolvent $(A - \lambda E)^{-1}$, the integral $\int_\gamma (A - \lambda E)^{-1} d\lambda$ exists on the contour γ, and

$$\left(\int_\gamma (A - \lambda E)^{-1} d\lambda h', h'' \right) = \int_\gamma ((A - \lambda E)^{-1} h', h'') \, d\lambda.$$

Since the closure of each of the sets $\mathfrak{H}'$ and $\mathfrak{H}''$ is the entire space $\mathfrak{H}$, we have $\int_\gamma (A - \lambda E)^{-1} d\lambda = 0$, and, according to a theorem of F. Riesz, there are no points of the spectrum of A within γ.

The theorem is proved.

Suppose that J is some linear operator operating in a separable Hilbert space $\mathfrak{G}$ and satisfying the conditions $J = J^*$ and $J^2 = E$. The operator-function of a complex variable $W(\lambda)$, whose values are bounded linear operators in $\mathfrak{G}$, will be assigned to the class $\Omega_J^{(q)}$ if:

($\mathrm{I}^{(q)}$) $W(\lambda)$ is holomorphic in a region G_W gotten by deleting from the extended complex plane some bounded set not having nonreal limit points;

$(\mathrm{II}^{(q)})$ $$\lim_{\lambda\to\infty} \| W(\lambda) - E \| = 0;$$

$(\mathrm{III}^{(q)})$ $$W^*(\lambda) J W(\lambda) - J \geqslant 0 \quad (\operatorname{Im}\lambda > 0,\ \lambda \in G_W);$$

$(\mathrm{IV}^{(q)})$ $$W^*(\lambda) J W(\lambda) - J = 0 \quad (\operatorname{Im}\lambda = 0,\ \lambda \in G_W);$$

$(\mathrm{V}^{(q)})$ All the operators $W(\lambda) - E$ $(\lambda \in G_W)$ are completely continuous.

From (6.1) and the equation

$$W_1(\lambda) W_2(\lambda) - E$$
$$= (W_1(\lambda) - E)(W_2(\lambda) - E) + (W_1(\lambda) - E) + (W_2(\lambda) - E), \tag{9.2}$$

it follows easily that the class $\Omega_J^{(q)}$ contains the product of any two functions $W_1(\lambda)$ and $W_2(\lambda)$ which it contains.

Obviously the COF of the quasi-Hermitian node

$$\Theta = \begin{pmatrix} A & K & J \\ \mathfrak{H} & & \mathfrak{G} \end{pmatrix}$$

lies in the class $\Omega_J^{(q)}$.

We will show that $\Omega_J^{(q)} \subset \Omega_J$. To this end we consider a function $W(\lambda) \in \Omega_J^{(q)}$ and denote by $G_W^{(0)}$ the set of those points of the region G_W at which the operator $W(\lambda) + E$ has a bounded inverse defined on the entire space $\mathfrak{G}$. The set $G_W^{(0)}$ is not empty: it follows from $(\mathrm{II}^{(q)})$ that it contains some neighborhood of the point at infinity. Since

$$W(\lambda) + E = 2(E + K(\lambda)),$$

where $K(\lambda) = (W(\lambda) - E)/2$ is a function holomorphic in G_W and taking on completely continuous values, by Theorem 1 of Appendix II the difference of the sets G_W and $G_W^{(0)}$ has no limit points in G_W.

Introduce the function

$$V(\lambda) = i(W(\lambda) + E)^{-1}(W(\lambda) - E)J \qquad (\lambda \in G_W^{(0)}).$$

Inasmuch as

$$\frac{V(\lambda) - V^*(\lambda)}{2i} = J(W^*(\lambda) + E)^{-1}(W^*(\lambda) J W(\lambda) - J)(W(\lambda) + E)^{-1} J,$$

we have

$$\frac{V(\lambda) - V^*(\lambda)}{2i} \geqslant 0 \quad (\operatorname{Im}\lambda > 0,\ \lambda \in G_W^{(0)}), \tag{9.3}$$

$$V(\lambda) = V^*(\lambda) \quad (\operatorname{Im}\lambda = 0,\ \lambda \in G_W^{(0)}). \tag{9.4}$$

Suppose that in the upper halfplane $V(\lambda)$ has an isolated singular point λ_0. Then for an appropriate vector $g_0 \in \mathfrak{G}$ the point λ_0 will be an isolated singular point for the function $(V(\lambda)g_0, g_0)$ as well. Accordingly the function $\mathrm{Im}(V(\lambda)g_0, g_0)$ will take on values of different signs in any neighborhood of the point λ_0, which contradicts inequality (9.3). Taking account of the properties of the sets G_W and $G_W^{(0)}$ indicated above, we arrive at the conclusion that $V(\lambda)$ is analytically continuable into the entire upper halfplane. In view of the symmetry principle and equation (9.4) it may also be analytically continued into the entire lower halfplane.

Theorem 9.4. *If $W(\lambda) \in \Omega_J^{(q)}$, then there exists a simple quasi-Hermitian node Θ such that $G_A \supseteq G_W$ and $W_\Theta(\lambda) = W(\lambda)$ $(\lambda \in G_W)$.*

Proof. Since $\Omega_J^{(q)} \subset \Omega_J$, from Theorem 5.2 it follows that there exists a simple node Θ for which $W_\Theta(\lambda) = W(\lambda)$ for all λ lying in some neighborhood of the point at infinity. In view of conditions $(\mathrm{II}^{(q)})$ and $(\mathrm{V}^{(q)})$ the function $\lambda(W(\lambda) - E)$ tends as $\lambda \to \infty$ in the sense of the uniform norm to some completely continuous operator. Denoting it by W_1 and taking account of the equation

$$\lambda(W(\lambda) - E) = -2i\lambda K^*(A - \lambda E)^{-1}KJ,$$

we get

$$W_1 = 2iK^*KJ.$$

Thus K is a completely continuous operator, which means that Θ is a quasi-Hermitian node. By the preceding theorem $G_A \supseteq G_W$, so that the equation $W_\Theta(\lambda) = W(\lambda)$ is valid in the entire region G_W.

§10. Volterra nodes

1. **Volterra operators.** A completely continuous operator is said to be a *Volterra operator* if its spectrum consists of only the single point zero.

It is useful to note that *if the operator A is a Volterra operator and if it is dissipative, then either the space $\mathfrak{H}$ on which it operates is infinite dimensional or else $A = 0$.* Indeed, if $\mathfrak{H}$ is finite dimensional, then $\mathrm{tr}\, A = \mathrm{tr}\, A_R + i\, \mathrm{tr}\, A_I = 0$, so that $\mathrm{tr}\, A_I = 0$, which along with the dissipativeness of A leads to the equation $A_I = 0$. The operator A, being at the same time a Volterra and a selfadjoint operator, is also equal to zero.

Theorem 10.1. *Suppose that the spectrum of the bounded linear operator A,* *operating in the Hilbert space* $\mathfrak{H}$, *does not contain points other than zero. If its imaginary component is completely continuous, then* A *is a Volterra operator.*

Proof. It suffices to prove that the operator A_R is completely continuous. Application of Theorem 1 of Appendix II to the function $K(\lambda) = -i(A - \lambda E)^{-1} A_I$ shows that the set of points other than zero at which the operator $E + K(\lambda)$ does not have a bounded inverse is a sequence $\lambda_1, \lambda_2, \cdots$ which either is finite or converges to zero. Since

$$A_R - \lambda E = (A - \lambda E)(E - i(A - \lambda E)^{-1} A_I) = (A - \lambda E)(E + K(\lambda)),$$

the spectrum of A_R consists of the points λ_j and possibly the point zero. Thus

$$\mathfrak{H} = \mathfrak{H}_0 \oplus \mathfrak{H}_1 \oplus \mathfrak{H}_2 \oplus \ldots,$$

where $A_R \mathfrak{H}_0 = 0$, $A_R h = \lambda_j h$ ($h \in \mathfrak{H}_j$, $j = 1, 2, \cdots$). It remains to verify that $\dim \mathfrak{H}_j < \infty$ ($j = 1, 2, \cdots$). But if $h \in \mathfrak{H}_j$ ($j \neq 0$), then

$$(E + K(\lambda_j))h = (A - \lambda_j E)^{-1}(A_R - \lambda_j E)h = 0,$$

i.e. $K(\lambda_j)h = -h$. Inasmuch as the eigenvectors of a completely continuous operator corresponding to a given nonzero eigenvalue form a finite-dimensional subspace, all the $\mathfrak{H}_j$, $j = 1, 2, \cdots$, are finite dimensional.

2. **Volterra nodes.** The node

$$\Theta = \begin{pmatrix} A & K & J \\ \mathfrak{H} & & \mathfrak{G} \end{pmatrix}$$

is said to be a *Volterra node* if the operator A is a Volterra operator and if K is completely continuous. Obviously, *the projection of a Volterra node onto any subspace which is semi-invariant relative to a basic operator is again a Volterra node.* Using Theorem 9.2, it is easy to show that *the product of two nodes is a Volterra node if and only if both factors are Volterra nodes.*

Theorem 10.2. *For the Volterra node* Θ *to be an excess node, it is necessary and sufficient that there exist a vector* $h \in \mathfrak{H}$ ($h \neq 0$) *orthogonal to* $\mathfrak{R}(K)$ *and annihilating the operator* A.

Proof. If the node Θ is excess, then the excess subspace $\mathfrak{E}_\Theta^{(0)}$ is distinct from zero, orthogonal to $\mathfrak{R}(K)$ and annihilates the operator A. Conversely, suppose that $Ah = 0$ ($h \neq 0$ and that $h \perp \mathfrak{R}(K)$. Then

$$(h, A^{*n}Kg) = (A^n h, Kg) = 0 \quad (n = 0, 1, \ldots; \ g \in \mathfrak{G})$$

so that $h \perp \mathfrak{G}_{\Theta^*}$. Since $\mathfrak{G}_{\Theta} = \mathfrak{G}_{\Theta^*}$, we have $h \perp \mathfrak{G}_{\Theta}$, i.e. Θ is an excess node.

Theorem 10.3. *For the dissipative Volterra node Θ to be excess, it is necessary and sufficient that there exist a vector $h \neq 0$ which annihilates the operator A.*

Since the simplicity of a dissipative node is equivalent to the complete nonselfadjointness of its basic operator, the assertion of the theorem follows from the remark preceding Theorem 7.1.

Theorem 10.4. *For the product Θ of two simple Volterra nodes*

$$\Theta_1 = \begin{pmatrix} A_1 & K_1 & J \\ \mathfrak{H}_1 & & \mathfrak{G} \end{pmatrix} \quad \text{and} \quad \Theta_2 = \begin{pmatrix} A_2 & K_2 & J \\ \mathfrak{H}_2 & & \mathfrak{G} \end{pmatrix}$$

to be excess, it is necessary and sufficient that $\mathfrak{H}_1$ and $\mathfrak{H}_2$ should contain one-dimensional subspaces $\mathfrak{L}_1$ and $\mathfrak{L}_2$ respectively, satisfying the conditions

$$A_1^*\mathfrak{L}_1 = 0, \quad A_2\mathfrak{L}_2 = 0, \quad K_1^*\mathfrak{L}_1 = K_2^*\mathfrak{L}_2. \tag{10.1}$$

Proof. If the node Θ is excess, then for any one-dimensional subspace $\mathfrak{L} \in \mathfrak{G}_{\Theta}^{(0)}$ the equations $A\mathfrak{L} = A^*\mathfrak{L} = 0$ hold. By Lemma 2.1 $\mathfrak{L}$ cannot lie in $\mathfrak{H}_1$ or $\mathfrak{H}_2$, so that the subspaces $\mathfrak{L}_1 = P_1\mathfrak{L}$ and $\mathfrak{L}_2 = P_2\mathfrak{L}$, where P_j is the orthoprojector onto $\mathfrak{H}_j$ $(j = 1, 2)$, are one-dimensional. Since $P_1A^* = A_1^*P_1$ and $P_2A = A_2P_2$, we have $A_1^*\mathfrak{L}_1 = A_2\mathfrak{L}_2 = 0$. Moreover,

$$(K_1^*P_1 + K_2^*P_2)\,\mathfrak{L} = K^*\mathfrak{L} = 0$$

so that $K_1^*\mathfrak{L}_1 = K_2^*\mathfrak{L}_2$.

Conversely, suppose that the one-dimensional subspaces $\mathfrak{L}_1 \subset \mathfrak{H}_1$ and $\mathfrak{L}_2 \subset \mathfrak{H}_2$ satisfy conditions (10.1). Then the operator $K^* = K_1^*P_1 + K_2^*P_2$ maps the two-dimensional subspace $\mathfrak{L}_1 \oplus \mathfrak{L}_2$ into a one-dimensional or zero-dimensional subspace, and therefore there exists a vector $h \in \mathfrak{L}_1 \oplus \mathfrak{L}_2$ which is nonzero and satisfies $K^*h = 0$. In view of the last equation

$$h \perp \mathfrak{R}(K),$$

$$K_1JK_2^*P_2h = -K_1JK_1^*P_1h = \frac{A_1^* - A_1}{2i}P_1h,$$

$$Ah = A_1P_1h + A_2P_2h + 2iK_1JK_2^*P_2h = A_1^*P_1h + A_2P_2h = 0$$

and it remains to apply Theorem 10.2.

Theorem 10.5. *The product of two simple dissipative Volterra nodes is a simple node.*

The proof follows easily from Theorems 10.3 and 10.4.

We note that Theorem 10.5 is a special case of Theorem 7.3.

3. **Characteristic operator-functions of Volterra nodes.** Suppose that J as before is an operator operating in the separable Hilbert space $\mathfrak{G}$ and satisfying the conditions $J = J^*$ and $J^2 = E$. The function $W(\lambda)$, whose values are bounded linear operators in $\mathfrak{G}$, will be assigned to the class $\Omega_J^{(0)}$ if:

($\mathrm{I}^{(0)}$) $W(\lambda)$ is an entire function of $\mu = 1/\lambda$;

($\mathrm{II}^{(0)}$)
$$\lim_{\lambda\to\infty} \| W(\lambda) - E \| = 0;$$

($\mathrm{III}^{(0)}$)
$$W^*(\lambda)\, JW(\lambda) - J \geqslant 0 \quad (\operatorname{Im}\lambda > 0);$$

($\mathrm{IV}^{(0)}$)
$$W^*(\lambda)\, JW(\lambda) - J = 0 \quad (\operatorname{Im}\lambda = 0,\ \lambda \neq 0);$$

($\mathrm{V}^{(0)}$) all the operators $W(\lambda) - E$ $(\gamma \neq 0)$ are completely continuous.

The COF of each Volterra node Θ obviously lies in $\Omega_J^{(0)}$.

Theorem 10.6. *If the function $W(\lambda)$ lies in the class $\Omega_J^{(0)}$, then there exists a simple Volterra node Θ for which it is characteristic.*

Proof. By Theorem 9.4 there exists a simple quasi-Hermitian node Θ such that A has no spectrum points other than zero and $W_\Theta(\lambda) = W(\lambda)$ $(\lambda \neq 0)$. In view of formula (1.2) and Theorem 10.1, A is a Volterra operator.

§11. Finite-dimensional nodes

1. **Characteristic operator-functions of finite-dimensional nodes.** The node
$$\Theta = \begin{pmatrix} A & K & J \\ \mathfrak{H} & & \mathfrak{G} \end{pmatrix}$$
is said to be finite dimensional if the spaces $\mathfrak{H}$ and $\mathfrak{G}$ are finite dimensional. Each operator A operating in a finite-dimensional space may be imbedded, as is obvious, in a simple finite-dimensional node.

Suppose that $\mathfrak{G}$ is a finite-dimensional Hilbert space and J is a linear operator in $\mathfrak{G}$ satisfying the conditions $J = J^*$ and $J^2 = E$. We will say that the function $W(\lambda)$ with values which are linear operators in $\mathfrak{G}$ lies in the class $\Omega_J^{(F)}$, if:

($\mathrm{I}^{(F)}$) $W(\lambda)$ is holomorphic in a region G_W which is gotten from the entire extended complex plane by deleting a finite number of points, each of which is a pole;

($\mathrm{II}^{(F)}$)
$$\lim_{\lambda\to\infty} \| W(\lambda) - E \| = 0;$$

($\mathrm{III}^{(F)}$)
$$W^*(\lambda)\, JW(\lambda) - J \geqslant 0 \quad (\operatorname{Im}\lambda > 0,\ \lambda \in G_W);$$

$$(\mathrm{IV}^{(F)}) \qquad W^*(\lambda)\, J W(\lambda) - J = 0 \quad (\operatorname{Im}\lambda = 0,\ \lambda \in G_W).$$

The class $\Omega_J^{(F)}$ contains the product of any two of its elements. It is not hard to see that the COF of a finite-dimensional node Θ lies in that class.

Theorem 11.1. *Each function $W(\lambda) \in \Omega_J^{(F)}$ is a characteristic operator-function for some simple finite-dimensional node.*

Proof. By Theorem 9.4 there exists a simple node Θ for which $G_A = G_W$ and $W_\Theta(\lambda) = W(\lambda)$ $(\lambda \in G_W)$. It remains to be proved that $\mathfrak{H}$ is finite dimensional.

In §3 it was proved that at the points of the set $G_A^{(0)} = G_A \cap G_{A_R}$ the operator $W_\Theta(\lambda) + E$ has an inverse, and

$$V_\Theta(\lambda) = K^*(A_R - \lambda E)^{-1} K = i\,(W_\Theta(\lambda) + E)^{-1}(W_\Theta(\lambda) - E)\,J.$$

At the same time, since all scalar products of the form $(W_\Theta(\lambda) f, g)$ are rational functions, the operator $W_\Theta(\lambda) + E$ can fail to have an inverse only at a finite number of points, and, in view of Lemma 9.1, only those points may be in the spectrum of the operator A_R. Accordingly, the orthogonal resolution of unity $E(s)$ corresponding to A_R is a piecewise constant function with a finite number of jumps, and therefore the closure $\mathfrak{H}_0$ of the linear envelope of vectors of the form $E(s)Kg$ $(-\infty < s < \infty,\ g \in \mathfrak{G})$ is finite dimensional. On the other hand, $\mathfrak{H}_0$ is invariant relative to A, since

$$AE(s)Kg = (A_R + iA_I)E(s)Kg = \int_{-\infty}^{s} t\,dE(t)\,Kg + iKJK^*E(s)\,Kg \in \mathfrak{H}_0$$

which means that $A^n Kg \in \mathfrak{H}_0$ $(n = 0, 1, \cdots;\ g \in \mathfrak{G})$. In view of the simplicity of Θ, $\mathfrak{H}_0 = \mathfrak{H}$.

2. Decomposition of functions of the class $\Omega_J^{(F)}$ into factors.

Lemma 11.1. *Suppose that $\mathfrak{H}$ is an n-dimensional space and that A is a linear operator operating in it. Then there exist subspaces*

$$0 = \mathfrak{H}_0 \subset \mathfrak{H}_1 \subset \mathfrak{H}_2 \subset \ldots \subset \mathfrak{H}_{n-1} \subset \mathfrak{H}_n = \mathfrak{H}, \qquad (11.1)$$
$$(\dim \mathfrak{H}_k = k, \quad k = 1, 2, \ldots, n-1),$$

invariant relative to A.

Proof. Construct in $\mathfrak{H}$ an orthonormal basis $e_1, \cdots, e_n$ such that the vector e_1 is an eigenvector for A, and the vector e_{k+1} $(k = 1, \cdots, n-1)$ is an eigenvector for the operator $P_k A$, where P_k is the orthoprojector onto the orthogonal complement to the linear envelope $\mathfrak{H}_k$ of vectors $e_1, \cdots, e_k$. Then

$$\begin{aligned} Ae_1 &= \lambda_1 e_1, \\ Ae_2 &= \alpha_{21}e_1 + \lambda_2 e_2, \\ &\dots\dots\dots\dots \\ Ae_n &= \alpha_{n1}e_1 + \alpha_{n2}e_2 + \dots + \lambda_n e_n, \end{aligned} \tag{11.2}$$

which means that the subspaces $\mathfrak{H}_k$, $k = 1, \cdots, n-1$ are invariant relative to A.

Theorem 11.2. *Suppose that the function $W(\lambda)$ lies in $\Omega_J^{(F)}$. Then it may be represented in the form*

$$W(\lambda) = \left(E + \frac{2i\sigma_1}{\lambda - \lambda_1} P_1 J\right)\left(E + \frac{2i\sigma_2}{\lambda - \lambda_2} P_2 J\right) \dots \left(E + \frac{2i\sigma_n}{\lambda - \lambda_n} P_n J\right), \tag{11.3}$$

where the λ_j are complex numbers, the σ_j are positive numbers, and the P_j are one-dimensional orthoprojectors satisfying the condition

$$P_j J P_j = \frac{\operatorname{Im} \lambda_j}{\sigma_j} P_j. \tag{11.4}$$

Proof. In Theorem 11.1 it was proved that the function $W(\lambda)$ is characteristic for some simple finite-dimensional node Θ. By Lemma 11.1 the space $\mathfrak{H}$ contains a subspace (11.1') invariant relative to A. In view of formula (3.3)

$$W(\lambda) = W_{\Theta_1}(\lambda) W_{\Theta_2}(\lambda) \dots W_{\Theta_n}(\lambda)$$
$$(\Theta_j = \operatorname{pr}_{\mathfrak{H}_j \ominus \mathfrak{H}_{j-1}} \Theta, \quad j = 1, 2, \dots, n).$$

Suppose that A_j and K_j are the basic and canal operators of the node Θ_j. Since the operator A_j operates in the one-dimensional space $\mathfrak{H}_j \ominus \mathfrak{H}_{j-1}$, we have

$$W_{\Theta_j}(\lambda) = E - 2iK_j^*(A_j - \lambda E)^{-1} K_j J = E + \frac{2i}{\lambda - \lambda_j} K_j^* K_j J.$$

In view of Theorem 2.2 the operator $K^* K_j$ is not zero. Accordingly $K_j^* K_j = \sigma_j P_j$, where $\sigma_j > 0$ and P_j is the orthoprojector onto the one-dimensional subspace. Moreover,

$$\begin{aligned} P_j J P_j &= \frac{1}{\sigma_j^2} K_j^* K_j J K_j^* K_j = \frac{1}{\sigma_j^2} K_j^* \frac{A_j - A_j^*}{2i} K_j \\ &= \frac{\operatorname{Im} \lambda_j}{\sigma_j^2} K_j^* K_j = \frac{\operatorname{Im} \lambda_j}{\sigma_j} P_j. \end{aligned}$$

The theorem is proved.

We note that each function of the form

$$W_0(\lambda) = E + \frac{2i\sigma_0}{\lambda - \lambda_0} P_0 J \quad \left(\sigma_0 > 0,\ P_0 J P_0 = \frac{\operatorname{Im} \lambda_0}{\sigma_0} P_0\right), \tag{11.5}$$

where P_0 is an orthoprojector onto a one-dimensional subspace, lies in the class $\Omega_J^{(F)}$. Indeed, from the easily verified equation

$$W_0^*(\lambda) J W_0(\lambda) - J = \frac{4\sigma_0 \operatorname{Im} \lambda}{|\lambda - \lambda_0|^2} J P_0 J$$

it follows that the function $W_0^*(\lambda) J W_0(\lambda) - J$ is positive in the upper halfplane and equal to zero on the real axis. Thus *the class* $\Omega_J^{(F)}$ *coincides with the collection of all possible products of simplest factors of the form* (11.5).

§12. Matrix nodes

1. **Matrix nodes and their characteristic matrix-functions.** An operator node is said to be $\mathfrak{G}$-finite dimensional if its exterior space is finite dimensional.

We consider the $\mathfrak{G}$-finite-dimensional node

$$\Theta = \begin{pmatrix} A & K & J \\ \mathfrak{H} & & \mathfrak{G} \end{pmatrix}$$

and denote the dimension of the space $\mathfrak{G}$ by r. Suppose that n^+, s and n^- are respectively the numbers of positive, zero and negative eigenvalues of the operator A_I, considered in the subspace $\Re(K)$ invariant for it. Thus the number $n = n^+ + s + n^-$ is the dimension of the subspace $\Re(K)$, and $\Re(K) \ominus \Re(A_I)$. Represent $\mathfrak{G}$ in the form of a sum $\mathfrak{G} = \mathfrak{G}^+ \oplus \mathfrak{G}^-$ such that

$$Jg = g \ (g \in \mathfrak{G}^+), \quad Jg = -g \ (g \in \mathfrak{G}^-),$$

and denote by r^+ and r^- respectively ($r = r^+ + r^-$) the dimensions of the subspaces $\mathfrak{G}^+$ and $\mathfrak{G}^-$.

Theorem 12.1. *The following inequalities hold*:

$$r^+ \geqslant n^+ + s, \ r^- \geqslant n^- + s. \tag{12.1}$$

Proof. In the spaces $\Re(K)$ and $\mathfrak{G}$ there exist orthonormalized bases $h_1, \ldots, h_n$ and $g_1, \ldots, g_n$ for which

$$\omega = \| (A_I h_i, h_j) \|_1^n = \begin{Vmatrix} \omega_1^+ & & & & & & & & \\ & \ddots & & & & & & & \\ & & \omega_{n^+}^+ & & & & & & \\ & & & 0 & & & & & \\ & & & & \ddots & & & & \\ & & & & & 0 & & & \\ & & & & & & \omega_1^- & & \\ & & & & & & & \ddots & \\ & & & & & & & & \omega_{n^-}^- \end{Vmatrix}$$

$$(\omega_k^+ > 0, \ \omega_k^- < 0),$$

$$j = \| (Jg_\alpha, g_\beta) \|_1^r = \begin{Vmatrix} I_{r^+} & \vdots & \\ \cdot & \cdot & \cdot \\ & \vdots & -I_{r^-} \end{Vmatrix}.$$

Since $A_I = KJK^*$, we have

$$(A_I h_i, h_j) = \sum_{\alpha, \beta = 1}^{r} (K^* h_i, g_\alpha)(Jg_\alpha, g_\beta)(Kg_\beta, h_j),$$

so that $\omega = \pi j \pi^*$, where $\pi = \|(K^* h_i, g_\alpha)\|$ is a rectangular matrix containing n rows and r columns.

Suppose that $r^+ < n^+ + s$. Then there exists a vector

$$\xi = \| \xi_1 \xi_2 \ldots \xi_{n^+ + s} \underbrace{0 \ldots 0}_{n^-} \|$$

such that at least one of the first of its $n^+ + s$ coordinates differs from zero and the first r^+ coordinates of the vector $\eta = \xi \pi$ are equal to zero. Also, in view of the linear independence of the rows of the matrix π at least one of the remaining r^- coordinates of the vector η will differ from zero. We have arrived at a contradiction, since the right side of the equation $\xi \omega \xi^* = \eta j \eta^*$ is negative, and the left is positive or equal to zero.

Analogously one obtains the second of inequalities (12.1).

The theorem is proved.

In the study of operators with finite-dimensional imaginary components it is sometimes convenient to operate with the concepts of matrix node and characteristic matrix-function instead of those of operator node and characteristic operator-function. We introduce the corresponding definitions.

The collection consisting of a bounded linear operator A, operating in a separable Hilbert space $\mathfrak{H}$, the vectors $k_\alpha \in \mathfrak{H}$ ($\alpha = 1, \cdots, r$; $r < \infty$) and the numerical matrix $j = \|j_{\alpha\beta}\|_1^r$ is said to be a *matrix node* and is denoted by the symbol

$$\theta = (A, k_1, \ldots, k_r, j),$$

if

$$\text{I)} \qquad j = j^*, \ j^2 = I; \tag{12.2}$$

$$\text{II)} \qquad A_I h = \sum_{\alpha, \beta = 1}^{r} (h, k_\alpha)\, j_{\alpha\beta} k_\beta \quad (h \in \mathfrak{H}). \tag{12.3}$$

We emphasize that in the definition of matrix node there is no requirement of linear independence of the vectors k_α. It is not excluded in particular that some of them can be equal to zero.

A matrix-function of the complex variable λ,

$$w_\theta(\lambda) = I - 2ij \left\| \left((A - \lambda E)^{-1} k_\alpha, k_\beta \right) \right\|_1^r \tag{12.4}$$

is called a *characteristic matrix-function* (CMF) of the matrix node θ.

Consider the $\mathfrak{G}$-finite-dimensional operator node

$$\Theta = \begin{pmatrix} A & K & J \\ \mathfrak{H} & & \mathfrak{G} \end{pmatrix},$$

and suppose that $g_1, \cdots, g_2$ is an orthonormal basis in $\mathfrak{G}$. Putting

$$k_\alpha = K g_\alpha \quad (\alpha = 1, 2, \ldots, r), \quad j = \| j_{\alpha\beta} \|_1^r = \| (J g_\alpha, g_\beta) \|_1^r, \tag{12.5}$$

we obtain the matrix node $\theta = (A, k_1, \cdots, k_r, j)$. Indeed, since $J = J^*$, $J^2 = E$ and $A_I = KJK^*$, we have $j = j^*$, $j^2 = I$ and

$$A_I h = \sum_{\alpha, \beta = 1}^{r} (K^* h, g_\alpha)(J g_\alpha, g_\beta)\, K g_\beta = \sum_{\alpha, \beta = 1}^{r} (h, k_\alpha)\, j_{\alpha\beta} k_\beta.$$

Also

$$\begin{aligned} \| (W_\Theta(\lambda) g_\alpha, g_\beta) \|_1^r &= I - 2i \left\| \left(K^* (A - \lambda E)^{-1} K J g_\alpha, g_\beta \right) \right\|_1^r \\ &= I - 2ij \left\| \left((A - \lambda E)^{-1} k_\alpha, k_\beta \right) \right\|_1^r = w_\theta(\lambda). \end{aligned} \tag{12.6}$$

Conversely, starting with a matrix node $\theta = (A, k_1, \cdots, k_r, j)$, we set up in any r-dimensional space $\mathfrak{G}$ an orthonormalized basis $g_1, \cdots, g_r$, and introduce the operators

$$Kg_\alpha = k_\alpha \quad (\alpha = 1, 2, \ldots, r), \quad Jg_\alpha = \sum_{\beta=1}^{r} j_{\alpha\beta} g_\beta \quad (\alpha = 1, 2, \ldots, r).$$

Then

$$\Theta = \begin{pmatrix} A & K & J \\ \mathfrak{H} & & \mathfrak{G} \end{pmatrix}$$

is a $\mathfrak{G}$-finite-dimensional operator node and

$$w_\theta(\lambda) = \| (W_\Theta(\lambda) g_\alpha, g_\beta) \|_1^r.$$

All the results of the preceding sections are valid for $\mathfrak{G}$-finite-dimensional nodes, and in this special case it is easy to reformulate them in terms of matrix nodes with the aid of formulas (12.5) and (12.6). The theorems resulting in this way are presented below without proof. Each of them is provided with the same index as that of the corresponding theorem on operator nodes.

2. Properties of matrix nodes. Suppose given a matrix node $\theta = (A, k_1, \cdots, k_r, j)$. The operator A will be called *basic*, the space $\mathfrak{H}_\theta$ in which it operates *interior*, each of the vectors k_α a *canal vector*, and the matrix j *directing*. The linear envelope of the vectors k_α $(\alpha = 1, 2, \cdots, r)$ and the range of the operator A_I will be denoted respectively by $\mathfrak{R}(k)$ and $\mathfrak{R}(A_I)$. Obviously

$$\mathfrak{R}(A_I) \subseteq \mathfrak{R}(k). \tag{12.7}$$

The subspace $\mathfrak{R}(A_I)$ is called *non-Hermitian*, and $\mathfrak{R}(k)$ a *canal subspace*.

By (12.1) the number r of canal vectors satisfies the inequality

$$r \geqslant m + 2s, \tag{12.8}$$

where m and s are the dimensions of the subspaces $\mathfrak{R}(A_I)$ and $\mathfrak{R}(k) \ominus \mathfrak{R}(A_I)$.

Theorem 1.1′. *If A is a bounded linear operator with a finite-dimensional imaginary component, and $\mathfrak{R}$ is any finite-dimensional subspace containing $\mathfrak{R}(A_I)$, then there exists a matrix node for which the operator A is basic and the subspace $\mathfrak{R}$ a canal subspace.*

The matrix node $\theta = (A, k_1, \cdots, k_r, j)$ constructed relative to a given operator A is called the imbedding of A into a matrix node. The operator A may be imbedded in a matrix node if and only if its imaginary component is finite dimensional.

Consider the matrix node $\theta = (A, k_1, \cdots, k_r, j)$ and denote by $\mathfrak{E}_\theta$ the closure of the linear envelope of vectors of the form

$$A^n k_\alpha \quad (n = 0, 1, \ldots; \alpha = 1, 2, \ldots, r).$$

The subspaces $\mathfrak{E}_\theta$ and $\mathfrak{E}_\theta^{(0)} = \mathfrak{H}_\theta \ominus \mathfrak{E}_\theta$ are called respectively the *principal* and *excess* subspaces. Each of them is invariant relative to A, while a selfadjoint operator is induced in $\mathfrak{E}_\theta^{(0)}$.

The principal subspaces of the nodes $\theta = (A, k_1, \cdots, k_r, j)$ and $\theta^* = (A^*, k_1, \cdots, k_r, -j)$ coincide.

The matrix node θ is said to be *simple* if $\mathfrak{E}_\theta = \mathfrak{H}_\theta$, and *excess* in the contrary case.

An operator A operating in the space $\mathfrak{H}$ admits imbedding into a simple matrix node if and only if its imaginary component is finite dimensional and there exists a finite-dimensional subspace $\mathfrak{R} \supseteq \mathfrak{R}(A_I)$ such that the linear envelope of vectors of the form $A^n h$ $(n = 0, 1, \cdots; h \in \mathfrak{R})$ is dense in $\mathfrak{H}$. In particular, every completely nonselfadjoint operator A with a finite-dimensional imaginary component may be imbedded in a simple matrix node.

Theorem 1.3′. *In order that a matrix node $\theta = (A, k_1, \cdots, k_r, j)$ be excess, it is necessary and sufficient that there exist a subspace $\mathfrak{H}_0 \subseteq \mathfrak{H}_\theta$ $(\mathfrak{H}_0 \neq 0)$ invariant relative to A and orthogonal to $\mathfrak{R}(k)$.*

Suppose that $\theta_i = (A_i, k_1^{(i)}, \cdots, k_r^{(i)}, j)$ $(i = 1, 2)$ are matrix nodes. Denoting by P_i the orthoprojector onto $\mathfrak{H}_{\theta_i}$ $(i = 1, 2)$ operating in the space $\mathfrak{H}_{\theta_1} \oplus \mathfrak{H}_{\theta_2}$, we obtain the matrix node

$$\theta = \left(A_1P_1 + A_2P_2 + 2i \sum_{\alpha, \beta=1}^{r} (\,\cdot\,, k_\alpha^{(2)})\, j_{\alpha\beta} k_\beta^{(1)},\ k_1^{(1)} + k_1^{(2)},\ \ldots,\ k_r^{(1)} + k_r^{(2)},\ j\right), \quad (12.9)$$

which is called the *product* of the nodes θ_1 and θ_2 $(\theta = \theta_1\theta_2)$. The operation of multiplication of matrix nodes has the associative property.

If $\theta = (A, k_1, \cdots, k_r, j)$ is a matrix node and P_0 is an orthoprojector onto some subspace $\mathfrak{H}_0$ of the interior space $\mathfrak{H}_\theta$, then $\theta_0 = (A_0, k_1^{(0)}, \cdots, k_r^{(0)}, j)$, where $A_0 h = P_0 A h$ $(h \in \mathfrak{H}_0)$, $k_\alpha^{(0)} = P_0 k_\alpha$, is also a matrix node, called the *projection* of the node θ onto the subspace $\mathfrak{H}_0$ and denoted by the symbol $\mathrm{pr}_{\mathfrak{H}_0}\theta$. The following formulas hold:

$$\mathrm{pr}_{\mathfrak{H}_0}\theta^* = (\mathrm{pr}_{\mathfrak{H}_0}\theta)^*, \quad (12.10)$$

$$\mathrm{pr}_{\mathfrak{H}_1}\theta = \mathrm{pr}_{\mathfrak{H}_1}(\mathrm{pr}_{\mathfrak{H}_2}\theta) \qquad (\mathfrak{H}_1 \subseteq \mathfrak{H}_2). \quad (12.11)$$

If the node $\theta = (A, k_1, \cdots, k_r, j)$ is the product of the nodes θ_1 and θ_2, then the subspace $\mathfrak{H}_{\theta_1}$ is invariant relative to A, and θ_1 and θ_2 are the projections of θ onto $\mathfrak{H}_{\theta_1}$ and $\mathfrak{H}_{\theta_2}$. Conversely, every matrix node $\theta = (A, k_1, \cdots, k_r, j)$

is the product of its projections onto any subspace invariant relative to A and its orthogonal complement.

The matrix node θ_0 is said to be a *left* (*right*) *divisor* of the node $\theta = (A, k_1, \cdots, k_r, j)$ if it is the projection of θ onto a subspace invariant relative to A (A^*). Suppose that the basic operator of the matrix node θ has the invariant subspaces

$$0 = \mathfrak{H}_0 \subset \mathfrak{H}_1 \subset \ldots \subset \mathfrak{H}_{n-1} \subset \mathfrak{H}_n = \mathfrak{H}_\theta.$$

Then

$$\theta = \mathrm{pr}_{\mathfrak{H}_1} \theta \; \mathrm{pr}_{\mathfrak{H}_2 \ominus \mathfrak{H}_1} \theta \ldots \mathrm{pr}_{\mathfrak{H}_n \ominus \mathfrak{H}_{n-1}} \theta. \tag{12.12}$$

Theorem 2.2′. *If $\theta = \theta_1 \theta_2 \cdots \theta_n$ is a simple matrix node, then all the nodes θ_j $(j = 1, 2, \cdots, n)$ are simple.*

The *spectrum* of a matrix node is the spectrum of its basic operator. Since the basic operator of a matrix node has a finite-dimensional imaginary component, by Theorem 9.1 the set of its regular points is connected.

Theorem 2.5′. *The spectrum of the product of two matrix nodes is equal to the sum of the spectra of the factors.*

Theorem 3.1′. *If the matrix node θ is the product of the matrix nodes θ_1 and θ_2, then at all the regular points of the basic operator of θ*

$$w_\theta(\lambda) = w_{\theta_2}(\lambda)\, w_{\theta_1}(\lambda). \tag{12.13}$$

Corollary. *If the basic operator of the matrix node θ has the invariant subspaces*

$$0 = \mathfrak{H}_0 \subset \mathfrak{H}_1 \subset \ldots \subset \mathfrak{H}_{n-1} \subset \mathfrak{H}_n = \mathfrak{H}_\theta,$$

then at all of its regular points

$$w_\theta(\lambda) = w_{\theta_n}(\lambda) \ldots w_{\theta_2}(\lambda)\, w_{\theta_1}(\lambda) \quad (\theta_j = \mathrm{pr}_{\mathfrak{H}_j \ominus \mathfrak{H}_{j-1}} \theta). \tag{12.14}$$

The matrix node $\theta_1 = (A_1, k_1^{(1)}, \cdots, k_r^{(1)}, j)$ will be called *unitarily equivalent* to the matrix node $\theta_2 = (A_2, k_1^{(2)}, \cdots, k_r^{(2)}, j)$ if there exists an isometric mapping U of the space $\mathfrak{H}_{\theta_1}$ onto $\mathfrak{H}_{\theta_2}$ such that

$$UA_1 = A_2 U, \qquad Uk_\alpha^{(1)} = k_\alpha^{(2)} \qquad (\alpha = 1, 2, \ldots, r). \tag{12.15}$$

By Theorem 9.3 the CMF of the matrix node $\theta = (A, k_1, \cdots, k_r, j)$ does not admit analytic continuation beyond the limits of the region G_A consisting of the regular points of the operator A. Hence it follows that if the CMF's of two matrix nodes

$$\theta_1 = (A_1, k_1^{(1)}, \dots, k_r^{(1)}, j) \text{ and } \theta_2 = (A_2, k_1^{(2)}, \dots, k_r^{(2)}, j)$$

coincide in the neighborhood of the point at infinity, then $G_{A_1} = G_{A_2}$ and $w_{\theta_1}(\lambda) \equiv w_{\theta_2}(\lambda)$ $(\lambda \in G_{A_1})$.

Theorem 3.2′. *Suppose that θ_1 and θ_2 are simple matrix nodes. If $w_{\theta_1}(\lambda) = w_{\theta_2}(\lambda)$, then θ_1 and θ_2 are unitarily equivalent.*

Theorem 3.3′. *Suppose that θ_1 and θ_2 are left divisors of the simple matrix node θ. If $w_{\theta_1}(\lambda) = w_{\theta_2}(\lambda)$, then $\theta_1 = \theta_2$.*

Suppose that the numerical matrix $j = \|j_{\alpha\beta}\|_1^r$ satisfies the conditions $j = j^*$ and $j^2 = I$. The matrix-function $w(\lambda) = \|w_{\alpha\beta}(\lambda)\|$ will be assigned to the class w_j, if:

(1) $w(\lambda)$ is holomorphic in a region G_w gotten from the entire extended complex plane by removing some bounded set which does not have any nonreal limit points;

(2) $\lim_{\lambda \to \infty} w(\lambda) = I$;

(3) $w(\lambda) j w^*(\lambda) - j \geqslant 0 \quad (\operatorname{Im} \lambda > 0,\ \lambda \in G_w)$;

(4) $w(\lambda) j w^*(\lambda) - j = 0 \quad (\operatorname{Im} \lambda = 0,\ \lambda \in G_w)$.

The CMF of the matrix node $\theta = (A, k_1, \cdots, k_r, j)$ lies in the class ω_j.

Theorem 9.4′. *If $w(\lambda) \in \omega_j$, then there exists a simple matrix node $\theta = (A, k_1, \cdots, k_r, j)$ such that $G_A \supseteq G_w$ and $w_\theta(\lambda) = w(\lambda)$ $(\lambda \in G_w)$.*

The function $w_1(\lambda) \in \omega_j$ is said to be a *right* (*left*) *divisor* of the function $w_2(\lambda) \in \omega_j$ if there exists a function $w_{21}(\lambda) \in \omega_j$ $(w_{12}(\lambda) \in \omega_j)$ such that in some neighborhood of the point at infinity

$$w_2(\lambda) = w_{21}(\lambda) w_1(\lambda) \qquad (w_2(\lambda) = w_1(\lambda) w_{12}(\lambda)).$$

A right divisor $w_1(\lambda)$ of the function $w_2(\lambda)$ is said to be *regular* if the product of the simple nodes $\theta_1 = (A_1, k_2^{(1)}, \cdots, k_r^{(1)}, j)$ and $\theta_{21} = (A_{21}, k_1^{(21)}, \cdots, k_r^{(21)}, j)$, for which $w_{\theta_1}(\lambda) = w_1(\lambda)$ and $w_{\theta_{21}}(\lambda) = w_{21}(\lambda)$, is a simple node.

Analogously one defines a regular left divisor.

Theorem 5.4′. *Suppose that $w(\lambda) \in \omega_j$ is a simple matrix node with directing matrix j such that $w_\theta(\lambda) = w(\lambda)$. For the function $w_1(\lambda)$ to be a regular right* *divisor of the function* $w(\lambda)$ *it is necessary and sufficient that it should be the CMF of some left divisor of the node* θ.

Suppose that $\theta = (A, k_1, \cdots, k_r, j)$ is a simple matrix node. Consider a mapping ϕ which assigns to each subspace $\mathfrak{H}_0 \subseteq \mathfrak{H}_\theta$ the matrix-function $w_{\theta_0}(\lambda)$, where $\theta_0 = \mathrm{pr}_{\mathfrak{H}_0}\theta$. The function $w_\theta(\lambda)$ lies in the class ω_j and $w_{\theta_0}(\lambda)$ is a regular right divisor of it.

Theorem 5.6'. *The mapping* ϕ *has the following properties:*

1. *Each regular right divisor of the function* $w_\theta(\lambda)$ *lies in the range of the mapping* ϕ.

2. ϕ *is one-to-one.*

3. *If* $\mathfrak{H}_1$ *and* $\mathfrak{H}_2$ *are invariant subspaces of the operator* A *and* $w_1(\lambda)$ *and* $w_2(\lambda)$ *are the regular right divisors of* $w(\lambda)$ *corresponding to them under* ϕ, *then* $w_1(\lambda)$ *is a regular right divisor of* $w_2(\lambda)$ *if and only if* $\mathfrak{H}_1 \subseteq \mathfrak{H}_2$.

A portion ω'_j of ω_j is said to be *ordered* if one of any two functions $w_1(\lambda) \in \omega'_j$ and $w_2(\lambda) \in \omega'_j$ is a right divisor of the other.

Theorem 5.7'. *Suppose that* $w(\lambda)$ *is the CMF of the simple node* $\theta = (A, k_1, \cdots, k_r, j)$. *For the operator* A *to be unicellular, it is necessary and sufficient that the set of regular right divisors of the function* $w(\lambda)$ *should be ordered.*

The matrix node $\theta = (A, k_1, \cdots, k_r, j)$ is said to be *dissipative* if $j = I$. Each dissipative operator with a finite-dimensional imaginary component may be imbedded in a dissipative matrix node. For the dissipative node $\theta = (A, k_1, \cdots, k_r, I)$ to be simple, it is necessary and sufficient that the operator A be completely nonselfadjoint.

We shall say that the matrix $w(\lambda) = \|w_{\alpha\beta}(\lambda)\|$ lies in the class ω_{I_r} if:

(1_I) $w(\lambda)$ is holomorphic in a region G_w which is gotten by removing from the extended complex plane some bounded set lying in the upper closed halfplane;

(2_I) $$\lim_{\lambda\to\infty} w(\lambda) = I;$$

(3_I) $$w(\lambda)w^*(\lambda) - I \leqslant 0 \quad (\mathrm{Im}\,\lambda < 0);$$

(4_I) $$w(\lambda)w^*(\lambda) - I = 0 \quad (\mathrm{Im}\,\lambda = 0,\ \lambda \in G_w).$$

The CMF of the dissipative node $\theta = (A, k_1, \cdots, k_r, I)$ lies in the class ω_{I_r}.

Theorem 7.4'. *If the function* $w(\lambda)$ *lies in the class* ω_{I_r}, *then there exists a simple dissipative node* $\theta = (A, k_1, \cdots, k_r, I)$ *such that* $w_\theta(\lambda) = w(\lambda)$ $(\lambda \in G_w)$.

We define a class $\omega_{I_r}^{(\exp)}$ of matrix-functions $w(\lambda) = \|w_{\alpha\beta}(\lambda)\|_1^r$ as follows:

($1_I^{(\exp)}$) all the elements of the matrix $w(\lambda)$ are entire functions of exponential type of $\mu = 1/\lambda$;

($2_I^{(\exp)}$)
$$\lim_{\lambda\to\infty} w(\lambda) = I;$$

($3_I^{(\exp)}$)
$$w(\lambda)\, w^*(\lambda) - I \geqslant 0 \quad (\operatorname{Im}\lambda > 0);$$

($4_I^{(\exp)}$)
$$w(\lambda)\, w^*(\lambda) - I = 0 \quad (\operatorname{Im}\lambda = 0,\ \lambda \neq 0).$$

If $\theta = (A, k_1, \cdots, k_r, I)$ and $A \in \Lambda^{(\exp)}$, then the CMF $w_\theta(\lambda)$ lies in the class $\omega_{I_r}^{(\exp)}$.

Theorem 8.3'. *If the function $w(\lambda)$ lies in the class $\omega_{I_r}^{(\exp)}$, then it is the characteristic function for some simple dissipative matrix node $\theta = (A, k_1, \cdots, k_r, I)$, where $A \in \Lambda^{(\exp)}$.*

The matrix node $\theta = (A, k_1, \cdots, k_r, j)$ is said to be a *Volterra node* if A is a Volterra operator.

Theorem 10.5'. *The product of two simple dissipative Volterra matrix nodes is a simple node.*

The matrix-function $w(\lambda) = \|w_{\alpha\beta}(\lambda)\|$ will be assigned to the class $\omega_j^{(0)}$ if all of its elements are entire functions of $\mu = 1/\lambda$ and if it satisfies conditions (2), (3), and (4).

The CMF of the Volterra node $\theta = (A, k_1, \cdots, k_r, j)$ lies in the class $\omega_j^{(0)}$.

Theorem 10.6'. *If the function $w(\lambda)$ lies in the class $\omega_j^{(0)}$, then there exists a simple Volterra node $\theta = (A, k_1, \cdots, k_r, j)$ for which it is characteristic.*

The matrix node θ is said to be finite dimensional if the space $\mathfrak{H}_\theta$ is finite dimensional.

We assign the matrix-function $w(\lambda) = \|w_{\alpha\beta}(\lambda)\|_1^r$ to the class $\omega_j^{(F)}$ if all of its elements are rational functions and if it satisfies conditions (2), (3), and (4).

The CMF of a finite-dimensional node $\theta = (A, k_1, \cdots, k_r, j)$ lies in the class $\omega_j^{(F)}$.

Theorem 11.1'. *Each function $w(\lambda) \in \omega_j^{(F)}$ is characteristic for some simple finite-dimensional matrix node.*

Theorem 11.2'. *The matrix $w(\lambda)$ lies in the class $\omega_j^{(F)}$ if and only if it is representable in the form*

$$w(\lambda) = \left(I + \frac{2i\sigma_n}{\lambda - \lambda_n} j\xi_n^*\xi_n\right) \ldots \left(I + \frac{2i\sigma_2}{\lambda - \lambda_2} j\xi_2^*\xi_2\right)\left(I + \frac{2i\sigma_1}{\lambda - \lambda_1} j\xi_1^*\xi_1\right), \quad (12.16)$$

where the λ_k *are complex numbers, the* σ_k *are positive numbers and the* ξ_k *are one-rowed numerical matrices satisfying the conditions*

$$\xi_k\xi_k^* = 1, \quad \xi_k j\xi_k^* = \frac{\operatorname{Im}\lambda_k}{\sigma_k} \quad (k = 1, 2, \ldots, n). \quad (12.17)$$

§13. Examples of matrix nodes

1. Suppose that the operator A operates in a one-dimensional space $\mathfrak{H}$. Then $Ah = \lambda_0 h$ $(h \in \mathfrak{H})$, where λ_0 does not depend on h, and $A_I h = \operatorname{Im}\lambda_0 h$.

In the case $\operatorname{Im}\lambda_0 \neq 0$ we put

$$k = |\operatorname{Im}\lambda_0|^{1/2} h_0 \qquad (\|h_0\| = 1), \qquad j = \operatorname{sign}\operatorname{Im}\lambda_0.$$

Inasmuch as $j = j^*$, $j^2 = I$ and

$$A_I h = (h, |\operatorname{Im}\lambda_0|^{1/2} h_0) \operatorname{sign}\operatorname{Im}\lambda_0 |\operatorname{Im}\lambda_0|^{1/2} h_0 = (h, k)\, jk,$$

we see that $\theta = (A, k, j)$ is a simple matrix node, and

$$w_\theta(\lambda) = 1 - 2ij\,((A - \lambda E)^{-1} k, k) = 1 - 2i\frac{\operatorname{Im}\lambda_0}{\lambda_0 - \lambda} = \frac{\lambda - \bar{\lambda}_0}{\lambda - \lambda_0} \quad (13.1)$$

is its CMF.

Now suppose that $\operatorname{Im}\lambda_0 = 0$. Inequality (12.8) shows that in this case the operator A already cannot be imbedded in a simple matrix node with one canal vector. Putting

$$k_1 = k_2 = k \neq 0, \; j = \begin{Vmatrix} 1 & 0 \\ 0 & -1 \end{Vmatrix},$$

we obtain a simple matrix node $\theta = (A, k_1, k_2, j)$ with the CMF

$$w_\theta(\lambda) = I - 2ij\left\|\left((A - \lambda E)^{-1} k_\alpha, k_\beta\right)\right\|_1^2 =$$

$$= I + \frac{2i\sigma}{\lambda - \lambda_0}\begin{Vmatrix} 1 & 1 \\ -1 & -1 \end{Vmatrix} = e^{\frac{2i\sigma}{\lambda - \lambda_0}\begin{Vmatrix} 1 & 1 \\ -1 & -1 \end{Vmatrix}} \quad (\sigma = \|k\|^2). \quad (13.2)$$

2. Consider a bounded selfadjoint operator A operating in Hilbert space $\mathfrak{H}$. If A has a simple spectrum, i.e. if there exists a vector h_0 such that the linear

envelope of the sequence $A^n h_0$ $(n = 0, 1, \cdots)$ is dense in $\mathfrak{H}$, then

$$\theta = (A,\ k_1,\ k_2,\ j) \qquad \left(k_1 = k_2 = h_0,\quad j = \left\| \begin{matrix} 1 & 0 \\ 0 & -1 \end{matrix} \right\|\right)$$

is a simple matrix node and

$$\begin{aligned} w_\theta(\lambda) &= I - 2ij\ \|((A - \lambda E)^{-1} k_\alpha,\ k_\beta)\|_1^2 \\ &= I - 2i\left((A - \lambda E)^{-1} h_0,\ h_0\right) \left\| \begin{matrix} 1 & 1 \\ -1 & -1 \end{matrix} \right\|. \end{aligned} \tag{13.3}$$

Put $p(\lambda) = ((A - \lambda E)^{-1} h_0,\ h_0)$ and note that

$$\operatorname{Im} p(\lambda) = \operatorname{Im} \lambda \|(A - \lambda E)^{-1} h_0\|^2 \geqslant 0 \qquad (\operatorname{Im} \lambda > 0).$$

It is easy to see that the function $p(\lambda)$ satisfies all the conditions of Theorem 4.9. Accordingly there exists a nondecreasing function $\sigma(t)$ $(-\infty < a \leq t \leq b < +\infty)$ such that

$$p(\lambda) = \int_a^b \frac{d\sigma(t)}{t - \lambda} \qquad (\lambda \notin [a,\ b]).$$

We introduce the space $L_2(a,\ b;\ \sigma)$ whose elements are σ-measurable functions $f(t)$ $(a \leq t \leq b)$ satisfying the condition $\int_a^b |f(t)|^2\, d\sigma(t) < \infty$. The scalar product in $L_2(a,\ b;\ \sigma)$ is defined by the formula

$$(f,\ g) = \int_a^b f(t)\, \overline{g(t)}\, d\sigma(t).$$

Define in $L_2(a,\ b;\ \sigma)$ the operator

$$Bf(t) = tf(t). \tag{13.4}$$

Since B is a selfadjoint operator and the linear envelope of the sequence $B^n f_0(t)$ $(n = 0, 1, \cdots;\ f_0(t) \equiv 1)$ is dense in $L_2(a,\ b;\ \sigma)$, we see that

$$\theta' = (B,\ k_1',\ k_2',\ j) \qquad \left(k_1' = k_2' = f_0(t),\ j = \left\| \begin{matrix} 1 & 0 \\ 0 & -1 \end{matrix} \right\|\right)$$

is a simple matrix node.

From the equation

$$w_{\theta'}(\lambda) = I - 2i \int_a^b \frac{d\sigma(t)}{t - \lambda} \left\| \begin{matrix} 1 & 1 \\ -1 & -1 \end{matrix} \right\| = I - 2ip(\lambda) \left\| \begin{matrix} 1 & 1 \\ -1 & -1 \end{matrix} \right\| = w_\theta(\lambda)$$

and Theorem 3.2′ of §11 it follows that the operators A and B are unitarily equivalent.

We have arrived at the known theorem that the collection of operators of the form (13.4) in the spaces $L_2(a, b; \sigma)$ is exhausted up to unitary equivalence by all bounded selfadjoint operators with simple spectra. The method of characteristic functions will yield an analogous result also for selfadjoint operators with a multiple spectrum.

3. Consider in the space $L_2(0, l)$ the operator

$$Af(x) = 2i \int_x^l f(t) K(t, x)\, dt, \tag{13.5}$$

where

$$\begin{gathered} K(t, x) = \overline{\varphi_1(t)}\, \varphi_1(x) + \ \dots \\ \dots + \overline{\varphi_p(t)}\, \varphi_p(x) - \overline{\varphi_{p+1}(t)}\, \varphi_{p+1}(x) - \dots - \overline{\varphi_{p+q}(t)}\, \varphi_{p+q}(x) \\ (\varphi_j(x) \in L_2(0, l),\ j = 1, 2, \dots, p+q). \end{gathered} \tag{13.6}$$

These are Volterra operators (see §24). Putting

$$\xi(x) = \| \overline{\varphi_1(x)}\, \overline{\varphi_2(x)} \dots \overline{\varphi_r(x)} \| \ (r = p+q), \ j = \| j_{\alpha\beta} \|_1^r = \begin{Vmatrix} I_p & 0 \\ 0 & -I_q \end{Vmatrix},$$

we rewrite (13.5) in the form

$$Af(x) = 2i \int_x^l f(t)\, \xi(t)\, dt\, j \xi^*(x).$$

Since

$$A^* f(x) = -\ 2i \int_0^x f(t)\, \xi(t)\, dt\, j \xi^*(x),$$

we have

$$\frac{A - A^*}{2i} f(x) = \int_0^l f(t) \xi(t)\, dt\, j \xi^*(x) = \sum_{\alpha=1}^r (f, \varphi_\alpha)\, j_{\alpha\alpha} \varphi_\alpha(x) = \sum_{\alpha, \beta=1}^r (f, \varphi_\alpha)\, j_{\alpha\beta} \varphi_\beta(x),$$

so that $\theta = (A, \phi_1, \dots, \phi_r, j)$ is a matrix node.

Suppose that $w(x, \lambda)$ is a matrix-function of the rth order, satisfying the conditions

$$\left. \begin{aligned} \frac{dw(x, \lambda)}{dx} &= \frac{2i}{\lambda}\, j \xi^*(x)\, \xi(x)\, w(x, \lambda), \\ w(0, \lambda) &\equiv I. \end{aligned} \right\} \tag{13.7}$$

Put

$$\eta(x, \lambda) = -\frac{1}{\lambda}\xi(x)\,w(x, \lambda).$$

Then

$$w(x, \lambda) = I + \frac{2i}{\lambda}j\int_0^x \xi^*(t)\,\xi(t)\,w(t, \lambda)\,dt$$

and

$$(A^* - \bar{\lambda}E)^{-1}\eta^*(x, \lambda) = -2i\int_0^x \eta^*(t, \lambda)\,\xi(t)\,dt\,j\xi^*(\lambda) - \bar{\lambda}\eta^*(x, \lambda)$$

$$= \left(\frac{2i}{\bar{\lambda}}\int_0^x w^*(t, \lambda)\,\xi^*(t)\,\xi(t)\,dt\,j + w^*(x, \lambda)\right)\xi^*(x) = \xi^*(x).$$

Accordingly

$$\|((A - \lambda E)^{-1}\varphi_\alpha, \varphi_\beta)\|_1^r = \int_0^l \xi^*(x)\,((A^* - \bar{\lambda}E)^{-1}\xi^*(x))^*\,dx$$

$$= \int_0^l \xi^*(x)\,\eta(x, \lambda)\,dx = -\frac{1}{\lambda}\int_0^l \xi^*(x)\,\xi(x)\,w(x, \lambda)\,dx = \frac{1}{2}ij\,(w(l, \lambda) - I),$$

i.e.

$$w(l, \lambda) = I - 2ij\,\|((A - \lambda E)^{-1}\varphi_\alpha, \varphi_\beta)\|_1^r = w_\theta(\lambda).$$

Thus, in order to obtain the CMF of the node θ we need to take the solution $w(x, \lambda)$ of the differential system (13.7) and put $x = l$ in it.

We shall establish criteria for the simplicity of the node θ, restricting ourselves to the dissipative case $q = 0$. It is easy to see that the excess subspace of the node $\theta = (A, \phi_1, \cdots, \phi_p, I)$ consists of those and only those functions $f(x)$ for which $Af(x) = 0$. Put

$$Af(x) = 2i\int_x^l f(t)\,\xi(t)\,dt\,\xi^*(x) = 0.$$

Putting $\zeta(x) = \int_x^l f(t)\,\xi(t)\,dt$, we get

$$\xi(x)\,\zeta^*(x) = \overline{\zeta(x)\,\xi^*(x)} = 0, \qquad \frac{d\zeta(x)}{dx}\,\zeta^*(x) = 0,$$

$$\frac{d}{dx}(\zeta(x)\,\zeta^*(x)) = \frac{d\zeta(x)}{dx}\,\zeta^*(x) + \zeta(x)\,\frac{d\zeta^*(x)}{dx} = 2\,\mathrm{Re}\left(\frac{d\zeta(x)}{dx}\,\zeta^*(x)\right) = 0.$$

Hence $\zeta(x) = 0$, which means that $f(x)\xi(x) = 0$.

Conversely, if the product $f(x)\xi(x)$ is equal to zero almost everywhere, then, obviously, $Af(x) = 0$.

We have arrived at the following conclusion. *For the matrix node* $\theta = (A, \phi_1, \cdots, \phi_p, I)$ *to be simple, it is necessary and sufficient that the vector* $\xi(x) = \|\phi_1(x)\phi_2(x)\cdots\phi_p(x)\|$ *should differ from zero almost everywhere.*

4. Suppose given in the space $L_2(0, l)$ the operator

$$Af(x) = \alpha(x)f(x) + 2i\int_x^l f(t)K(t, x)\,dt, \tag{13.8}$$

where $K(t, x)$ is defined by formula (13.6) and $\alpha(x)$ is a bounded measurable function. One may imbed the operator (13.8) into a matrix node and construct the CMF of this node according to the schema indicated in the previous example, with the only difference that in place of the differential system (13.7) one has to choose the system

$$\left.\begin{aligned} \frac{dw(x, \lambda)}{dx} &= \frac{2i}{\lambda - \alpha(x)} j\xi^*(x)\xi(x)w(x, \lambda), \\ w(0, \lambda) &\equiv I. \end{aligned}\right\} \tag{13.9}$$

5. Define in $L_2(0, l)$ the operator

$$Af(x) = 2\int_x^l f(t)\left(\overline{\varphi_1(t)}\varphi_2(x) - \overline{\varphi_2(t)}\varphi_1(x)\right)dt \tag{13.10}$$

$$(\varphi_\alpha(x) \in L_2(0, l),\ \alpha = 1, 2)$$

and introduce the notation

$$\xi(x) = \|\varphi_1(x)\varphi_2(x)\|, \qquad j = \|j_{\alpha\beta}\| = \begin{Vmatrix} 0 & -i \\ i & 0 \end{Vmatrix}.$$

Inasmuch as

$$Af(x) = 2i\int_x^l f(t)\xi(t)\,dt\,j\xi^*(x),$$

$$A^*f(x) = -2i\int_0^x f(t)\xi(t)\,dt\,j\xi^*(x),$$

$$\frac{A - A^*}{2i}f(x) = \int_0^l f(t)\xi(t)\,j\xi^*(x) = \sum_{\alpha,\beta=1}^{2}(f, \varphi_\alpha)\,j_{\alpha\beta}\varphi_\beta(x),$$

the node $\theta = (A, \phi_1, \phi_2, j)$ is a matrix node. As in example 3, the solution $w(x, \lambda)$ of the differential system

$$\left.\begin{aligned} \frac{dw(x, \lambda)}{dx} &= \frac{2i}{\lambda} j\xi^*(x)\,\xi(x)\,w(x, \lambda), \\ w(0, \lambda) &\equiv I \end{aligned}\right\} \tag{13.11}$$

coincides for $x = l$ with the CMF of θ.

CHAPTER II

TRIANGULAR REPRESENTATIONS OF VOLTERRA OPERATORS AND MULTIPLICATIVE REPRESENTATIONS OF CHARACTERISTIC FUNCTIONS OF VOLTERRA NODES

§14. Chains of orthoprojectors

Suppose that $\mathfrak{H}$ is a separable Hilbert space and $\mathfrak{P}_{\mathfrak{H}}$ the collection of orthoprojectors onto all possible subspaces in $\mathfrak{H}$. A portion π of the set $\mathfrak{P}_{\mathfrak{H}}$ containing at least two orthoprojectors is said to be a *chain* if it follows from the conditions $P_1, P_2 \in \pi$, $P_1 \neq P_2$ that either $P_1 < P_2$ or $P_2 < P_1$.

In the set of all chains it is possible to introduce a natural ordering relation: the chain π_1 is said to *precede* the chain π_2 if every orthoprojector entering into the structure of π_1 also lies in π_2. A chain π preceding itself only is said to be *maximal*. It easily follows from Zorn's lemma that every chain precedes some maximal chain.

We present two examples.

1. Suppose that P_x is an orthoprojector operating in $L_2(0, l)$ onto the subspace consisting of all functions equal to zero almost everywhere in the interval (x, l). The collection π of all the orthoprojectors $P_x (0 \leq x \leq l)$ is a maximal chain.

2. Denote by P_n an operator operating in l_2 onto the subspace formed by all sequences of the form

$$\{\xi_1, \xi_2, \ldots, \xi_n, 0, 0, \ldots\}.$$

It is easy to see that the set π consisting of the orthoprojectors $P_0 = 0$, P_n $(n = 1, 2, \cdots)$ and $P_\infty = E$ is a maximal chain.

If a chain contains orthoprojectors P^- and P^+ $(P^- < P^+)$ such that every orthoprojector $P \in \pi$ distinct from them satisfies one of the inequalities $P < P^-$, $P^+ < P$, then the pair (P^-, P^+) is said to be a *jump* in the chain π, and the dimension of the subspace $P^+\mathfrak{H} \ominus P^-\mathfrak{H}$ is the *dimension of the jump*. A chain without jumps is said to be *continuous*.

The orthoprojector is said to be *limiting* for the chain π if there exists a sequence $P_n \in \pi$ $(m = 1, 2, \cdots)$ converging strongly to P_0. We recall that weak convergence of a sequence of orthoprojectors to an orthoprojector implies strong convergence.

If one adjoins the limiting orthoprojector P_0 to a chain π, then the resulting collection of orthoprojectors will again be a chain. Indeed, suppose that $P_n \to P_0$ $(P_n \in \pi)$. If P is any orthoprojector of the chain π, then there exists either a subsequence P_n' of the sequence P_n such that $P_n' \leq P$ or else a subsequence P_n'' such that $P \leq P_n''$. In the first case $P_0 \leq P$, and in the second $P \leq P_0$.

A chain is said to be *closed* if it contains all of its limiting orthoprojectors. The orthoprojector Q is said to be an *upper* (*lower*) *bound* of the chain π if $P \leq Q$ $(Q \leq P)$ for all $P \in \pi$. It is easy to see that the orthoprojectors onto the subspaces $\bigcup_{P\in\pi} P\mathfrak{H}$ and $\bigcap_{P\in\pi} P\mathfrak{H}$ are respectively the least among the upper bounds and the greatest among the lower. We agree to denote them by $\sup \pi$ and $\inf \pi$.

If the chain π is closed, then $\sup \pi$ *and* $\inf \pi$ *lie in π.* Indeed, suppose that ϵ_j $(j = 1, 2, \cdots)$ is a sequence of positive numbers tending to zero, e_n $(n = 1, 2, \cdots)$ an orthonormal basis in $\mathfrak{H}$, and $a_n = \sup_{P\in\pi}(Pe_n, e_n)$. π contains a nondecreasing sequence of orthoprojectors P_j $(j = 1, 2, \cdots)$ such that

$$a_n - (P_j e_n, e_n) < \varepsilon_j \qquad (n \leqslant j).$$

As is known, a monotone sequence of orthoprojectors converges to some orthoprojector (see Ahiezer and Glazman [1]). The limit Q of the sequence P_j lies in π. Also

$$(Qe_n, e_n) = \lim_{j\to\infty} (P_j e_n, e_n) = a_n,$$

and therefore $Q = \sup \pi$. Analogously one proves that $\inf \pi \in \pi$.

Theorem 14.1. *For the chain π to be maximal, it is necessary and sufficient that it be closed, that it contain the orthoprojectors O and E and that its jumps, if there are any, be one-dimensional.*

Proof. The necessity is obvious. For the proof of sufficiency we suppose that it is possible to extend the chain π by adjoining to it some orthoprojector

P_0. Denote by π' and π'' chains consisting of the orthoprojectors $P \in \pi$ for which $P < P_0$ and $P_0 < P$ respectively. Obviously $O \in \pi'$ and $E \in \pi''$. Since π' and π'' are closed, we have $\sup \pi' \in \pi'$, $\inf \pi'' \in \pi''$ and $\sup \pi' < P_0 < \inf \pi''$. Thus $\sup \pi'$ and $\inf \pi''$ constitute a jump in the chain π whose dimension exceeds unity.

§15. Maximal chains belonging to completely continuous operators

1. Existence of a maximal chain belonging to a given completely continuous operator. Suppose in the space $\mathfrak{H}$ we are given an operator A and a chain π. We say that π *belongs to the operator* A if all the subspaces $P\mathfrak{H}$ for $P \in \pi$ are invariant relative to A. The expression "A possesses the chain π" has the same meaning.

Lemma 15.1. *Suppose that P_n $(n = 1, 2, \cdots)$ is a sequence of orthoprojectors operating in $\mathfrak{H}$. If all the subspaces $\mathfrak{H}_n = P_n \mathfrak{H}$ have the same dimension k and the sequence P_n converges strongly to the orthoprojector P_0, then the dimension of the subspace $\mathfrak{H}_0 = P_0 \mathfrak{H}$ does not exceed k.*

Proof. It is necessary only to consider the case when k is finite. Suppose that $e_1^{(n)}, e_2^{(n)}, \cdots, e_k^{(n)}$ is an orthonormalized basis in $\mathfrak{H}_n$. Selecting from the sequences $\{e_1^{n}\}, \{e_2^{n}\}, \cdots, \{e_k^{(n)}\}$ weakly converging subsequences $\{e_1^{(n_j)}\}$, $\{e_2^{(n_j)}\}, \cdots, \{e_k^{(n_j)}\}$ and denoting their limits by $h_1, \cdots, h_k$ respectively, we get

$$\begin{aligned} P_0 h &= \lim_{j \to \infty} P_{n_j} h \\ &= \lim_{j \to \infty} \left[\left(h, e_1^{(n_j)}\right) e_1^{(n_j)} + \left(h, e_2^{(n_j)}\right) e_2^{(n_j)} + \ldots + \left(h, e_k^{(n_j)}\right) e_k^{(n_j)} \right] \\ &= (h, h_1) h_1 + (h, h_2) h_2 + \ \ldots \ + (h, h_k) h_k \qquad (h \in \mathfrak{H}). \end{aligned}$$

Thus $\mathfrak{H}_0$ is contained in the linear envelope of the vectors $h_1, \cdots, h_k$, which means that it is not more than k-dimensional.

Theorem 15.1. *Every completely continuous operator A operating in a separable Hilbert space $\mathfrak{H}$ whose dimension exceeds unity possesses an invariant subspace distinct from O and $\mathfrak{H}$.*

Proof. Suppose that $e_1, e_2, \cdots$ is an orthonormalized basis in $\mathfrak{H}$ and that $\mathfrak{H}_n$ is the linear envelope of the vectors $e_1, \cdots, e_n$. Suppose that P_n is the orthoprojector onto $\mathfrak{H}_n$. By Lemma 11.1 there exist subspaces

$$O = \mathfrak{H}_n^{(0)} \subset \mathfrak{H}_n^{(1)} \subset \ldots \subset \mathfrak{H}_n^{(n-1)} \subset \mathfrak{H}_n^{(n)} = \mathfrak{H}_n \qquad (\dim \mathfrak{H}_n^{(k)} = k),$$

invariant relative to the operator $A_n = P_n A P_n$. Since

$$A_n P_n^{(k)} = P_n^{(k)} A_n P_n^{(k)}, \quad P_n P_n^{(k)} = P_n^{(k)} P_n = P_n^{(k)},$$

where $P_n^{(k)}$ is the orthoprojector onto $\mathfrak{H}_n^{(k)}$, we have

$$P_n A P_n^{(k)} = P_n^{(k)} A P_n^{(k)}. \tag{15.1}$$

In view of the inequalities

$$0 = (P_n^{(0)} e_1, e_1) \leqslant (P_n^{(1)} e_1, e_1) \leqslant \ldots \leqslant (P_n^{(n)} e_1, e_1) = 1$$

there exist orthoprojectors $Q_n' = P_n^{(k_n)}$ and $Q_n'' = P_n^{(k_n+1)}$, satisfying the condition

$$(Q_n' e_1, e_1) < \frac{1}{2} \leqslant (Q_n'' e_1, e_1).$$

Selecting from the sequences Q_n' and Q_n'' weakly converging subsequences Q_{n_j}' and Q_{n_j}'' and denoting their limits by F' and F'' respectively, we get

$$(F' e_1, e_1) \leqslant \frac{1}{2} \leqslant (F'' e_1, e_1). \tag{15.2}$$

By (15.1)

$$P_{n_j} A Q_{n_j}' = Q_{n_j}' A Q_{n_j}', \quad P_{n_j} A Q_{n_j}'' = Q_{n_j}'' A Q_{n_j}''.$$

Since the sequences AQ_{n_j}', AQ_{n_j}'' and P_{n_j} strongly converge to AF', AF'' and E respectively,

$$AF' = F'AF', \quad AF'' = F''AF''.$$

It follows from (15.2) that $F' \neq E$ and $F'' \neq O$. Suppose that simultaneously $F' = O$ and $F'' = E$. Then the sequence $Q_{n_j}'' - Q_{n_j}'$ of orthoprojectors onto one-dimensional subspaces converges weakly (and so strongly) to E, contradicting Lemma 15.1. Thus at least one of the operators F' or F'' differs both from zero and from E. Denoting it by F, we arrive at the equation

$$AF = FAF \quad (F \neq O, \ F \neq E).$$

Suppose that $\mathfrak{H}_0$ is the subspace consisting of all vectors $f \in \mathfrak{H}$ for which $Ff = f$. There exists a vector $h_0 \neq 0$, such that $Fh_0 \neq 0$. If $AFh_0 = 0$, then the one-dimensional subspace generated by the vector Fh_0 is invariant relative to A. If now $AFh_0 \neq 0$, then $\mathfrak{H}_0 \neq O$. At the same time, $\mathfrak{H}_0 \neq \mathfrak{H}$, since in the contrary case the equation $F = E$ would be satisfied. It remains to be noted that $\mathfrak{H}_0$ is invariant relative to A. Indeed, if $f \in \mathfrak{H}_0$, then

$$FAf = FAFf = AFf = Af$$

so that $Af \in \mathfrak{H}_0$.

Theorem 15.2. *Every completely continuous operator A operating in a separable Hilbert space $\mathfrak{H}$ possesses a maximal chain.*

Proof. Denote by $\mathfrak{A}$ the collection of all chains belonging to A. Since each ordered subset $\mathfrak{A}_0 \subset \mathfrak{A}$ has an upper bound, by Zorn's lemma there exists a chain $\pi^* \in \mathfrak{A}$ which is maximal in $\mathfrak{A}$. This last fact means that if π^* precedes the chain $\pi_1^* \neq \pi^*$, then π_1^* does not lie in $\mathfrak{A}$. It remains to be proved that π^* is maximal in the set of all chains.

Obviously $0 \in \pi^*$ snd $E \in \pi^*$. Suppose that P_0 is a limiting orthoprojector for π^*. Then, as was shown in the preceding section, the collection π^{**} formed by the chain π^* and the orthoprojector P_0 is also a chain. If $P_n \in \pi^*$ $(n = 1, 2, \cdots)$ and $P_n \to P_0$, then $AP_n = P_nAP_n$, so that $AP_0 = P_0AP_0$, which means that the subspace $P_0\mathfrak{H}$ is invariant relative to A. Thus $\pi^{**} \in \mathfrak{A}$, which means that $\pi^* = \pi^{**}$. This proves that the chain π^* is closed.

Now suppose that the chain π^* has a jump (P^-, P^+) whose dimension is larger than unity. This would imply that π^* is not maximal in $\mathfrak{A}$. In fact, if there were such a jump, then in view of Theorem 15.1 the operator $(P^+ - P^-)A$ would have an invariant subspace $\mathfrak{H}^{(0)}$ $(\mathfrak{H}^{(0)} \neq 0, \mathfrak{H}^{(0)} \neq P^+\mathfrak{H} \ominus P^-\mathfrak{H})$ in the space $P^+\mathfrak{H} \ominus P^-\mathfrak{H}$. But then the inequality

$$P^- < P^- + P^{(0)} < P^+$$

would be satisfied, where $P^{(0)}$ is an orthoprojector from $\mathfrak{H}^{(0)}$, and the subspace $(P^- + P^{(0)})\mathfrak{H}$ would be invariant relative to A, since

$$\begin{aligned} Ah &= AP^-h + AP^{(0)}h \\ &= P^-AP^-h + P^-AP^{(0)}h + (P^+ - P^-)AP^{(0)}h \in P^-\mathfrak{H} \oplus P^{(0)}\mathfrak{H} \\ &\qquad (h \in P^-\mathfrak{H} \oplus P^{(0)}\mathfrak{H}). \end{aligned}$$

The maximality of π^* in the set of all chains now follows from Theorem 14.1.

Theorem 15.3. *If the chain π belongs to the completely continuous operator A, then it precedes some maximal chain also belonging to A.*

For the proof it suffices to consider the set $\mathfrak{A}$ of all chains belonging to A which π precedes, and repeat the arguments applied for the proof of Theorem 15.2.

2. **Maximal chains of operators not having spectral points other than zero.**

Theorem 15.4. *If the operator A does not have any point other than zero in its spectrum, and (P^-, P^+) is a jump of a maximal chain belonging to it, then*

$$(P^+ - P^-) A (P^+ - P^-) = 0. \tag{15.3}$$

Proof. Suppose that A induces an operator A_0 in the subspace $P^+\mathfrak{H}$. Since $P^-\mathfrak{H}$ is invariant relative to A_0, the one-dimensional subspace $(P^+ - P^-)\mathfrak{H}$ is invariant relative to A_0^*. Since the spectrum of A_0^* consists of the point zero only, we get

$$A_0^* (P^+ - P^-) = 0.$$

Hence

$$(P^+ - P^-) A (P^+ - P^-) = (P^+ - P^-) AP^+ = (P^+ - P^-) A_0 P^+ = 0.$$

Theorem 15.5. *Suppose that the maximal chain π belongs to the completely nonselfadjoint operator A, and that the spectrum of A has no points other than zero. If A is dissipative, then the chain π is continuous.*

Proof. Suppose that the chain π has a jump (P^-, P^+). Then, from Theorem 15.4,

$$(P^+ - P^-) A_I (P^+ - P^-) = 0$$

and so $A_I(P^+ - P^-) = 0$. Thus

$$A (P^+ - P^-) = A^* (P^+ - P^-).$$

On the other hand, again in view of Theorem 15.4,

$$A (P^+ - P^-) = P^- A (P^+ - P^-),$$
$$A^* (P^+ - P^-) = (E - P^+) A^* (P^+ - P^-).$$

Since the orthoprojectors P^- and $E - P^+$ are mutually orthogonal,

$$A (P^+ - P^-) = A^* (P^+ - P^-) = 0,$$

which contradicts the complete nonselfadjointness of A.

§16. Triangular representations of Volterra operators

1. **The finite-dimensional case.** Suppose that A is a linear operator operating in an n-dimensional space $\mathfrak{H}$. In Lemma 1.1 it was proved that there exist subspaces

$$0 = \mathfrak{H}_0 \subset \mathfrak{H}_1 \subset \ldots \subset \mathfrak{H}_n = \mathfrak{H} \quad (\dim \mathfrak{H}_k = k),$$

invariant relative to A. The orthoprojectors P_k onto the subspaces $\mathfrak{H}_k$ form a maximal chain belonging to A. Since $AP_k = P_k AP_k$, we have

$$A\,\Delta P_k = \Delta P_k AP_{k-1} + P_k A\,\Delta P_k \quad (\Delta P_k = P_k - P_{k-1})$$

so that $\Delta P_j A \Delta P_k = 0$ $(k < j)$.

In particular, if A is a Volterra operator, then $\Delta P_k A \Delta P_k = 0$ by Theorem 15.4, and therefore

$$\Delta P_j A\,\Delta P_k = 0 \quad (k \leqslant j), \quad \Delta P_j A^*\,\Delta P_k = 0 \quad (j \leqslant k).$$

Taking account of the equations

$$\sum_{j=1}^{n} \Delta P_j = E, \; A = \sum_{j,\,k=1}^{n} \Delta P_j A\,\Delta P_k = \sum_{j<k} \Delta P_j A\,\Delta P_k,$$

we get

$$A = 2\sum_{j<k} \Delta P_j A_R \Delta P_k = 2i \sum_{j<k} \Delta P_j A_I\,\Delta P_k. \tag{16.1}$$

Formulas (16.1) may obviously be rewritten in the form

$$A = 2\sum_{k=1}^{n} P_{k'} A_R\,\Delta P_k = 2i \sum_{k=1}^{n} P_{k'} A_I\,\Delta P_k, \tag{16.2}$$

where $P_{k'}$ is either of the orthoprojectors P_{k-1}, P_k.

In relations (16.2) the Volterra operator A is expressed in terms of a maximal chain belonging to it and one of its Hermitian components. It is natural to call such a representation of A *triangular*.

We shall give equations (16.2) another form, which admits generalization to the case of a Volterra operator operating in an infinite-dimensional space. To this end we introduce the following definitions, relating to both finite-dimensional and infinite-dimensional spaces.

Suppose that π is any closed chain in the space $\mathfrak{H}$. The collection of orthoprojectors

$$\inf \pi = Q_0 < Q_1 < \ldots < Q_{m-1} < Q_m = \sup \pi \quad (Q_k \in \pi) \tag{16.3}$$

is said to be a *subdivision* of the chain π. The subdivision

$$P_0 < P_1 < \ldots < P_{n-1} < P_n \quad (P_j \in \pi)$$

is said to be a *continuation* of the subdivision (16.3) if every orthoprojector Q_k coincides with one of the orthoprojectors P_j.

Let $F(P)$ and $G(P)$ $(P \in \pi)$ be functions on the chain, with values which are bounded linear operators acting in $\mathfrak{H}$. The operator $\mathfrak{J}$ is called the *integral* of $F(P)$ relative to $G(P)$ in the sense of S. O. Šatunovskiĭ, and denoted by $\int_\pi F(P)dG(P)$, if for each $\epsilon > 0$ there exists a subdivision of the chain π such that for each of its continuations $P_0 < P_1 < \cdots < P_n$ and for any orthoprojectors $Q_j \in \pi$ satisfying the conditions $P_{j-1} \leq Q_j \leq P_j$, the inequality

$$\left\| \mathcal{J} - \sum_{j=1}^{n} F(Q_j)(G(P_j) - G(P_{j-1})) \right\| < \varepsilon$$

is satisfied.

Analogously one defines the integral $\int_\pi dF(P)G(P)$.

If A is a Volterra operator in finite-dimensional space and π is a maximal chain belonging to it, then formulas (16.2) may be given the form

$$A = 2\int_\pi PA_R\, dP = 2i\int_\pi PA_I\, dP. \tag{16.4}$$

2. **Triangular representation of an arbitrary Volterra operator.**

Theorem 16.1. *If π is a closed chain and one of the integrals $\int_\pi F(P)dG(P)$, $\int_\pi dF(P)G(P)$ exists, then the second exists as well, while*

$$\int_\pi F(P)\,dG(P) + \int_\pi dF(P)\,G(P) = F(\overline{P})\,G(\overline{P}) - F(\underline{P})\,G(\underline{P}) \tag{16.5}$$

$$(\underline{P} = \inf \pi,\ \ \overline{P} = \sup \pi).$$

Proof. Suppose for example that the integral $\int_\pi F(P)dG(P)$ exists. Choosing a number $\epsilon > 0$, we find a subdivision $\mathfrak{z}$ of the chain π such that for each of its continuations $P_0 < P_1 < \cdots < P_m$ and any $Q_j \in \pi$ satisfying the conditions $P_{j-1} \leq Q_j \leq P_j$ the inequality

$$\left\| \int_\pi F(P)\,dG(P) - \sum_{j=1}^{m} F(Q_j)(G(P_j) - G(P_{j-1})) \right\| < \varepsilon$$

will be satisfied. In particular, the subdivision

$$P_0 \leqslant Q_1 \leqslant P_1 \leqslant Q_2 \leqslant \ldots \leqslant P_{m-1} \leqslant Q_m \leqslant P_m$$

is a continuation of $\mathfrak{z}$. Since

$$\sum_{j=1}^{m} F(P_{j-1})(G(Q_j)-G(P_{j-1}))+\sum_{j=1}^{m} F(P_j)(G(P_j)-G(Q_j))$$
$$=F(\bar{P})\,G(\bar{P})-F(\underline{P})\,G(\underline{P})-\sum_{j=1}^{m}(F(P_j)-F(P_{j-1}))\,G(Q_j),$$

we have

$$\left\| F(\bar{P})\,G(\bar{P})-F(\underline{P})\,G(\underline{P})-\int_{\pi} F(P)\,dG(P)-\sum_{j=1}^{m}(F(P_j)-F(P_{j-1}))\,G(Q_j)\right\|<\varepsilon.$$

The theorem is proved.

It is not hard to see that the existence of one of the integrals $\int_{\pi} F(P)dG(P)$ or $\int_{\pi} dG^*(P)F^*(P)$ implies the existence of the other. In addition,

$$\left(\int_{\pi} F(P)\,dG(P)\right)^*=\int_{\pi} dG^*(P)\,F^*(P). \tag{16.6}$$

Lemma 16.1. *Suppose that π is a closed chain in the space $\mathfrak{H}$. For given $h \in \mathfrak{H}$ and $\epsilon > 0$ there exists a subdivision of the chain π such that each of its continuations $P_0 < P_1 < \cdots < P_m$ will satisfy the following condition: if $((P_j - P_{j-1})h, h) \geq \epsilon$, then (P_{j-1}, P_j) is a jump of the chain π.*

Proof. It is easy to see that the set $\mathfrak{N}$ of numbers of the form (Ph, h), $P \in \pi$, is closed. Therefore there exists a subdivision

$$x_0<x_1<\ldots<x_n \quad (x_0=\min\mathfrak{N},\ x_n=\max\mathfrak{N},\ x_k\in\mathfrak{N})$$

such that if $x_k - x_{k-1} \geq \epsilon$, then (x_{k-1}, x_k) is an interval adjacent to the set $\mathfrak{N}$. Suppose that π_j $(j = 0, \cdots, n)$ consists of those orthoprojectors $P \in \pi$ for which $(Ph, h) = x_j$. Then the subdivision

$$P_0'\leqslant P_0''<P_1'\leqslant P_1''<\ldots<P_n'\leqslant P_n''$$
$$(P_j'=\inf\pi_j,\ P_j''=\sup\pi_j)$$

satisfies the requirements of the lemma.

Theorem 16.2. *Suppose that A is a completely continuous operator and π a closed chain for which every jump (P^-, P^+) satisfies the condition*

$$(P^+-P^-)\,A\,(P^+-P^-)=0. \tag{16.7}$$

Then to each $\epsilon > 0$ one may assign a subdivision $\mathfrak{z}_\epsilon$ of the chain π such that for each of its continuations

$$P_0 < P_1 < \ldots < P_m \tag{16.8}$$

the inequality

$$\left\| \sum_{k=1}^{m} \Delta P_k A \, \Delta P_k \right\| \leqslant \varepsilon \qquad (\Delta P_k = P_k - P_{k-1}) \tag{16.9}$$

will be satisfied.

Proof. First we prove the theorem for the case when A is a selfadjoint operator. Denote by $e_1, e_2, \cdots$ an orthonormalized basis in the space $\overline{A\mathfrak{H}}$ such that $Ae_j = \omega_j e_j$. Since $\omega_j \to 0$, for a given $\epsilon > 0$ there exists an integer N for which

$$|\omega_j| < \frac{\varepsilon}{2} \qquad (j > N).$$

Moreover, in view of Lemma 16.1 there exists a subdivision of the chain π such that any of its continuations (16.8) still satisfies the following condition: if for some $k = 1, 2, \cdots, m$ and some $j = 1, 2, \cdots, N$

$$((P_k - P_{k-1}) e_j, e_j) \geqslant \frac{\varepsilon}{2 \sum_{r=1}^{N} |\omega_r|},$$

then (P_{k-1}, P_k) is a jump of the chain π. From the equation

$$\left(\sum_{k=1}^{m} \Delta P_k A \, \Delta P_k h, \, h \right) = \left(\sum_{k=k_1}^{k_p} \Delta P_k A \, \Delta P_k h, \, h \right)$$
$$= \sum_{k=k_1}^{k_p} \sum_{j=1}^{\infty} |(\Delta P_k h, e_j)|^2 \, \omega_j \qquad (h \in \mathfrak{H}),$$

where the summation is carried out only over those values $k_1, \cdots, k_p$ of the index k for which $\Delta P_k A \Delta P_k \neq 0$, and the estimates

$$\sum_{k=k_1}^{k_p} \sum_{j=1}^{N} |(\Delta P_k h, e_j)|^2 |\omega_j| \leqslant \sum_{k=k_1}^{k_p} \sum_{j=1}^{N} \|\Delta P_k h\|^2 \, \|\Delta P_k e_j\|^2 \, |\omega_j|$$
$$\leqslant \frac{\varepsilon}{2 \sum_{r=1}^{N} |\omega_r|} \sum_{k=k_1}^{k_p} \sum_{j=1}^{N} \|\Delta P_k h\|^2 \, |\omega_j| \leqslant \frac{\varepsilon}{2} \|h\|^2$$

and

$$\sum_{k=k_1}^{k_p}\sum_{j=N+1}^{\infty}|(\Delta P_k h, e_j)|^2|\omega_j| \leqslant \frac{\varepsilon}{2}\sum_{k=k_1}^{k_p}\sum_{j=N+1}^{\infty}|(\Delta P_k h, e_j)|^2$$

$$\leqslant \frac{\varepsilon}{2}\sum_{k=k_1}^{k_p}\|\Delta P_k h\|^2 \leqslant \frac{\varepsilon}{2}\|h\|^2$$

(16.9) follows.

We are now in a position to obtain a complete proof of the theorem, using the representation $A = A_R + iA_I$ and noting that the selfadjoint operators A_R and A_I satisfy the condition (16.7).

Theorem 16.3. *Suppose that A is a completely continuous operator and that π is a maximal chain belonging to A. If for each jump (P^-, P^+) of the chain π condition (16.7) is satisfied, then*

$$\int_{\pi} PA\,dP = A, \quad \int_{\pi} PA^*\,dP = 0 \tag{16.10}$$

so that

$$A = 2\int_{\pi} PA_R\,dP = 2i\int_{\pi} PA_I\,dP. \tag{16.11}$$

Proof. Take an $\epsilon > 0$. By the preceding theorem there exists a subdivision of the chain π such that any of its continuations $P_0 < P_1 < \cdots < P_m$ satisfies inequality (16.9).

If $P_{k-1} \leq Q_k \leq P_k$ $(Q_k \in \pi)$, then $P_k - Q_k \leq \Delta P_k$. Accordingly,

$$\left(\sum_{k=1}^{m}(P_k - Q_k)A\,\Delta P_k\right)^*\sum_{k=1}^{m}(P_k - Q_k)A\,\Delta P_k$$

$$= \sum_{k=1}^{m}\Delta P_k A^*(P_k - Q_k)A\,\Delta P_k \leqslant \sum_{k=1}^{m}\Delta P_k A^*\,\Delta P_k A\,\Delta P_k$$

$$= \left(\sum_{k=1}^{m}\Delta P_k A\,\Delta P_k\right)^*\sum_{k=1}^{m}\Delta P_k A\,\Delta P_k,$$

and therefore

$$\left\|\sum_{k=1}^{m}(P_k - Q_k)A\,\Delta P_k\right\| \leqslant \varepsilon.$$

Using the relations $\Delta P_k A \Delta P_r = 0$ $(r < k)$, which follow from the invariance of the subspaces $P\mathfrak{H}$ $(P \in \pi)$ relative to A, we obtain

$$A = \sum_{k,\, r=1}^{m} \Delta P_k A\, \Delta P_r = \sum_{k \leqslant r} \Delta P_k A\, \Delta P_r = \sum_{k=1}^{m} P_k A\, \Delta P_k,$$

$$\left\| A - \sum_{k=1}^{m} Q_k A\, \Delta P_k \right\| = \left\| \sum_{k=1}^{m} (P_k - Q_k)\, A\, \Delta P_k \right\| \leqslant \varepsilon.$$

Thus the integral $\int_\pi PAdP$ exists and the formula $A = \int_\pi PAdP$ holds. In view of (16.6) and (16.5)

$$A^* = \int_\pi dPA^*P = A^* - \int_\pi PA^*\, dP$$

so that $\int_\pi PA^* dP = 0$.

Theorem 16.4. *If A is a Volterra operator and π is a maximal chain belonging to it, then*

$$A = 2\int_\pi PA_R\, dP = 2i\int_\pi PA_I\, dP.$$

Proof. By Theorem 15.4 the jumps of the chain π satisfy condition (16.7). Thus Theorem 16.3 can be applied to the operator A and its maximal chain π.

Corollary. *There exists not more than one Volterra operator possessing a given maximal chain and a given real and imaginary component.*

3. **Maximal chains of a completely nonselfadjoint Volterra operator.**

Theorem 16.5. *If A is a completely nonselfadjoint Volterra operator operating in the space $\mathfrak{H}$, and π is a maximal chain belonging to it, then the closure $\mathfrak{H}_1$ of the linear envelope of vectors of the form PA_Ih $(P \in \pi,\ h \in \mathfrak{H})$ coincides with $\mathfrak{H}$.*

Proof. We suppose that $\mathfrak{H}_0 = \mathfrak{H} \ominus \mathfrak{H}_1 \neq 0$. Since each vector $g \in \mathfrak{H}_0$ satisfies the condition

$$(PA_I h, g) = 0 \quad (P \in \pi,\ h \in \mathfrak{H}),$$

we have $A_I P g = 0$ $(P \in \pi)$. Accordingly,

$$Ag = 2i\int_\pi PA_I\, dPg = 0$$

and

$$A^*g = -2i \int_{\pi} dPA_I Pg = 0,$$

which contradicts the complete nonselfadjointness of the operator A.

Theorem 16.6. *Suppose that A is a dissipative Volterra operator operating in the space $\mathfrak{H}$, and π a maximal chain belonging to it. For the operator A to be completely nonselfadioint, it is necessary and sufficient that the closure of the linear envelope of vectors of the form $PA_I h$ $(P \in \pi,\ h \in \mathfrak{H})$ coincide with $\mathfrak{H}$.*

Proof. The necessity was established in the preceding theorem. If A is not completely nonselfadjoint, then there exists a vector $g \neq 0$ such that $A^* g = 0$. Using the equation $PAP = AP$ $(P \in \pi)$, we find that

$$\| A_I^{1/2} Pg \|^2 = (PA_I Pg,\ g) = \operatorname{Im}(PAPg,\ g) = \operatorname{Im}(Pg,\ A^*g) = 0.$$

Therefore $A_I Pg = 0$, so that g is orthogonal to all vectors of the form $PA_I h$ $(P \in \pi,\ h \in \mathfrak{H})$.

The theorem is proved.

We note that the criterion obtained in §13 for the operator

$$Af(x) = 2i \int_x^l f(t)\, [\overline{\varphi_1(t)}\, \varphi_1(x) + \overline{\varphi_2(t)}\, \varphi_2(x) + \ \dots\ + \overline{\varphi_p(t)}\, \varphi_p(x)]\, dt \tag{16.12}$$

operating in $L_2(0,\ l)$ to be completely nonselfadjoint follows easily also from Theorem 16.6. It suffices to note that the operator (16.12) possesses a maximal chain consisting of the orthoprojectors P_x introduced in example 1 of §14.

§17. Integral of triangular truncation

1. **Integral of triangular truncation and Volterra operators.** The integral $\int_\pi PT\, dP$, where T is a bounded operator and π is a maximal chain, is said to be the *integral of triangular truncation* of the operator T relative to the chain π. These terms express the fact that for any subdivision $P_0 < P_1 < \cdots < P_m$ of the chain π

$$T = \sum_{j,\ k=1}^{m} \Delta P_j T \Delta P_k \qquad (\Delta P_j = P_j - P_{j-1}),$$

while at the same time the integral $\int_\pi PT\, dP$ (if it exists) is the limit of sums of the form $\Sigma_{j<k} \Delta P_j T \Delta P_k$.

In Theorem 16.4 it was proved that every Volterra operator is representable in the form of an integral of triangular truncation. In the present subsection we shall prove the validity of the converse assertion.

If the sequence A_n $(n = 1, 2, \cdots)$ consists of completely continuous operators and $\|A_n - A\| \to 0$, then the operator A, as is known, is also completely continuous. In the case when all the A_n are Volterra operators, the following assertion holds.

Lemma 17.1. *If the sequence A_n of Volterra operators converges in norm to A, then A is a Volterra operator.*

Proof. Recall that the spectrum of a completely continuous operator is either a finite set or else consists of zero and a sequence tending to zero. Suppose that the spectrum of A contains a point $\lambda_0 \neq 0$. Suppose that γ is a circumference with center λ_0 and with radius so small that zero and all the points other than λ_0 of the spectrum lie outside γ. Since the function $\|(A - \lambda E)^{-1}\|$ is bounded for $\lambda \in \gamma$ and

$$\|(A_n - \lambda E)^{-1}\| - \|(A - \lambda E)^{-1}\| \leqslant \|(A_n - \lambda E)^{-1} - (A - \lambda E)^{-1}\| \leqslant \|(A_n - \lambda E)^{-1}\| \, \|A - A_n\| \, \|(A - \lambda E)^{-1}\|,$$

for sufficiently large n we have

$$\|(A_n - \lambda E)^{-1}\| \leqslant \frac{\|(A - \lambda E)^{-1}\|}{1 - \|A - A_n\| \, \|(A - \lambda E)^{-1}\|}.$$

Accordingly the functions $\|A_n - \lambda E)^{-1}\|$ $(n = 1, 2, \cdots)$ are uniformly bounded on γ, which means that

$$\int_\gamma (A - \lambda E)^{-1} d\lambda = \int_\gamma (A - \lambda E)^{-1} d\lambda - \int_\gamma (A_n - \lambda E)^{-1} d\lambda = \int_\gamma (A - \lambda E)^{-1} (A_n - A)(A_n - \lambda E)^{-1} d\lambda \to 0,$$

i.e. $\int_\gamma (A - \lambda E)^{-1} d\lambda = 0$, which contradicts a theorem of F. Riesz (see Riesz and Sz.-Nagy [1]).

Lemma 17.2. *Suppose that π is a closed chain and T a bounded operator. If* *the integral* $\int_\pi PT\,dP$ *exists and* $Q \in \pi$, *then*

$$\int_\pi PT\,dPQ = \int_{\pi_1} PT\,dP, \tag{17.1}$$

$$Q\int_\pi PT\,dP = \int_{\pi_1} PT\,dP + QT(\bar{P} - Q) \quad (\bar{P} = \sup \pi), \tag{17.2}$$

where π_1 *is the chain consisting of all the orthoprojectors* $P \in \pi$ *satisfying the condition* $P \leq Q$.

Proof. For any $\epsilon > 0$ there exists a subdivision $\mathfrak{z}$ of the chain π such that

$$\left\| \int_\pi PT\,dP - S \right\| \leqslant \varepsilon$$

whenever S is an integral sum of the integral $\int_\pi PT\,dP$, constructed using any continuation of the subdivision $\mathfrak{z}$. Denote by π_2 the chain consisting of the orthoprojectors $P \in \pi$ for which $Q \leq P$. Choose subdivisions $\mathfrak{z}_1$ and $\mathfrak{z}_2$ of the chains π_1 and π_2 so that they jointly constitute a continuation of the subdivision $\mathfrak{z}$. Let

$$\underline{P} = P_0 < P_1 < \ldots < P_m = Q \quad (\underline{P} = \inf \pi),$$
$$Q = P_m < P_{m+1} < \ldots < P_n = \bar{P}$$

be continuations of the subdivisions $\mathfrak{z}_1$ and $\mathfrak{z}_2$. Putting

$$S_1 = \sum_{j=1}^m Q_j T\,\Delta P_j, \quad S_2 = \sum_{j=m+1}^n Q_j T\,\Delta P_j$$
$$(P_{j-1} \leqslant Q_j \leqslant P_j, \quad \Delta P_j = P_j - P_{j-1}, \quad Q_j \in \pi)$$

and taking account of the fact that $S_1 Q = S_1$ and $S_2 Q = 0$, we get

$$\left\| \int_\pi PT\,dPQ - S_1 \right\| = \left\| \int_\pi PT\,dPQ - (S_1 + S_2)\,Q \right\|$$
$$\leqslant \left\| \int_\pi PT\,dP - (S_1 + S_2) \right\| \leqslant \varepsilon.$$

Thus we have proved the existence of the integral $\int_{\pi_1} PT\,dP$ and also equation (17.1). The second assertion of the theorem may be obtained analogously, using the relations

$$QS_1 = S_1, \qquad QS_2 = QT(\overline{P} - Q).$$

Theorem 17.1. *If the integral*

$$A = 2i \int_{\pi} PH\, dP,$$

where π is a maximal chain and H is a completely continuous selfadjoint operator, then A is a Volterra operator, π belongs to A and $H = A_I$.

Proof. For a given $\epsilon > 0$ there exists a subdivision $P_0 < P_1 < \cdots < P_m$ of the chain π such that

$$\left\| A - \sum_{j=1}^{m} P_{j-1} H\, \Delta P_j \right\| \leqslant \varepsilon \qquad (\Delta P_j = P_j - P_{j-1}).$$

The operator

$$S = \sum_{j=1}^{m} P_j H\, \Delta P_j = \sum_{k<j} \Delta P_k H\, \Delta P_j$$

is a Volterra operator, since $S^m = 0$. By Lemma 17.1 A is a Volterra operator.

Suppose that $Q \in \pi$ and that π_1 is the collection of those orthoprojectors $P \in \pi$ for which $P \leq Q$. By formula (17.1) $AQ = 2i \int_{\pi_1} PHdP$, from which it follows that $AQ = QAQ$. This last equation means that the subspaces $Q\mathfrak{H}$ are invariant relative to A.

Using formulas (16.5) and (16.6), we get

$$A = 2i \int_{\pi} PH\, dP = 2iH - 2i \int_{\pi} dPHP = 2iH + A^*,$$

$$H = \frac{A - A^*}{2i} = A_I.$$

Corollary 1. *Suppose that A is a completely continuous operator and that π is a maximal chain belonging to it. If for each jump (P^-, P^+) of the chain π the equation*

$$(P^+ - P^-)\, A\, (P^+ - P^-) = 0$$

holds, then A is a Volterra operator. In particular, each completely continuous operator possessing a continuous maximal chain is a Volterra operator.

The proof follows directly from Theorem 16.3 and 17.1.

Corollary 2. *If the Volterra operators A, B and the bounded operator C have a common maximal chain π, then $\lambda A+\mu B$, AC and CA are Volterra operators, with arbitrary numbers λ, μ.*

Indeed, the operators $\lambda A+\mu B$, AC and CA are completely continuous and the chain π belongs to each of them. If (P^-, P^+) is a jump of the chain π, then by Theorem 15.4

$$(P^+-P^-)A(P^+-P^-)=0,\ (P^+-P^-)B(P^+-P^-)=0, \tag{17.3}$$

so that

$$(P^+-P^-)(\lambda A+\mu B)(P^+-P^-)=0. \tag{17.4}$$

The first of equations (17.3) is equivalent to the relation $AP^+\mathfrak{H}\subseteq P^-\mathfrak{H}$. Since

$$AC(P^+-P^-)\mathfrak{H}\subseteq AP^+\mathfrak{H}\subseteq P^-\mathfrak{H},$$
$$CA(P^+-P^-)\mathfrak{H}\subseteq CAP^+\mathfrak{H}\subseteq CP^-\mathfrak{H}\subseteq P^-\mathfrak{H},$$

we have

$$(P^+-P^-)AC(P^+-P^-)=0,\quad (P^+-P^-)CA(P^+-P^-)=0. \tag{17.5}$$

The corollary now follows from (17.4), (17.5) and Corollary 1.

Theorem 17.2. *For there to exist a Volterra operator A having a given maximal chain π and a given imaginary component H, it is necessary and sufficient that the integral $\int_\pi PH\,dP$ should exist. If this condition is fulfilled, then A is uniquely defined by the formula $A=2i\int_\pi PH\,dP$.*

The proof follows from Theorems 16.4 and 17.1.

2. **Conditions for the existence of the integral of triangular truncation.** Suppose that π is a closed chain and T a bounded operator.

Theorem 17.3. *If the integral $\mathfrak{I}=\int_\pi PT\,dP$ exists, then for each jump (P^-, P^+) of the chain π the equation*

$$(P^+-P^-)T(P^+-P^-)=0 \tag{17.6}$$

holds.

Proof. Consider some jump (P^-, P^+) of π. For a given $\epsilon>0$ there exists a subdivision $\mathfrak{z}$ of π having the following properties: 1) P^- and P^+ enter into $\mathfrak{z}$, 2) for any integral sum S of the integral $\int_\pi PT\,dP$, constructed using the subdivision $\mathfrak{z}$, the inequality $\|\mathfrak{I} - S\| \leq \epsilon$ holds. Construct two integral sums S_1 and S_2 differing only in that the term $P^- T(P^+ - P^-)$ contained in the first of them is replaced in the second by the term $P^+ T(P^+ - P^-)$. Since $\|\mathfrak{I} - S_1\| \leq \epsilon$ and $\|\mathfrak{I} - S_2\| \leq \epsilon$, we have $\|(P^+ - P^-)T(P^+ - P^-)\| \leq 2\epsilon$, so that (17.6) follows in view of the arbitrary choice of ϵ.

Theorem 17.4. *Suppose that π is a closed chain and that T is a finite-dimensional operator. If equation* (17.6) *is satisfied for each jump (P^-, P^+) of π then the integral $\int_\pi PT\,dP$ exists.*

Proof. It suffices to prove that for a given $\epsilon > 0$ there exists a subdivision $\mathfrak{z}$ of the chain π such that for any integral sums S and S' of the integral $\int_\pi PT\,dP$, constructed using the subdivision $\mathfrak{z}$ and any of its continuations the estimate $\|S - S'\| \leq \epsilon$ is satisfied. Here we may obviously restrict the considerations to the case when T is a selfadjoint operator.

Suppose that $e_1, e_2, \cdots, e_r$ is an orthonormalized basis in the space $T\mathfrak{H}$ and that $Te_k = \omega_k e_k$ $(k = 1, 2, \cdots, r)$. By Lemma 16.1 the chain π admits a subdivision

$$P_0 < P_1 < \ldots < P_m \tag{17.7}$$

such that if for each $j = 1, 2, \cdots, m$ and some $k = 1, \cdots, r$

$$\|(P_j - P_{j-1})e_k\| > \frac{\varepsilon}{\sum_{i=1}^{r} |\omega_i|},$$

then (P_{j-1}, P_j) is a jump of the chain π.

Consider some continuation $P'_0 < P'_1 < \cdots < P'_n$ of the subdivision (17.7) and set up the integral sums

$$S = \sum_{j=1}^{m} Q_j T\,\Delta P_j \qquad (P_{j-1} \leqslant Q_j \leqslant P_j),$$

$$S' = \sum_{j=1}^{n} Q'_j T\,\Delta P'_j \qquad (P'_{j-1} \leqslant Q'_j \leqslant P'_j).$$

The first of them may be rewritten in the form $S = \sum_{j=1}^n Q''_j\, T\Delta P'_j$, where each orthoprojector Q''_j coincides with one of the Q_j. Thus

$$S - S' = \sum_{j=1}^{n} (Q''_j - Q'_j)\,T\,\Delta P'_j, \tag{17.8}$$

while according to condition (17.6) the only terms which can be different from zero are those for which

$$\|(Q''_j - Q'_j)\,e_k\| \leqslant \frac{\varepsilon}{\sum_{i=1}^{r} |\omega_i|} \qquad (k = 1,\ 2,\ \ldots,\ r).$$

Denoting the indices of these terms by $n_1, n_2, \cdots, n_q$, we get

$$|((S-S')f,\ g)| \leqslant \sum_{j=n_1}^{n_q} \sum_{k=1}^{r} |(\Delta P'_j f,\ e_k)|\,|\omega_k|\,|((Q''_j - Q'_j)\,e_k,\ g)|$$

$$\leqslant \frac{\varepsilon}{\sum_{i=1}^{r} |\omega_i|} \|g\| \sum_{k=1}^{r} \sum_{j=1}^{n} \|\Delta P'_j f\|\,\|\Delta P'_j e_k\|\,|\omega_k|.$$

Since

$$\sum_{j=1}^{n} \|\Delta P'_j f\|\,\|\Delta P'_j e_k\| \leqslant \left(\sum_{j=1}^{n} \|\Delta P'_j f\|^2\right)^{1/2} \left(\sum_{j=1}^{n} \|\Delta P'_j e_k\|^2\right)^{1/2} = \|f\|,$$

we have $|((S - S')f,\ g)| \leq \epsilon \|f\|\ \|g\|$, so that $\|S - S'\| \leq \epsilon$.

The following assertion follows from Theorems 17.4 and 17.2.

Theorem 17.5. *Suppose that* π *is a maximal chain and that* H *is a finite-dimensional selfadjoint operator. If for each jump* $(P^-,\ P^+)$ *of* π *the equation* $(P^+ - P^-)H(P^+ - P^-) = 0$ *holds, then there exists one and only one Volterra operator* A *such that* 1) π *belongs to* A, and 2) $H = A_I$.

§18. Spectral functions

Suppose given some chain π in the space $\mathfrak{H}$. The subspace $\mathfrak{H}_0 \subset \mathfrak{H}$ is said to be *reproducing* for π if the linear envelope of all vectors of the form $Ph\,(P \in \pi,\ h \in \mathfrak{H}_0)$ is dense in $\mathfrak{H}$. For a chain to have a reproducing subspace it is necessary and sufficient that the unit operator enter into its structure. The smallest of the dimensions of all reproducing subspaces is called the *rank* of the chain π. A system of vectors is called *reproducing* for the chain π if the closed linear envelope of this system is a reproducing subspace.

Consider some bounded closed set $\mathfrak{M}$ of real numbers, containing at least two elements. A function $P(x)$ $(x \in \mathfrak{M})$ is said to be a *spectral* function if it has the following properties:

I) $P(x) \in \mathfrak{P}_{\mathfrak{H}}$ $(x \in \mathfrak{M})$;

II) if $x_1 < x_2$ $(x_1,\ x_2 \in \mathfrak{M})$, then $P(x_1) < P(x_2)$;

III) if $x_n \to x_0$ $(x_n \in \mathfrak{M})$, then $P(x_n) \to P(x_0)$ in the sense of strong convergence;

IV) $P(\alpha) = 0$ and $P(\beta) = E$, where $\alpha = \min \mathfrak{M}$, $\beta = \max \mathfrak{M}$;

V) for any interval (a, b) with $a \in \mathfrak{M}, b \in \mathfrak{M}$ but such that no point $x \in (a, b)$ lies in $\mathfrak{M}$ ("interval adjacent to $\mathfrak{M}$")

$$\dim \{P(b)\mathfrak{H} \ominus P(a)\mathfrak{H}\} = 1.$$

The range π of the spectral function $P(x)$ $(x \in \mathfrak{M})$ is obviously a chain, while the pair (P^-, P^+) is a jump of the chain if and only if $P^- = P(a)$ and $P^+ = P(b)$, where (a, b) is an interval adjacent to $\mathfrak{M}$. Thus the chain π is continuous if and only if $\mathfrak{M} = [\alpha, \beta]$.

A subspace $\mathfrak{H}_0$ is said to be *reproducing* for the spectral function $P(x)$ if it is reproducing for the chain of its values. Analogously one introduces for a spectral function the concepts of rank and reproducing system.

Theorem 18.1. *Suppose that π is a maximal chain and that $h_1, h_2, \cdots, h_s$ $(s \leq \infty)$ is a reproducing system for it, normalized by the condition $\sum_{k=1}^{\infty} \|h_k\|^2 < \infty$. Then there exists a spectral function $P(x)$ $(x \in \mathfrak{M})$ such that*

1) $$\sum_{k=1}^{s} (P(x) h_k, h_k) = x \quad (x \in \mathfrak{M}); \tag{18.1}$$

2) *the range of the function $P(x)$ $(x \in \mathfrak{M})$ coincides with π.*

Proof. We assign to each orthoprojector $P \in \pi$ the number

$$x(P) = \sum_{k=1}^{s} (Ph_k, h_k).$$

We denote by $\mathfrak{M}$ the set of all numbers obtained in this way. The mapping $x(P)$ of the chain π onto the set $\mathfrak{M}$ has the following properties:

a) *If $P_1 < P_2$ $(P_1, P_2 \in \pi)$, then $x(P_1) < x(P_2)$.* Indeed, at least one of the vectors $(P_2 - P_1)h_k$ $(k = 1, 2, \cdots, s)$ is distinct from zero, for in the contrary case we would have the equations

$$(Ph_k, h) = (Ph_k, (P_2 - P_1)h) = (P(P_2 - P_1)h_k, h) = 0$$
$$(P \in \pi, \ k = 1, 2, \ldots, s)$$

for any vector $h \in (P_2 - P_1)\mathfrak{H}$, so that the system $h_1, \cdots, h_s$ would not be reproducing. Accordingly,

$$x(P_2) - x(P_1) = \sum_{k=1}^{s} ((P_2 - P_1) h_k, h_k) = \sum_{k=1}^{s} \| (P_2 - P_1) h_k \|^2 > 0.$$

b) *If* $P_n \to P_0$, *then* $x(P_n) \to x(P_0)$. This assertion is obvious for $s < \infty$. In the case $s = \infty$ it follows from the inequality

$$| x(P_n) - x(P_0) | \leqslant \sum_{k=1}^{N} | ((P_n - P_0) h_k, h_k) | + \sum_{k=N+1}^{\infty} \| h_k \|^2,$$

inasmuch as one may make the second term arbitrarily small by taking N sufficiently large, and then the first term as well by choosing a sufficiently large n.

c) *If the sequence* $x(P_n)$ *converges, then the sequence* P_n *converges as well.* Indeed, suppose that P_{n_k} is some monotone subsequence of the sequence P_n. Then P_{n_k} converges strongly to some orthoprojector $P_0 \in \pi$, and therefore $x(P_{n_k}) \to x(P_0)$. It is easy to see that any monotone subsequences selected from sequence P_n converges to the limit P_0. But this means that $P_n \to P_0$.

Consider the function $P(x)$ $(x \in \mathfrak{M})$ inverse to the function $x(P)$ $(P \in \pi)$. From the properties of the function $x(P)$ established above it follows that the set $\mathfrak{M}$ is closed, and $P(x)$ satisfies condition (18.1) and all the enumerated points of the definition of a spectral function.

Theorem 18.2. *For the chain* π *to be maximal, it is necessary and sufficient that it should coincide with the range of some spectral function.*

Proof. From the definition of a spectral function it easily follows that the chain π of all of its values satisfies the conditions of Theorem 14.1. Therefore π is a maximal chain.

The necessity was proved in Theorem 18.1.

Theorem 18.3. *If a spectral function* $P(x)$ *is given in the space* $\mathfrak{H}$, $x \in \mathfrak{M}$, *and* $\mathfrak{M} \neq [\alpha, \beta]$ $(\alpha = \min \mathfrak{M}, \beta = \max \mathfrak{M})$, *then in some space* $\tilde{\mathfrak{H}} \supset \mathfrak{H}$ *there exists a spectral function* $\tilde{P}(x)$ $(\alpha \leq x \leq \beta)$ *having the following properties:*

1) $\tilde{P}(x)h = P(x)h$ $(x \in \mathfrak{M}, h \in \mathfrak{H})$;

2) $P(x)$ *and* $\tilde{P}(x)$ *have the same rank;*

3) *every reproducing system for* $P(x)$ *is also reproducing for* $\tilde{P}(x)$;

4) *if the system of vectors* $h_1, h_2, \cdots, h_s$ $(s \leq \infty)$ *is such that*

$$\sum_{k=1}^{s} (P(x) h_k, h_k) = x \qquad (x \in \mathfrak{M}),$$

then

$$\sum_{k=1}^{s} (\tilde{P}(x) h_k, h_k) = x \quad (\alpha \leqslant x \leqslant \beta).$$

Proof. We define a function $F(x)$ on the interval $[\alpha, \beta]$ by putting

$$F(x) = \begin{cases} P(x), \text{ if } x \in \mathfrak{M}, \\ P(a) + \dfrac{x-a}{b-a}(P(b) - P(a)), \text{ if } x \text{ lies in an interval } (a, b) \text{ adjacent to } \mathfrak{M}. \end{cases}$$

It is not hard to see that $F(x)$ is continuous and strictly increasing. By Theorem 2 of Appendix I there exists a space $\tilde{\mathfrak{H}} \supset \mathfrak{H}$ and a spectral function $\tilde{P}(x)$ in it $(\alpha \leq x \leq \beta)$ such that the convex hull of the set of vectors of the form $\tilde{P}(x)h$ $(\alpha \leq x \leq \beta,\ h \in \mathfrak{H})$ is dense in $\tilde{\mathfrak{H}}$ and

$$Q\tilde{P}(x) h = F(x) h \qquad (\alpha \leqslant x \leqslant \beta,\ h \in \mathfrak{H}),$$

where Q is an orthoprojector onto $\mathfrak{H}$. Inasmuch as

$$Q\tilde{P}(x) h = P(x) h \qquad (x \in \mathfrak{M},\ h \in \mathfrak{H})$$

and

$$\| Q\tilde{P}(x) h \|^2 = \| P(x) h \|^2 = (P(x) h, h) = (Q\tilde{P}(x) h, h)$$
$$= (\tilde{P}(x) h, h) = \| \tilde{P}(x) h \|^2,$$

we have $Q\tilde{P}(x)h = \tilde{P}(x)h$, so that

$$\tilde{P}(x) h = P(x) h \qquad (x \in \mathfrak{M},\ h \in \mathfrak{H}). \tag{18.2}$$

Consider the sequence (a_j, b_j) of all intervals adjacent to the set $\mathfrak{M}$. Denote by e_j the unit vector of the one-dimensional subspace $P(b_j)\mathfrak{H} \ominus P(a_j)\mathfrak{H}$. Obviously

$$P(x) e_j = \begin{cases} 0 & (x \leqslant a_j), \\ e_j & (x > a_j). \end{cases} \tag{18.3}$$

Moreover, in view of (18.2)

$$Q\tilde{P}(x) \tilde{h} = \tilde{P}(x) Q\tilde{h} = P(x) Q\tilde{h} \qquad (x \in \mathfrak{M},\ \tilde{h} \in \tilde{\mathfrak{H}}). \tag{18.4}$$

Thus

$$Q(\tilde{P}(b_j)-\tilde{P}(a_j))\tilde{h}=(P(b_j)-P(a_j))Q\tilde{h},$$

$$Q(\tilde{P}(x)-\tilde{P}(a_j))\tilde{\mathfrak{H}}\subseteq Q(\tilde{P}(b_j)-\tilde{P}(a_j))\tilde{\mathfrak{H}}$$
$$=(P(b_j)-P(a_j))\mathfrak{H}\qquad (a_j<x<b_j),$$

and therefore

$$Q\tilde{P}(x)\tilde{h}=P(a_j)Q\tilde{h}+\lambda_j(x,\ \tilde{h})e_j$$
$$(a_j<x<b_j,\ \tilde{h}\in\tilde{\mathfrak{H}},\ \lambda_j(x,\ \tilde{h})=((\tilde{P}(x)-\tilde{P}(a_j))\tilde{h},\ e_j)). \tag{18.5}$$

If $\tilde{g}_1, \tilde{g}_2, \cdots, \tilde{g}_n$ $(n<\infty)$ is a reproducing system of the function $\tilde{P}(x)$, then the system

$$g_k=Q\tilde{g}_k-\sum_j(\tilde{g}_k,\ e_j)e_j+\sum_j\frac{e_j}{2^j}\quad (k=1,\ 2,\ \ldots,\ n)\cdot \tag{18.6}$$

is reproducing for $P(x)$. Indeed, suppose that for some fixed $h\in\mathfrak{H}$

$$(P(x)g_k,\ h)=0\qquad (x\in\mathfrak{M},\ k=1,\ 2,\ \ldots,\ n). \tag{18.7}$$

Then

$$0=((P(b_j)-P(a_j))g_k,\ h)=(g_k,\ e_j)(e_j,\ h),$$

and since $(g_k, e_j)\neq 0$,

$$(e_j,\ h)=0\qquad (j=1,\ 2,\ \ldots). \tag{18.8}$$

Using relations (18.3)–(18.8), we obtain

$$(\tilde{P}(x)\tilde{g}_k,\ h)=(P(x)Q\tilde{g}_k,\ h)$$
$$=(P(x)g_k,\ h)+\sum_j(\tilde{g}_k,\ e_j)(P(x)e_j,\ h)-\sum_j\frac{(P(x)e_j,\ h)}{2^j}=0$$
$$(x\in\mathfrak{M}),$$
$$(\tilde{P}(x)\tilde{g}_k,\ h)=(P(a_j)Q\tilde{g}_k,\ h)+\lambda_j(x,\ \tilde{g}_k)(e_j,\ h)=0$$
$$(a_j<x<b_j).$$

Therefore $(\tilde{P}(x)\tilde{g}_k, h)=0$ $(\alpha\le x\le\beta,\ k=1,2,\cdots,n)$, i.e. $h=0$. Thus we have proved that the rank $\tilde{r}$ of $\tilde{P}(x)$ is not less than the rank r of $P(x)$. On the other hand, $\tilde{r}\le r$, since any reproducing systems $g_1, g_2, \cdots, g_n$ of $P(x)$ is a reproducing system for $\tilde{P}(x)$ as well. Indeed, if for a fixed $\tilde{h}\in\tilde{\mathfrak{H}}$

$$(\tilde{P}(x)\, g_k,\ \tilde{h}) = 0 \quad (\alpha \leqslant x \leqslant \beta,\ k = 1, 2, \ldots, n),$$

then

$$(\tilde{P}(x)\, P(y)\, g_k,\ \tilde{h}) = 0 \quad (\alpha \leqslant x \leqslant \beta,\ y \in \mathfrak{M},\ k = 1, 2, \ldots, n),$$

which means that $(\widetilde{P}(x)h,\ \widetilde{h}) = 0$ $(\alpha \leq x \leq \beta,\ h \in \mathfrak{H})$. Since the collection of vectors of the form $\widetilde{P}(x)h$ is complete in $\widetilde{\mathfrak{H}}$ it follows that $\widetilde{h} = 0$.

Suppose finally that

$$\sum_{k=1}^{s} (P(x)\, h_k,\ h_k) = x \quad (x \in \mathfrak{M},\ h_k \in \mathfrak{H}).$$

Then

$$\sum_{k=1}^{s} (\tilde{P}(x)\, h_k,\ h_k) = x \quad (x \in \mathfrak{M})$$

and

$$\sum_{k=1}^{s} (\tilde{P}(x)\, h_k,\ h_k) = \sum_{k=1}^{s} (Q\tilde{P}(x)\, h_k,\ h_k) = \sum_{k=1}^{s} (F(x)\, h_k,\ h_k)$$

$$= \sum_{k=1}^{s} (P(a_j)\, h_k,\ h_k) + \frac{x - a_j}{b_j - a_j} \left(\sum_{k=1}^{s} (P(b_j)\, h_k,\ h_k) - \sum_{k=1}^{s} (P(a_j)\, h_k,\ h_k) \right) = x$$

$$(a_j < x < b_j).$$

§19. Integral of triangular truncation relative to a spectral function. Inessential extensions of Volterra operators

1. **Stieltjes integral on a closed set.** Suppose that on a bounded closed set of real numbers $\mathfrak{M}$ there are given functions $F(x)$ and $G(x)$ whose values are bounded linear operators operating in a Hilbert space $\mathfrak{H}$. For each subdivision

$$\alpha = x_0 < x_1 < \ldots < x_m = \beta$$
$$(x_k \in \mathfrak{M},\ \alpha = \min \mathfrak{M},\quad \beta = \max \mathfrak{M}) \tag{19.1}$$

we construct sums of the form

$$\sum_{k=1}^{m} F(\zeta_k)[G(x_k) - G(x_{k-1}) \quad (\zeta_k \in \mathfrak{M},\ x_{k-1} \leqslant \zeta_k \leqslant x_k). \tag{19.2}$$

We agree that by the *weight* of the subdivision (19.1) we will mean the largest of the lengths of all the intervals (x_{k-1}, x_k) which are not adjacent to $\mathfrak{M}$. If all the intervals (x_{k-1}, x_k) are adjacent to $\mathfrak{M}$, then all subdivisions will be regarded as equal to zero.

The operator $\mathfrak{J}$ is said to be the *integral* of $F(x)$ relative to $G(x)$, and is denoted by the symbol $\int_{\mathfrak{M}} F(x)dG(x)$, [1] if for each $\epsilon > 0$ there exists a $\delta > 0$ such that any sum S of the form (19.2) constructed using a subdivision whose weight is less than δ satisfies the inequality $\|\mathfrak{J} - S\| \leq \epsilon$.

Analogously one defines the integrals $\int_{\mathfrak{M}} dF(x)\, G(x)$ and $\int_{\mathfrak{M}} F_1(x)dG(x)F_2(x)$.

From the existence of one of the integrals $\int_{\mathfrak{M}} F(x)dG(x)$ *and* $\int_{\mathfrak{M}} dF(x)G(x)$ *follows the existence of the other, along with the formula*

$$\int_{\mathfrak{M}} F(x)\, dG(x) + \int_{\mathfrak{M}} dF(x)\, G(x) = F(\beta)\, G(\beta) - F(\alpha)\, G(\alpha). \tag{19.3}$$

The proof of this assertion differs little from that which was employed in the derivation of formula (16.5).

We note also that the existence of one of the integrals $\int_{\mathfrak{M}} F(x)dG(x)$ and $\int_{\mathfrak{M}} dG^*(x)F^*(x)$ implies the existence of the second and the equation

$$\left(\int_{\mathfrak{M}} F(x)\, dG(x)\right)^* = \int_{\mathfrak{M}} dG^*(x)\, F^*(x). \tag{19.4}$$

Suppose that the functions $F(P)$ and $G(P)$ are defined for all P lying in some maximal chain π. If $P(x)$ $(x \in \mathfrak{M})$ is a spectral function the range of which coincides with π, and the integral $\int_{\mathfrak{M}} F(x)dG(x)$ exists, where $F(x) = F(P(x))$ and $G(x) = G(P(x))$, then obviously the integral in the sense of S. O. Šatunovskiĭ $\int_{\pi} F(P)dG(P)$ exists, and the equation

$$\int_{\mathfrak{M}} F(x)\, dG(x) = \int_{\pi} F(P)\, dG(P) \tag{19.5}$$

holds.

In the following theorem the restrictions will be given under which the converse assertion is valid.

Theorem 19.1. *If* $F(P)$ *is continuous in norm and the function* $\|G(P)\|$ *is bounded, then the existence of the integral* $\mathfrak{J} = \int_{\pi} F(P)dG(P)$ *implies the existence of the integral* $\int_{\mathfrak{M}} F(x)dG(x)$.

1) In the case $\mathfrak{M} = [\alpha, \beta]$ we shall use the notation $\int_{\alpha}^{\beta} F(x)dG(x)$.

Proof. For a given $\epsilon > 0$ one can produce a subdivision $\alpha = x_0 < x_1 < \cdots < x_m = \beta$ of the set $\mathfrak{M}$ such that the integral sum of the integral $\int_{\mathfrak{M}} F(x)dG(x)$, formed for any one of its refinements, will differ in norm from $\mathfrak{J}$ by less than $\epsilon/2$. Moreover, there exists a $\delta > 0$ such that if $|x' - x''| < \delta$ $(x', x'' \in \mathfrak{M})$, then

$$\| F(x') - F(x'') \| < \frac{\varepsilon}{8(m-1)M},$$

where M is the upper bound of the function $\|G(x)\|$.

Denote by $\mathfrak{z}$ some subdivision $\alpha = y_0 < y_1 < \cdots < y_n = \beta$ of the set $\mathfrak{M}$, whose weight is less than δ, and by $\mathfrak{z}'$ the refinement of $\mathfrak{z}$ obtained by adjoining to $\mathfrak{z}$ the points $x_1, x_2, \cdots, x_{m-1}$. If $y_{j-1} < x_i < y_j$, then

$$\| F(\zeta_j)(G(y_j) - G(y_{j-1})) - [F(\eta_j')(G(x_i) - G(y_{j-1})) + F(\eta_j'')(G(y_j) - G(x_i))] \|$$
$$\leqslant \|(F(\zeta_j) - F(\eta_j'))(G(x_i) - G(y_{j-1}))\|$$
$$+ \|(F(\zeta_j) - F(\eta_j''))(G(y_j) - G(x_i))\| < \frac{\varepsilon}{2(m-1)}$$
$$(y_{j-1} \leqslant \zeta_j \leqslant y_j, \quad y_{j-1} \leqslant \eta_j' \leqslant x_i \leqslant \eta_j'' \leqslant y_j),$$

so that for any integral sum S of the integral $\int_{\mathfrak{M}} F(x)dG(x)$, constructed relative to the subdivision $\mathfrak{z}$, there exists an integral sum S' of the same integral constructed relative to the subdivision $\mathfrak{z}'$ and satisfying $\|S - S'\| < \epsilon/2$. Inasmuch as $\|\mathfrak{J} - S'\| < \epsilon/2$, we have $\|\mathfrak{J} - S\| < \epsilon$.

2. Integral of triangular truncation relative to a spectral function.

Lemma 19.1. *If a sequence F_n $(n = 1, 2, \cdots)$ of bounded linear operators converges strongly to F_0, and K is a completely continuous operator, then the sequence F_nK converges to F_0K in norm.*

Proof. Supposing the contrary, we find a positive number ϵ and a subsequence F_{n_j} such that for all $j = 1, 2, \cdots$ the inequality

$$\|(F_{n_j} - F_0)K\| \geqslant \varepsilon \tag{19.6}$$

holds. There exists a sequence of unit vectors h_j for which

$$\|(F_{n_j} - F_0)K\| = \|(F_{n_j} - F_0)Kh_j\| \quad (j = 1, 2, \ldots).$$

Without loss of generality we may suppose that the sequence h_j converges weakly to some vector h_0. Then Kh_j converges strongly to Kh_0, so that

$$\|(F_{n_j} - F_0)K\| \leqslant (\|F_{n_j}\| + \|F_0\|)\|Kh_j - Kh_0\| + \|(F_{n_j} - F_0)Kh_0\| \to 0,$$

which contradicts inequality (19.6).

The lemma is proved.

Suppose that $P(x)$ ($x \in \mathfrak{M}$) is a spectral function and that π is the chain of all its values. If K is a completely continuous operator, then in view of Lemma 19.1 the function $P(x)K$ is continuous in norm. Taking account of what was said in the preceding subsection, we arrive at the conclusion that from the existence of one of the integrals $\int_{\mathfrak{M}} P(x)K dP(x)$ or $\int_{\pi} PK\,dP$ the existence of the other follows, along with the equality

$$\int_{\mathfrak{M}} P(x)\,K\,dP(x) = \int_{\pi} PK\,dP. \tag{19.7}$$

We shall say that the spectral function $P(x)$ ($x \in \mathfrak{M}$) belongs to the operator A if all of the subspaces $P(x)\mathfrak{H}$ ($x \in \mathfrak{M}$) are invariant relative to A.

We may now reformulate Theorems 15.4, 15.5, 16.4, 17.1 and 17.4 as follows.

Theorem 19.2. *If the spectral function $P(x)$ ($x \in \mathfrak{M}$) belongs to the Volterra operator A and (a, b) is an interval adjacent to the set $\mathfrak{M}$, then*

$$(P(b) - P(a))\,A\,(P(b) - P(a)) = 0.$$

Theorem 19.3. *If the spectral function $P(x)$ ($x \in \mathfrak{M}$) belongs to a completely nonselfadjoint dissipative Volterra operator, then $\mathfrak{M} = [\alpha, \beta]$.*

Theorem 19.4. *If A is a Volterra operator and $P(x)$ ($x \in \mathfrak{M}$) is the spectral function belonging to it, then*

$$A = 2 \int_{\mathfrak{M}} P(x)\,A_R\,dP(x) = 2i \int_{\mathfrak{M}} P(x)\,A_I\,dP(x). \tag{19.8}$$

Theorem 19.5. *Suppose that the integral $A = 2i \int_{\mathfrak{M}} P(x)H\,dP(x)$ exists, where $P(x)$ ($x \in \mathfrak{M}$) is a spectral function and H is a completely continuous selfadjoint operator. Then A is a Volterra operator, $P(x)$ belongs to A and $H = A_I$.*

Theorem 19.6. *Suppose that $P(x)$ ($x \in \mathfrak{M}$) is a spectral function and T is a finite-dimensional operator. If for each interval (a, b) adjacent to $\mathfrak{M}$ the equation*

$$(P(b) - P(a))\,T\,(P(b) - P(a)) = 0$$

is satisfied, then the integral $\int_{\mathfrak{M}} P(x)T\,dP(x)$ exists.

3. **Inessential extensions of Volterra operators.** Suppose given in the space $\mathfrak{H}$ a Volterra operator A. An operator $\tilde{A}$ operating in a space $\tilde{\mathfrak{H}} \supset \mathfrak{H}$ is said to

be an *inessential extension* of the operator A, if

$$\tilde{A}h = \begin{cases} Ah & (h \in \mathfrak{H}), \\ 0 & (h \perp \mathfrak{H}). \end{cases}$$

Theorem 19.7. *Suppose that in the spaces* $\mathfrak{H}$ *and* $\tilde{\mathfrak{H}} \supset \mathfrak{H}$ *there are given respectively spectral functions* $P(x)$ $(x \in \mathfrak{M})$ *and* $\tilde{P}(x)$ $(\alpha \le x \le \beta,\ \alpha = \min \mathfrak{M},\ \beta = \max \mathfrak{M})$. *If*

$$\tilde{P}(x)\,h = P(x)\,h \quad (x \in \mathfrak{M},\ h \in \mathfrak{H}) \tag{19.9}$$

and $P(x)$ $(x \in \mathfrak{M})$ *belongs to the Volterra operator* A, *then* $\tilde{P}(x)$ $(\alpha \le x \le \beta)$ *belongs to its inessential extension* $\tilde{A}$.

Proof. By (19.9) the subspaces $\mathfrak{H}$ and $\tilde{\mathfrak{H}} \ominus \mathfrak{H}$ are invariant relative to $\tilde{P}(x)$ for $x \in \mathfrak{M}$. Therefore

$$\tilde{A}\tilde{P}(x)\,\tilde{\mathfrak{H}} = AP(x)\,\mathfrak{H} \subseteq P(x)\,\mathfrak{H} = \tilde{P}(x)\,\mathfrak{H} \subseteq \tilde{P}(x)\,\tilde{\mathfrak{H}} \quad (x \in \mathfrak{M}).$$

Moreover, if (a, b) is an interval adjacent to $\mathfrak{M}$, then in view of the relation $AP(b)\mathfrak{H} \subseteq P(a)\mathfrak{H}$, which follows from Theorem 15.4, we have

$$\begin{aligned}\tilde{A}\tilde{P}(x)\,\tilde{\mathfrak{H}} \subseteq \tilde{A}\tilde{P}(b)\,\tilde{\mathfrak{H}} = AP(b)\,\mathfrak{H} \subseteq P(a)\,\mathfrak{H} \subseteq \tilde{P}(a)\,\tilde{\mathfrak{H}} \subseteq \\ \subseteq \tilde{P}(x)\,\tilde{\mathfrak{H}} \quad (a < x < b).\end{aligned}$$

Thus all the subspaces $\tilde{P}(x)\tilde{\mathfrak{H}}$ $(\alpha \le x \le \beta)$ are invariant relative to $\tilde{A}$.

The following theorems follow easily from Theorems 18.3 and 19.7.

Theorem 19.8. *Every Volterra operator admits an inessential extension having a continuous maximal chain.*

Theorem 19.9. *Suppose that there are given in the spaces* $\mathfrak{H}$ *and* $\tilde{\mathfrak{H}} \supset \mathfrak{H}$ *respectively operators* H *and* $\tilde{H}$ *and spectral functions* $P(x)$ $(x \in \mathfrak{M})$ *and* $\tilde{P}(x)$ $(\alpha \le x \le \beta,\ \alpha = \min \mathfrak{M},\ \beta = \max \mathfrak{M})$, *while*:

1) H *is a completely continuous selfadjoint operator*;

2) $$\tilde{H}h = \begin{cases} Hh & (h \in \mathfrak{H}), \\ 0 & (h \perp \mathfrak{H}); \end{cases} \tag{19.10}$$

3) $P(x)$ *and* $\tilde{P}(x)$ *are connected by the relation* (19.9);

4) *for each interval* (a, b) *adjacent to* $\mathfrak{M}$ *the equation*

$$(P(b) - P(a))\,H\,(P(b) - P(a)) = 0. \tag{19.11}$$

is satisfied.

For there to exist a Volterra operator possessing the spectral function $P(x)$ *and imaginary component* H, *it is necessary and sufficient that there should exist a Volterra operator possessing the spectral function* $\widetilde{P}(x)$ *and the imaginary component* $\widetilde{H}$.

Proof. If A is a Volterra operator possessing the spectral function $P(x)$ $(x \in \mathfrak{M})$ and imaginary component H, then its inessential extension $\widetilde{A}$ in $\widetilde{\mathfrak{H}}$ obviously has the imaginary component $\widetilde{H}$, and, according to Theorem 19.7, the spectral function $\widetilde{P}(x)$ $(\alpha \leq x \leq \beta)$.

For the proof of the converse assertion we choose a $\delta > 0$ and consider a subdivision $\alpha = x_0 < x_1 < \cdots < x_n = \beta$ satisfying the following conditions: 1) $x_j - x_{j-1} < \delta$ $(j = 1, 2, \cdots, n)$; 2) if either of the points x_{j-1}, x_j does not lie in $\mathfrak{M}$, then the interval (x_{j-1}, x_j) is entirely contained in some interval adjacent to $\mathfrak{M}$.

Suppose that

$$\widetilde{S} = 2i \sum_{j=1}^{n} \widetilde{P}(x_{j-1}) \widetilde{H} (\widetilde{P}(x_j) - \widetilde{P}(x_{j-1})).$$

If the interval (x_{l-1}, x_l) is a portion of an interval (a, b) adjacent to $\mathfrak{M}$, then there exist integers $k \leq l$ and $m \geq l$ such that

$$a = x_{k-1} < \ldots < x_{l-1} < x_l < \ldots < x_m = b.$$

In view of the equation

$$(\widetilde{P}(b) - \widetilde{P}(a)) \widetilde{H} (\widetilde{P}(b) - \widetilde{P}(a)) = 0,$$

which follows from (19.11), we have

$$\begin{aligned}\sum_{j=k}^{m} \widetilde{P}(x_{j-1}) \widetilde{H} (\widetilde{P}(x_j) - \widetilde{P}(x_{j-1})) - \widetilde{P}(a) \widetilde{H} (\widetilde{P}(b) - \widetilde{P}(a)) \\ = \sum_{j=k}^{m} (\widetilde{P}(x_{j-1}) - \widetilde{P}(a)) \widetilde{H} (\widetilde{P}(x_j) - \widetilde{P}(x_{j-1})) \\ = \sum_{j=k}^{m} (\widetilde{P}(x_{j-1}) - \widetilde{P}(a)) (\widetilde{P}(b) - \widetilde{P}(a)) \widetilde{H} \\ \times (\widetilde{P}(b) - \widetilde{P}(a)) (\widetilde{P}(x_j) - \widetilde{P}(x_{j-1})) = 0.\end{aligned}$$

Thus

$$\tilde{S} = 2i \sum_{j=1}^{r} \tilde{P}(y_{j-1}) \tilde{H} (\tilde{P}(y_j) - \tilde{P}(y_{j-1})),$$

where all the points y_j lie in $\mathfrak{M}$. Denote by $\tilde{A}$ a Volterra operator having the spectral function $\tilde{P}(x)$ and imaginary component $\tilde{H}$. It follows from (19.9) and (19.10) that $\tilde{S}$ carries $\mathfrak{H}$ into itself and annihilates $\tilde{\mathfrak{H}} \ominus \mathfrak{H}$. The operator $\tilde{A}$ operates in the same way, inasmuch as it is the limit of sums of the form (19.12). It is now easy to see that the operator A induced by $\tilde{A}$ in $\mathfrak{H}$ has the spectral function $P(x)$ $(x \in \mathfrak{M})$ and imaginary component H.

§20. Volterra operators with one-dimensional imaginary components

1. **Unicellularity.** Suppose given in the space $\mathfrak{H}$ a nonselfadjoint Volterra operator A with a one-dimensional imaginary component:

$$A_I h = (h, e) le \qquad (h \in \mathfrak{H}, \ \| e \| = 1, \ l \neq 0).$$

The existence of such an operator follows from Theorems 17.4 and 17.1. If $l > 0$, then the operator A is dissipative; in the contrary case the operator $-A$ is dissipative. In this section we shall assume that $l > 0$.

Consider some maximal chain π of the operator A. By Theorem 15.6 it is continuous, and in view of Theorem 16.5 the vector e is reproducing for it. By Theorem 18.1 the chain π is the range of a spectral function $P(x)$ $(0 \leq x \leq l)$ satisfying the condition

$$(P(x) k_0, k_0) = x \qquad (0 \leqslant x \leqslant l, \ k_0 = l^{1/2} e). \tag{20.1}$$

Theorem 20.1. *Each completely nonselfadjoint Volterra operator A with a one-dimensional imaginary component is unicellular.* [1)]

Proof. We will suppose that

$$A_I h = (h, e) le \qquad (\| e \| = 1, \ l > 0). \tag{20.2}$$

We denote by P and P' orthoprojectors onto certain invariant subspaces of the operator A, and construct spectral functions $P(x)$ $(0 \leq x \leq l)$ and $P'(x)$ $(0 \leq x \leq l)$, having the following properties:

1) $P(x)$ and $P'(x)$ belong to A;

1) The concept of a unicellular operator was introduced in §5.

2) the ranges of the functions $P(x)$ and $P'(x)$ contain P and P' respectively;

3) $(P(x)k_0, k_0) = (P'(x)k_0, k_0) = x$ $(0 \le x \le l,\ k_0 = l^{1/2} e)$.

Consider the linear mapping U_0 which assigns to each vector of the form $P(x)k_0$ the vector $P'(x)k_0$. From property 3) and Theorem 16.5 it follows that U_0 may be extended to an isometric mapping U of the entire space $\mathfrak{H}$ onto itself. Since for each fixed x

$$UP(x)P(y)k_0 = P'(x)P'(y)k_0 = P'(x)UP(y)k_0 \quad (0 \leqslant y \leqslant l),$$

it follows that

$$UP(x) = P'(x)U. \tag{20.3}$$

Moreover, $Uk_0 = k_0$, so that

$$UA_I = A_I U. \tag{20.4}$$

In view of formulas (20.3) and (20.4)

$$UA = 2i\int_0^l UP(x)A_I\,dP(x) = 2i\int_0^l P'(x)A_I\,dP'(x)U = AU,$$
$$UA^n k_0 = A^n U k_0 = A^n k_0 \quad (n = 0, 1, \ldots).$$

Since the linear envelope of the vectors $A^n k_0$ $(n = 0, 1, \cdots)$ is dense in $\mathfrak{H}$, we have $U = E$, which means that $P(x) = P'(x)$ $(0 \le x \le l)$.

Theorem 20.2. *A completely nonselfadjoint Volterra operator A with a one-dimensional imaginary component (20.2) is unitarily equivalent to the operator $\mathcal{J}f(x) = 2i\int_x^l f(t)dt$, operating in the space $L_2(0, l)$.*

Proof. Suppose that $P(x)$ $(0 \le x \le l)$ is a spectral function of the operator A satisfying the condition (20.1). Consider the mapping U_0 which assigns to each piecewise constant function $f(x) \in L_2(0, l)$ the vector

$$f = \int_0^l f(x)\,dP(x)\,k_0.$$

This is isometric, since

$$(f, g) = \int_0^l f(x)\overline{g(x)}\, d(P(x)k_0, k_0) = \int_0^l f(x)\overline{g(x)}\, dx$$

$$\left(g = \int_0^l g(x)\, dP(x)\, k_0\right),$$

and puts into correspondence with the functions

$$f_t(x) = \begin{cases} 1, & x \leqslant t, \\ 0, & x > t \end{cases} \qquad (0 \leqslant t \leqslant l)$$

vectors $P(t)k_0$ whose linear envelope, according to Theorem 16.5, is dense in $\mathfrak{H}$. Consequently U_0 may be extended to an isometric mapping U of the entire space $L_2(0, l)$ onto the entire space $\mathfrak{H}$. Since the formula

$$Uf(x) = \int_0^l f(x)\, dP(x)\, k_0$$

is valid not only for piecewise constant functions, but also for arbitrary continuous functions,

$$AUf(x) = 2i\int_0^l P(x) A_l\, dP(x) f = 2i\left[A_l - \int_0^l dP(x) A_l P(x)\right] f$$

$$= 2i\int_0^l dP(x) A_l (E - P(x)) f = 2i \int_0^l (f, (E - P(x))k_0)\, dP(x)\, k_0$$

$$= 2i\int_0^l \int_x^l f(t)\, dt\, dP(x)\, k_0 = U\mathcal{J}f(x).$$

Corollary 1. *The operator* $\mathcal{J}f(x) = 2i\int_x^l f(t)dt$, *operating in the space* $L_2(0, l)$, *is Volterra and unicellular. Denote by* $L_2(0, \sigma, l)$ *the invariant subspace of the operator* $\mathcal{J}$ *consisting of all possible functions which are equal to zero almost everywhere on the segment* $[\sigma, l]$. *Any invariant subspace of the operator* $\mathcal{J}$ *is comparable to each of the subspaces* $L_2(0, \sigma, l)$, *and so coincides with one of them.*

Corollary 2. *For the linear envelope of the sequence of functions*

$$\mathcal{J}^n f(x) \left(f(x) \in L_2(0, l),\ \mathcal{J}f(x) = 2i\int_x^l f(t)\, dt,\ n = 0, 1, \dots\right)$$

to be dense in $L_2(0, l)$, *it is necessary and sufficient that the set of points of the segment* $[l-\epsilon, l]$ *at which* $f(x) \neq 0$ *should have positive measure for each* $\epsilon > 0$.

Corollary 3. *If the imaginary components of the completely nonselfadjoint Volterra operators* A *and* A' *are one-dimensional and have the same nonzero eigenvalue, then* A *and* A' *are unitarily equivalent.*

As an application of the results obtained in this section we present a proof of the following theorem of Titchmarsh (see E. C. Titchmarsh [11]).

Theorem 20.3. *If the convolution*

$$F * G = \int_0^x F(x-y)\,G(y)\,dy \tag{20.5}$$

of the functions $F(x)$ *and* $G(x)$, *lying in* $L(0, l)$, *is almost everywhere equal to zero, then there exist nonnegative numbers* α *and* β *with* $\alpha + \beta = l$ *such that* $F(x)$ *and* $G(x)$ *are respectively equal to zero almost everywhere on the intervals* $[0, \alpha]$ *and* $[0, \beta]$.

Proof. Suppose that the convolution (20.5) is equal to zero almost everywhere. Since the convolution operation is commutative and associative, the function

$$\int_0^x f(x-y_n)\int_0^{y_n}\cdots\int_0^{y_1} g(y)\,dy\,dy_1\ldots dy_n = f*(1*\ldots(1*(1*g))\ldots) \tag{20.6}$$

$$\left(f(x) = 1 * F(x) = \int_0^x F(y)\,dy,\ g(x) = 1 * G(x) = \int_0^x G(y)\,dy\right),$$

being continuous, is equal to zero for all $x \in [0, l]$. *In particular,*

$$\int_0^l f(l-y_n)\int_0^{y_n}\cdots\int_0^{y_1} g(y)\,dy\,dy_1\ldots dy_n = 0\ (n=0, 1, \ldots). \tag{20.7}$$

Using the integration operator

$$\mathcal{J}f = 2i\int_x^l f(t)\,dt \qquad (f(t) \in L_2(0, l))$$

and noting that the adjoint operator is defined by the formula

$$\mathcal{J}^* f = -2i \int_0^x f(t)\,dt,$$

we rewrite (20.7) in the form

$$\int_0^l f(l-x)\left[\mathcal{J}^{*n} g(x)\right] dx = (\mathcal{J}^n f(l-x),\ \overline{g(\overline{x})}) = 0$$
$$(n = 0,\ 1,\ \ldots).$$

The closure of the linear envelope of the functions $\mathcal{J}^n f(l-x)$ $(n = 0, 1, \cdots)$ is invariant relative to the operator $\mathcal{J}$. By Corollary 1 of the preceding theorem it coincides with some subspace $L_2(0, \beta, l)$. Thus $f(l-x) \in L_2(0, \beta, l)$ and $g(x) \perp L_2(0, \beta, l)$, so that $f(x)$ and $g(x)$ are equal to zero almost everywhere respectively on intervals $[0, \alpha]$ and $[0, \beta]$, where $\beta = l - \alpha$. The assertion of the theorem on the functions $F(x)$ and $G(x)$ follows from formula (20.6).

2. **Eigenvectors and eigenvalues of real components.**

Theorem 20.4. *In the completely nonselfadjoint Volterra operator A, operating in the space $\mathfrak{H}$, has a one-dimensional imaginary component*

$$A_I h = (h, e) l e \qquad (\|e\| = 1,\ l > 0),$$

then the vectors

$$h_j = \int_0^l \exp\left(\frac{(2j-1)\pi i x}{l}\right) dP(x) e \ (j = 0,\ \pm 1,\ \pm 2, \ldots), \tag{20.8}$$

where $P(x)$ $(0 \le x \le l)$ is a spectral function of the operator A satisfying condition (20.1), *form an orthonormalized basis in $\mathfrak{H}$, while*

$$A_R h_j = -\frac{2l}{(2j-1)\pi} h_j \quad (j = 0,\ \pm 1,\ \pm 2,\ \ldots). \tag{20.9}$$

Proof. The system (20.8) is orthonormalized, since

$$(h_j,\ h_k) = \int_0^l \exp\left(\frac{2(j-k)\pi i x}{l}\right) d(P(x) e,\ e)$$
$$= \frac{1}{l} \int_0^l \exp\left(\frac{2(j-k)\pi i x}{l}\right) dx = \begin{cases} 1, & j = k, \\ 0, & j \ne k. \end{cases}$$

Moreover,

$$A_R = iA_I + A^* = iA_I - 2i \int_0^l dP(x)\, A_I\, dP = i \int_0^l dP(x)\, A_I (E - 2P(x)),$$

$$A_R h = i \int_0^l [(h, k_0) - 2(P(x)h, k_0)]\, dP(x)\, k_0, \tag{20.10}$$

so that

$$A_R h_j = i \int_0^l \left[\int_0^l \exp\left(\frac{(2j-1)\pi i t}{l}\right) dt - 2 \int_0^x \exp\left(\frac{(2j-1)\pi i t}{l}\right) dt \right] dP(x)\, e \tag{20.11}$$

$$= -\frac{2l}{(2j-1)\pi} \int_0^l \exp\left(\frac{(2j-1)\pi i x}{l}\right) dP(x)\, e = -\frac{2l}{(2j-1)\pi} h_j. \tag{20.12}$$

It remains to be proved that the operator A_R does not have eigenvectors distinct from (20.8). If $A_R h = 0$, then

$$0 = (P(x) A_R h, k_0) = i \int_0^x [(h, k_0) - 2(P(t)h, k_0)]\, dt,$$

so that $(P(x)h, k_0) = 0$. In view of Theorem 16.5 $h = 0$. If now $A_R h = \lambda h$ $(\lambda \neq 0,\ h \neq 0)$, then by formula (20.10) the vector h is representable in the form $h = \int_0^l h(x) dP(x) k_0$, where $h(x)$ is a continuous function for which

$$i \int_0^l \left[\int_0^l h(t)\, dt - 2 \int_0^x h(t)\, dt \right] dP(x)\, k_0 = \lambda \int_0^l h(t)\, dP(t)\, k_0.$$

Applying the operator $P(s)$ to each side of this last equation and multiplying the result by k_0, we get

$$i \int_0^s \left[\int_0^l h(t)\, dt - 2 \int_0^x h(t)\, dt \right] dx = \lambda \int_0^s h(t)\, dt,$$

$$i \int_0^l h(t)\, dt - 2i \int_0^x h(t)\, dt = \lambda h(x),$$

$$-2ih(x) = \lambda \frac{dh(x)}{dx}, \qquad h(0) + h(l) = 0.$$

Accordingly λ is equal to one of the numbers

$$\lambda_j = -\frac{2l}{(2j-1)\pi} \quad (j = 0, \pm 1, \pm 2, \ldots),$$

and $h(x)$, up to a constant factor, coincides with the function $e^{-2ix/\lambda}$.

§21. Two-sided ideals $\mathfrak{S}_\pi$ of the ring of bounded linear operators

1. **Eigenvalues of positive completely continuous operators.** We recall that the nonzero eigenvalues of a positive completely continuous operator H, operating in Hilbert space $\mathfrak{H}$, are positive and form a sequence $\omega_1 > \omega_2 > \omega_3 > \cdots$ which is either finite or infinite. In the latter case $\lim_{j\to\infty} \omega_j = 0$. The collection $\mathfrak{H}_j$ of all eigenvectors of the operator H corresponding to the eigenvalue ω_j is a finite-dimensional space, while

$$\mathfrak{H} = \mathfrak{H}_0 \oplus \mathfrak{H}_1 \oplus \mathfrak{H}_2 \oplus \ldots,$$

where $H\mathfrak{H}_0 = 0$. Denote by n_j the dimension of $\mathfrak{H}_j$ and construct the sequence $s_j = s_j(H)$, putting

$$s_1 = s_2 = \ldots = s_{n_1} = \omega_1,$$
$$s_{n_1+1} = s_{n_1+2} = \ldots = s_{n_1+n_2} = \omega_2,$$
$$\cdots\cdots\cdots\cdots$$

In the case when the dimension r of the subspace $\mathfrak{H} \ominus \mathfrak{H}_0$ is finite we put $0 = s_{r+1} = s_{r+2} = \cdots$. Thus the sequence $s_j(H)$ is always infinite, and $\lim_{j\to\infty} s_j(H) = 0$.

The numbers $s_j(H)$ are called the *singular numbers* of the operator H. For their calculation one uses the formula (see Riesz and Sz.-Nagy [1], §95)

$$s_{j+1} = \min_{\mathfrak{G}_j} \max_{\substack{\|e\|=1 \\ e \perp \mathfrak{G}_j}} (He, e), \tag{21.1}$$

where $\mathfrak{G}_j$ is any j-dimensional subspace in $\mathfrak{H}$.

If $0 < H_1 < H_2$, then in view of (21.1)

$$s_j(H_1) \leqslant s_j(H_2) \quad (j = 1, 2, \ldots).$$

2. **Singular numbers of completely continuous operators.** Suppose that A is any completely continuous operator. Its *singular numbers* are those of the non-

negative operator $H = (A^* A)^{1/2}$:

$$s_j(A) = s_j((A^*A)^{1/2}).$$

Lemma 21.1. *For each completely continuous operator A it is possible to indicate orthonormalized sequences $\{g_j\}_1^\infty$ and $\{h_j\}_1^\infty$ such that for any $f \in \mathfrak{H}$*

$$Af = \sum_{j=1}^{\infty} s_j(A)(f, g_j) h_j \tag{21.2}$$

and

$$A^*f = \sum_{j=1}^{\infty} s_j(A)(f, h_j) g_j. \tag{21.3}$$

Proof. Consider first the case when all the numbers $s_j(A)$ are distinct from zero. Suppose that $g_1, g_2, \cdots$ is an orthonormalized system for which

$$(A^*A)^{1/2} g_j = s_j(A) g_j \quad (j = 1, 2, \ldots).$$

Then, obviously, the system $h_j = Ag_j/s_j(A)$ $(j = 1, 2, \cdots)$ is orthonormalized as well. Noting that the operator A annihilates the orthogonal complement to the linear envelope of the vectors g_j, we get

$$Af = \sum_{j=1}^{\infty} (f, g_j) Ag_j = \sum_{j=1}^{\infty} s_j(A)(f, g_j) h_j.$$

If only the numbers $s_1(A), \cdots, s_r(A)$, $r < \infty$, are nonzero, an analogous argument leads to the equation $Af = \Sigma_{j=1}^r s_j(A)(f, g_j)h_j$, which may be rewritten in the form (21.2), arbitrarily supplementing the orthogonal systems $g_1, \cdots, g_r$ and $h_1, \cdots, h_r$.

From the relation

$$(A^*f, g) = \left(f, \sum_{j=1}^{\infty} s_j(A)(g, g_j) h_j\right) = \left(\sum_{j=1}^{\infty} s_j(A)(f, h_j) g_j, g\right)$$

there results (21.3). The lemma is proved.

We note the following properties of singular numbers:

1) $$s_j(\lambda A) = |\lambda| s_j(A) \quad (j = 1, 2, \cdots); \tag{21.4}$$

2) $$s_j(A) = s_j(A^*) \quad (j = 1, 2, \cdots); \tag{21.5}$$

3) If A is a bounded operator and B is completely continuous, then

$$s_j(AB) \leqslant \| A \| s_j(B), \quad s_j(BA) \leqslant \| A \| s_j(B). \tag{21.6}$$

The first of these is a direct consequence of the definition. The second is easily obtained from formulas (21.2) and (21.3) on noting that

$$AA^* f = \sum_{j=1}^{\infty} s_j^2(A)(f, h_j) h_j, \quad (AA^*)^{1/2} h_j = s_j(A) h_j.$$

For the proof of the third we write down the inequalities

$$(B^*A^*ABf, f) = \| ABf \|^2 \leqslant \| A \|^2 (B^*Bf, f),$$
$$B^*A^*AB \leqslant \| A \|^2 B^*B.$$

Since

$$s_j(B^*A^*AB) \leqslant \| A \|^2 s_j(B^*B),$$

we have

$$s_j(AB) = s_j((B^*A^*AB)^{1/2}) = (s_j(B^*A^*AB))^{1/2} \leqslant \| A \| (s_j(B^*B))^{1/2}$$
$$= \| A \| s_j((B^*B)^{1/2}) = \| A \| s_j(B)$$

and

$$s_j(BA) = s_j(A^*B^*) \leqslant \| A^* \| s_j(B^*) = \| A \| s_j(B).$$

Lemma 2.21. *Suppose that A is a completely continuous operator and π_j ($j = 1, \cdots, n$; $\pi_1 \geq \cdots \geq \pi_n > 0$) are nonnegative numbers. Then*

$$\max \sum_{j=1}^{n} \pi |(Ag_j, h_j)| = \sum_{j=1}^{n} \pi_j s_j(A), \tag{21.7}$$

where the maximum is taken over all possible orthonormalized systems $g_1, \cdots, g_n$ and $h_1, \cdots, h_n$.

Proof. Suppose that $g_1, \cdots, g_n$ and $h_1, \cdots, h_n$ are certain orthonormalized systems, P an orthoprojector onto the linear envelope $\mathfrak{L}$ of the vectors $g_1, \cdots, g_n$ and U a unitary operator satisfying the conditions

$$Uh_j = e^{-i\theta_j} g_j \quad (\theta_j = \arg(Ag_j, h_j)).$$

Then

$$|(Ag_j, h_j)| = e^{-i\theta_j}(Ag_j, h_j) = (UAg_j, g_j) = (PUAPg_j, g_j).$$

Inasmuch as the operator $H = (B^* B)^{1/2}$, where $B = PUAP$, carries $\mathfrak{L}$ into itself and annihilates the orthogonal complement to $\mathfrak{L}$, in $\mathfrak{L}$ there exists an orthonormalized basis $f_1, \cdots, f_n$ such that $Hf_j = s_j(B)f_j$ $(j = 1, 2, \cdots, n)$. In view of (21.6)

$$|(Bf_j, f_j)|^2 \leqslant \|Bf_j\|^2 = (B^*Bf_j, f_j) = s_j^2(B) \leqslant s_j^2(A)$$

so that

$$\sum_{j=1}^{n} |(Ag_j, h_j)| = \sum_{j=1}^{n} (Bg_j, g_j) = \sum_{j=1}^{n} (Bf_j, f_j) \leqslant \sum_{j=1}^{n} s_j(A),$$

$$\sum_{j=1}^{n} \pi_j |(Ag_j, h_j)| = (\pi_1 - \pi_2)|(Ag_1, h_1)| + (\pi_2 - \pi_3) \sum_{j=1}^{2} |(Ag_j, h_j)| + \dots$$
$$\dots + (\pi_{n-1} - \pi_n) \sum_{j=1}^{n-1} |(Ag_j, h_j)| + \pi_n \sum_{j=1}^{n} |(Ag_j, h_j)|$$
$$\leqslant (\pi_1 - \pi_2)\, s_1(A) + (\pi_2 - \pi_3) \sum_{j=1}^{2} s_j(A) + \dots$$
$$\dots + (\pi_{n-1} - \pi_n) \sum_{j=1}^{n-1} s_j(A) + \pi_n \sum_{j=1}^{n} s_j(A) = \sum_{j=1}^{n} \pi_j s_j(A).$$

It remains to be proved that under an appropriate choice of the vectors g_j and h_j we will have the equation

$$\sum_{j=1}^{n} \pi_j |(Ag_j, h_j)| = \sum_{j=1}^{n} \pi_j s_j(A).$$

But this follows from Lemma 21.1, in which we established the existence of orthonormalized sequences $g_1, g_2, \cdots$ and $h_1, h_2, \cdots$ for which $(Ag_j, h_j) = s_j(A)$ $(j = 1, 2, \cdots)$.

3. The ideals $\mathfrak{S}_\pi$. Fix a nonincreasing sequence of positive numbers $\pi_1 = 1, \pi_2, \pi_3, \cdots$ and denote by $\mathfrak{S}_\pi$ the class of all completely continuous operators satisfying the condition

$$\sum_{j=1}^{\infty} \pi_j s_j(A) < \infty.$$

The set $\mathfrak{S}_\pi$ *is linear.* Indeed, suppose that A and B lie in $\mathfrak{S}_\pi$ and that λ is some number. Then, according to Lemma 21.2,

$$\sum_{j=1}^{n} \pi_j s_j(A+B) = \max \sum_{j=1}^{n} \pi_j |((A+B)g_j, h_j)|$$
$$\leqslant \max \sum_{j=1}^{n} \pi_j |(Ag_j, h_j)| + \max \sum_{j=1}^{n} \pi_j |(Bg_j, h_j)|$$
$$= \sum_{j=1}^{n} \pi_j s_j(A) + \sum_{j=1}^{n} \pi_j s_j(B)$$

and, moreover,

$$\sum_{j=1}^{n} \pi_j s_j(\lambda A) = |\lambda| \sum_{j=1}^{n} \pi_j s_j(A).$$

Accordingly,

$$\sum_{j=1}^{\infty} \pi_j s_j(A+B) \leqslant \sum_{j=1}^{\infty} \pi_j s_j(A) + \sum_{j=1}^{\infty} \pi_j s_j(B), \tag{21.8}$$

$$\sum_{j=1}^{\infty} \pi_j s_j(\lambda A) = |\lambda| \sum_{j=1}^{\infty} \pi_j s_j(A), \tag{21.9}$$

which means that $A+B$ and λA lie in $\mathfrak{S}_\pi$.

The function assigning to each operator $A \in \mathfrak{S}_\pi$ *a number*

$$\| A \|_\pi = \sum_{j=1}^{\infty} \pi_j s_j(A) \tag{21.10}$$

is a norm in $\mathfrak{S}_\pi$. Indeed, since $\|A\|_\pi \geq s_1(A)$, and $s_1(A) = \|A\|$,

$$\| A \| \leqslant \| A \|_\pi. \tag{21.11}$$

Therefore the equation $\|A\|_\pi = 0$ holds if and only if $A = 0$. That the remaining axioms for a norm are satisfied follows from (21.8) and (21.9).

We shall show that *the space* $\mathfrak{S}_\pi$ *is complete in the sense of the norm* (21.10). Suppose that the sequence $A_1, A_2, \cdots$ of operators of $\mathfrak{S}_\pi$ has the following property: for any $\epsilon > 0$ there exists an integer N_ϵ such that

$$\| A_m - A_n \|_\pi = \sum_{j=1}^{\infty} \pi_j s_j(A_m - A_n) \leqslant \varepsilon$$

for $m, n > N_\epsilon$. According to (21.11) $\|A_m - A_n\| < \epsilon$, and therefore the sequence $A_1, A_2, \cdots$ converges in the uniform norm to some completely continuous operator A. Fixing the number n in the inequality

$$\sum_{j=1}^{k} \pi_j s_j(A - A_n) \leqslant \sum_{j=1}^{k} \pi_j s_j(A - A_m) + \sum_{j=1}^{k} \pi_j s_j(A_m - A_n$$

$$\leqslant \sum_{j=1}^{k} \pi_j s_j(A - A_n) + \varepsilon \leqslant \| A - A_m \| \sum_{j=1}^{k} \pi_j + \varepsilon \qquad (m,\ n > N_\varepsilon),$$

and letting first m and then k tend to infinity, we get

$$\sum_{j=1}^{\infty} \pi_j s_j(A - A_n) \leqslant \varepsilon \qquad (n > N_\varepsilon). \tag{21.12}$$

With this it is proved that

$$A-A_n\in\mathfrak{S}_\pi \quad (n>N_\varepsilon), \qquad A=(A-A_n)+A_n\in\mathfrak{S}_\pi.$$

Meanwhile the inequality (21.12) means that $\|A-A_n\|_\pi \to 0$ as $n\to\infty$.

We have proved that in the norm (21.10) $\mathfrak{S}_\pi$ is a Banach space.

The following assertions follow from (21.5) and (21.6).

1. *If* $A\in\mathfrak{S}_\pi$, *then* $A^*\in\mathfrak{S}_\pi$, *while* $\|A\|_\pi=\|A^*\|_\pi$.

2. *If* A *is any bounded operator and* $B\in\mathfrak{S}_\pi$, *then* AB *and* BA *lie in* $\mathfrak{S}_\pi$.

Thus $\mathfrak{S}_\pi$ is a selfadjoint two-sided ideal in the space of all bounded linear operators.

Suppose that H *is a selfadjoint operator of the class* $\mathfrak{S}_\pi$ *and* e_α $(\alpha=1,2,\cdots)$ *an orthonormalized sequence for which*

$$He_\alpha=\omega_\alpha e_\alpha \qquad (|\omega_\alpha|=s_\alpha(H)).$$

Then $\|H-P_nHP_n\|_\pi\to 0$, *where* P_n *denotes the orthoprojector onto the linear envelope of the vectors* $e_1,\cdots,e_n$.

Indeed, fixing on a number $\epsilon>0$, we choose N so that the inequality

$$\sum_{j=N}^{\infty}\pi_js_j\leqslant\frac{\varepsilon}{2} \qquad (s_j=s_j(H))$$

holds. Since

$$\|H-P_nHP_n\|_\pi=\pi_1s_{n+1}+\pi_2s_{n+2}+\ \ldots+\pi_{N-1}s_{n+N-1}+\pi_Ns_{n+N}+\ldots$$

$$=\pi_1s_1\frac{s_{n+1}}{s_1}+\pi_2s_2\frac{s_{n+2}}{s_2}+\ldots+\pi_{N-1}s_{N-1}\frac{s_{n+N-1}}{s_{N-1}}+\pi_Ns_N\frac{s_{n+N}}{s_N}+\ldots$$

$$\leqslant\frac{s_{n+1}}{s_{N-1}}(\pi_1s_1+\pi_2s_2+\ldots+\pi_{N-1}s_{N-1})+\sum_{j=N}^{\infty}\pi_js_j\ \leqslant\frac{s_{n+1}}{s_{N-1}}\|H\|_\pi+\frac{\varepsilon}{2},$$

we have $\|H-P_nHP_n\|_\pi<\epsilon$ for sufficiently large n.

The result just obtained shows that *the ideal* $\mathfrak{S}_\pi$ *is the closure relative to the norm* (21.10) *of the set of all finite-dimensional operators.*

In this chapter we shall apply the ideal $\mathfrak{S}_\pi$ constructed with respect to the sequence $\pi_j=1/(2j-1)$. We shall denote this ideal by $\mathfrak{S}_\pi$, and the corresponding norm by the symbol $\|\ \|_\omega$.

We denote by $\mathfrak{S}_p$ $(p \geq 1)$ the collection of completely continuous operators for which

$$\sum_{j=1}^{\infty} s_j^p(A) < \infty.$$

It is possible to show that for each fixed $p \geq 1$ the set $\mathfrak{S}_p$ is a Banach space with the norm

$$\| A \|_p = \left(\sum_{j=1}^{\infty} s_j^p(A) \right)^{1/p}$$

and is a two-sided selfadjoint ideal in the space of all bounded linear operators (see Gohberg and Kreĭn [4], Chapter III).

In particular, $\mathfrak{S}_1$ coincides with the ideal $\mathfrak{S}_\pi$ constructed relative to the sequence $\pi_j = 1$ $(j = 1, 2, \cdots)$. The operators of the class $\mathfrak{S}_1$ are called *nuclear*.

If there exists a $p \geq 1$ such that $A \in \mathfrak{S}_p$, then $A \in \mathfrak{S}_\omega$. This is obvious for $p = 1$ and follows from the inequality

$$\sum_{j=1}^{\infty} \frac{s_j(A)}{2j-1} \leqslant \left(\sum_{j=1}^{\infty} s_j^p(A) \right)^{1/p} \left(\sum_{j=1}^{\infty} \left(\frac{1}{2j-1} \right)^q \right)^{1/q} \quad \left(\frac{1}{p} + \frac{1}{q} = 1 \right)$$

for $p > 1$.

§22. The problem of existence of a Volterra operator having a given maximal chain and given imaginary component

Suppose that the Volterra operators $A^{(1)}$ and $A^{(2)}$ have one-dimensional imaginary components

$$A_I^{(1)} h = (h, e_1) l_1 e_1 \ (\| e_1 \| = 1), \qquad A_I^{(2)} h = (h, e_2) l_2 e_2 \ (\| e_2 \| = 1)$$

and a common spectral function $P(x)$ $(\alpha \leq x \leq \beta)$. Then

$$(A^{(1)} e_2, e_2) = 2i \left(\int_\alpha^\beta P(x) A_I^{(1)} dP(x) e_2 e_2 \right) = 2il_1 \int_\alpha^\beta (P(x) e_1, e_2) \, d(P(x) e_2, e_1),$$

$$(A^{(2)} e_1, e_1) = 2i \left(\int_\alpha^\beta P(x) A_I^{(2)} dP(x) e_1, e_1 \right) = 2il_2 \int_\alpha^\beta (P(x) e_2, e_1) \, d(P(x) e_1, e_2),$$

so that

$$l_2(A^{(1)}e_2,\ e_2) + l_1(A^{(2)^*}e_1,\ e_1) = 0 \tag{22.1}$$

and accordingly

$$l_2(A_R^{(1)}e_2,\ e_2) + l_1(A_R^{(2)}e_1,\ e_1) = 0. \tag{22.2}$$

Lemma 22.1. *If the Volterra operator A has a finite-dimensional imaginary component, then*

$$\| A_R \| \leqslant \frac{4}{\pi} \| A_I \|_\omega. \tag{22.3}$$

Proof. In view of Theorem 19.8 we may without loss of generality suppose that one of the maximal chains belonging to A is continuous. We denote this one by π. Let B be a Volterra operator having the chain π and imaginary component $B_I h = (h,\ e)e$, where e is an arbitrarily chosen unit vector. By Theorem 20.4, B_R is the difference of two mutually orthogonal positive operators B_R^+ and B_R^- such that

$$s_j(B_R^+) = s_j(B_R^-) = \frac{2}{(2j-1)\pi} \qquad (j = 1,\ 2,\ \ldots).$$

Suppose

$$A_I h = \sum_{j=1}^{n} (h,\ e_j)\,\omega_j e_j$$

$$\left(|\omega_1| \geqslant |\omega_2| \geqslant \ldots \geqslant |\omega_n| > 0;\ (e_j,\ e_k) = \begin{cases} 1, & j = k, \\ 0, & j \neq k \end{cases} \right).$$

Obviously $A = \sum_{j=1}^{n} A^{(j)}$, where $A^{(j)}$ is a Volterra operator having the chain π and imaginary component $A_I^{(j)} h = (h, e_j)\omega_j e_j$. Using formula (22.2) and Lemma 21.2, we get

$$(A_R e,\ e) = \sum_{j=1}^{n} (A_R^{(j)} e,\ e) = -\sum_{j=1}^{n} \omega_j (B_R e_j,\ e_j)$$

$$= -\sum_{j=1}^{n} \omega_j (B_R^+ e_j,\ e_j) + \sum_{j=1}^{n} \omega_j (B_R^- e_j,\ e_j),$$

$$|(A_R e, e)| \leqslant \sum_{j=1}^{n} |\omega_j| \, |(B_R^+ e_j,\ e_j)| + \sum_{j=1}^{n} |\omega_j| \, |(B_R^- e_j,\ e_j)|$$

$$\leqslant \frac{4}{\pi} \sum_{j=1}^{n} \frac{|\omega_j|}{2j-1} = \frac{4}{\pi} \| A_I \|_\omega.$$

Accordingly,

$$\| A_R \| = \sup_{\|e\|=1} |(A_R e, e)| \leqslant \frac{4}{\pi} \| A_I \|_\omega .$$

Theorem 22.1. *Suppose that π is a maximal chain and H a selfadjoint operator lying in the ideal $\mathfrak{S}_\omega$. If for each jump (P^-, P^+) of the chain π the equation*

$$(P^+ - P^-) H (P^+ - P^-) = 0$$

is satisfied, then there exists a Volterra operator A such that π belongs to A and $H = A_I$. In addition

$$\| A_R \| \leqslant \frac{4}{\pi} \| A_I \|_\omega . \tag{22.4}$$

Proof. We consider first the case when the chain π is continuous. In the preceding section it was proved that there exists a sequence $H_1, H_2, \cdots$ of finite-dimensional selfadjoint operators such that $\|H - H_n\|_\omega \to 0$. Denote by $A^{(n)}$ a Volterra operator having the chain π and imaginary component H_n. The sequence $A^{(n)}$ converges in the uniform norm to some operator A, since in view of (22.3) and (21.11)

$$\begin{aligned} \| A^{(m)} - A^{(n)} \| &\leqslant \| (A^{(m)} - A^{(n)})_R \| + \| H_m - H_n \| \\ &\leqslant \frac{4}{\pi} \| H_m - H_n \|_\omega + \| H_m - H_n \|_\omega \to 0 \end{aligned}$$

as $m, n \to \infty$. It is clear that π belongs to A and $H = A_I$. In view of (22.3) $\|A_R^{(n)}\| \le (4/\pi) \|A_I^{(n)}\|_\omega$. Passing to the limit in this inequality as $n \to \infty$, we obtain (22.4).

The case when the chain π has jumps reduces in view of Theorems 18.3 and 19.9 to the case already considered.

According to Theorem 17.2, this result is equivalent to the following assertion.

Theorem 22.1$'$. *Suppose that $P(x)$ $(x \in \mathfrak{M})$ is a spectral function and H a selfadjoint operator lying in the ideal $\mathfrak{S}_\omega$. If for each interval (a, b) adjacent to $\mathfrak{M}$ the equation*

$$(P(b) - P(a)) H (P(b) - P(a)) = 0$$

is satisfied, then the integral $\int_{\mathfrak{M}} P(x) H \, dP(x)$ exists.

Lemma 22.2. *For each completely continuous selfadjoint operator H not lying in the ideal $\mathfrak{S}_\omega$, it is possible to produce a spectral function $P(x)$ $(0 \le x \le 1)$*

such that the integral $\int_0^1 P(x)H\,dP(x)$ *does not exist.*

Proof. Consider first the case when the operator H has a special structure, as follows. Suppose that

$$Hf = \sum_{j=-\infty}^{+\infty} (f,\ h_j)\,\omega_j h_j$$

$$\begin{pmatrix} \omega_j > 0 \quad (j = 1,\ 2,\ \ldots), & \omega_j < 0 \quad (j = 0,\ -1,\ \ldots), \\ \omega_1 \geqslant \omega_2 \geqslant \ldots & \omega_0 \leqslant \omega_{-1} \leqslant \ldots \end{pmatrix},$$

where $\{h_j\}_{-\infty}^{+\infty}$ is an orthonormalized basis. Inasmuch as H lies in $\mathfrak{S}_\omega$,

$$\sum_{j=-\infty}^{+\infty} \frac{\omega_j}{2j-1} = \infty.$$

It follows easily from Theorem 20.4 that there exist a unit vector e and a spectral function $P(x)$ $(0 \leq x \leq 1)$ for which

$$(P(x)\,e,\ e) = x \qquad (0 \leqslant x \leqslant 1)$$

and

$$h_j = \int_0^1 \exp\left((2j-1)\,\pi i x\right) dP(x)\,e.$$

We break the unit interval $[0,\ 1]$ into n equal pieces and construct the integral sum

$$S_n = \sum_{k=1}^{n} P(x_k)\,H\,(P(x_k) - P(x_{k-1})) \qquad \left(x_k = \frac{k}{n}\right).$$

Since

$$(S_n e,\ e) = \sum_{k=1}^{n} \sum_{j=-\infty}^{+\infty} \left(\left(P(x_k) - P(x_{k-1})\,e,\ h_j\right) \omega_j \left(P(x_k)\,h_j,\ e\right)\right)$$

$$= \sum_{k=1}^{n} \sum_{j=-\infty}^{+\infty} \omega_j \int_{x_{k-1}}^{x_k} e^{-(2j-1)\,\pi i x}\,dx \int_0^{x_k} e^{(2j-1)\,\pi i x}\,dx$$

$$= \sum_{j=-\infty}^{+\infty} \frac{\omega_j}{(2j-1)^2\,\pi^2} \left[n\left(1 - e^{\frac{(2j-1)\,\pi i}{n}}\right) + 2\right],$$

we have

$$\operatorname{Im}(S_n e,\ e) = -\sum_{j=-\infty}^{+\infty} \frac{n\omega_j}{(2j-1)^2\pi^2} \sin\frac{(2j-1)\pi}{n}. \tag{22.5}$$

We denote by n_1 the integral part of the number $n/4$. From the inequalities

$$0 < \frac{(2j-1)\pi}{n} < \frac{\pi}{2} \quad (j = 1, 2, \ldots, n_1 - 1),$$
$$-\frac{\pi}{2} < \frac{(2j-1)\pi}{n} < 0 \quad (j = 0, -1, \ldots, -n_1 + 1)$$

it follows that the terms of (22.5) with indices $j = -n_1 + 1, -n_1 + 2, \cdots, n_1 - 1$ are positive. Taking into account the fact that

$$\frac{\sin x}{x} \geqslant \frac{2}{\pi} \quad \left(0 \leqslant x \leqslant \frac{\pi}{2}\right),$$

we arrive at the estimate

$$\sum_{j=-n_1+1}^{n_1-1} \frac{n\omega_j}{(2j-1)^2\pi^2} \sin\frac{(2j-1)\pi}{n}$$
$$= \frac{n}{\pi^2} \sum_{j=-n_1+1}^{n_1-1} \frac{|\omega_j|}{(2j-1)^2} \sin\frac{|2j-1|\pi}{n} \geqslant \frac{2}{\pi^2} \sum_{j=-n_1+1}^{n_1-1} \frac{\omega_j}{2j-1} \to \infty. \tag{22.6}$$

At the same time

$$\left|\sum_{j=n_1}^{+\infty} \frac{n\omega_j}{(2j-1)^2\pi^2} \sin\frac{(2j-1)\pi}{n}\right| \leqslant \frac{n\omega_1}{\pi^2} \sum_{j=n_1}^{+\infty} \frac{1}{(2j-1)^2}$$
$$\leqslant \frac{n\omega_1}{2\pi^2} \sum_{j=n_1}^{+\infty} \left(\frac{1}{2j-3} - \frac{1}{2j-1}\right) = \frac{n\omega_1}{2\pi^2(2n_1-3)} \leqslant \frac{n\omega_1}{\pi^2(n-10)} \to \frac{\omega_1}{\pi^2}, \tag{22.7}$$

and analogously

$$\left|\sum_{j=-\infty}^{-n_1} \frac{n\omega_j}{(2j-1)^2\pi^2} \sin\frac{(2j-1)\pi}{n}\right| \leqslant \frac{n|\omega_0|}{\pi^2(n-6)} \to \frac{|\omega_0|}{\pi^2}. \tag{22.8}$$

Relations (22.6), (22.7) and (22.8) show that

$$\lim_{n\to\infty} [\operatorname{Im}(S_n e,\ e)] = -\infty.$$

Thus the integral $\int_0^1 P(x)H\,dP(x)$ does not exist even in the sense of weak convergence.

Passing to the general case, we consider any completely continuous selfadjoint operator H not lying in the ideal $\mathfrak{S}_\omega$. It is easy to see that it is possible to choose a selfadjoint operator $H_1 \in \mathfrak{S}_\omega$ in such a way that the sum $H + H_1$ has the properties indicated at the beginning of the proof. Suppose that $P(x)$ $(0 \leq x \leq 1)$ is a spectral function for which the integral $\int_0^1 P(x)\,(H + H_1)dP(x)$ does not exist. In view of Theorem 22.1 the integral $\int_0^1 P(x)H\,dP(x)$ does not exist either.

The following is an equivalent formulation of this theorem. *For a given selfadjoint completely continuous operator H not lying in the ideal $\mathfrak{S}_\omega$, there exists a continuous maximal chain such that any Volterra operator having this chain will have an imaginary component distinct from H.*

Comparing Theorem 22.1 and Lemma 22.2, we arrive at the following conclusion.

Theorem 22.2. *Suppose that H is a completely continuous selfadjoint operator. For there to exist a Volterra operator with imaginary component H and arbitrary continuous maximal chain given in advance, it is necessary and sufficient that $H \in \mathfrak{S}_\omega$.*

Theorem 22.3. *Every completely continuous selfadjoint operator H operating in an infinite-dimensional space $\mathfrak{H}$ is the imaginary component of some Volterra operator.*

Proof. It is easy to see that $\mathfrak{H}$ may be represented as an orthogonal sum $\mathfrak{H} = \mathfrak{H}_1 \oplus \mathfrak{H}_2 \oplus \cdots$ in such a way that the following conditions are satisfied:

1. All the subspaces $\mathfrak{H}_n$ are infinite dimensional and invariant relative to H.

2. $\|H_n\|_\omega \leq 2s_n(H)$ $(n = 1, 2, \cdots)$, where H_n is the operator induced by H in $\mathfrak{H}_n$.

Suppose that $\tilde{A}_n$ is an inessential extension in $\mathfrak{H}$ of a Volterra operator A_n operating in $\mathfrak{H}_n$ and having imaginary component H_n. In view of (22.4) and (22.11)

$$\begin{aligned}\|\tilde{A}_n\| = \|A_n\| &\leqslant \|(A_n)_R\| + \|H_n\| \\ &\leqslant \frac{4}{\pi}\|H_n\|_\omega + \|H_n\| \leqslant \left(\frac{4}{\pi} + 1\right)\|H_n\|_\omega \leqslant 2\left(\frac{4}{\pi} + 1\right)s_n(H),\end{aligned}$$

so that

$$\begin{aligned}&\|\tilde{A}_m + \tilde{A}_{m+1} + \ldots + \tilde{A}_n\| \\ &\quad = \max\{\|A_m\|, \|A_{m+1}\|, \ldots, \|A_n\|\} \leqslant 2\left(\frac{4}{\pi} + 1\right)s_m(H).\end{aligned}$$

With this it is proved that $\tilde{A}_1 + \tilde{A}_2 + \cdots$ converges in the uniform norm. This sum is obviously a Volterra operator whose imaginary component is equal to H.

§23. Absolutely continuous and canonical spectral functions

1. **Computation of the rank of a spectral function.** The spectral function $P(x)$ $(\alpha \leq x \leq \beta)$ is said to be *absolutely continuous* if all the scalar functions of the form $(P(x)f, g)$ are absolutely continuous.

Lemma 23.1. *If $h_1, \cdots, h_s$ $(s \leq \infty)$ is a reproducing system of the spectral function $P(x)$ $(0 \leq x \leq l)$ and*

$$\sum_{k=1}^{s} (P(x) h_k, h_k) = x \qquad (0 \leqslant x \leqslant l),$$

then $P(x)$ is absolutely continuous.

Proof. We note first that if the functions $(P(x)f, f)$ and $(P(x)g, g)$ are absolutely continuous, then the function $(P(x)f, g)$ is too, as is shown by the estimate

$$\sum_{j=1}^{n} |(\Delta P_j f, g)| \leqslant \sum_{j=1}^{n} (\Delta P_j f, f)^{1/2} (\Delta P_j g, g)^{1/2}$$
$$\leqslant \left(\sum_{j=1}^{n} (\Delta P_j f, f) \right)^{1/2} \left(\sum_{j=1}^{n} (\Delta P_j g, g) \right)^{1/2}$$

$$(0 \leqslant x_1 < x_1' \leqslant x_2 < x_2' \leqslant \ldots \leqslant x_n < x_n' \leqslant l, \quad \Delta P_j = P(x_j') - P(x_j)).$$

In view of the relation

$$\sum_{j=1}^{n} (\Delta P_j h_k, h_k) \leqslant \sum_{j=1}^{n} \sum_{k=1}^{s} (\Delta P_j h_k, h_k) = \sum_{j=1}^{n} (x_j' - x_j)$$

all the functions $(P(x)h_k, h_k)$ are absolutely continuous. Inasmuch as

$$(P(x) P(t) h_k, P(t) h_k) = \begin{cases} (P(x) h_k, h_k), & x \leqslant t, \\ (P(t) h_k, h_k), & x > t, \end{cases}$$

a function of the form $(P(x)g, g)$, where g is in the linear envelope $\mathfrak{H}_0$ of vectors of the form $P(t)h_k$ $(0 \leq t \leq l,\ k = 1, 2, \cdots, s)$ is also absolutely continuous. In view of the density of $\mathfrak{H}_0$ in $\mathfrak{H}$ and the inequality

$$\sum_{j=1}^{n} (\Delta P_j h, h) \leqslant \sum_{j=1}^{n} (\Delta P_j g, g) + \sum_{j=1}^{n} |(\Delta P_j (h-g), g)| + \sum_{j=1}^{n} |(\Delta P_j h, h-g)|$$

$$\leqslant \sum_{j=1}^{n} (\Delta P_j g, g) + \| h - g \| (\| h \| + \| g \|) \quad (h \in \mathfrak{H},\ g \in \mathfrak{H}_0),$$

the functions $(P(x)h, h)$ are absolutely continuous for any h.

The lemma is proved.

The reproducing system $h_1, \cdots, h_s$ $(s \leq \infty)$ of the spectral function $P(x)$ $(x \in \mathfrak{M})$ is said to be *fully orthogonal* if

$$(P(x)h_j, h_k) = 0 \quad (x \in \mathfrak{M},\ j \neq k). \tag{23.1}$$

Lemma 23.2. *Every spectral function $P(x)$ $(x \in \mathfrak{M})$ of rank $r \leq \infty$ has a fully orthogonal reproducing system consisting of r vectors.*

Proof. Suppose that $g_1, \cdots, g_r$ is a reproducing system of the function $P(x)$. Denote by $\mathfrak{G}_k$ the closure of the linear envelope of vectors of the form $P(x)g_j$ $(x \in \mathfrak{M},\ j = 1, \cdots, k)$, and put

$$\mathfrak{H}_1 = \mathfrak{G}_1, \quad \mathfrak{H}_k = \mathfrak{G}_k \ominus \mathfrak{G}_{k-1} \quad (k = 2, 3, \ldots, r).$$

Since

$$\mathfrak{H} = \mathfrak{H}_1 \oplus \mathfrak{H}_2 \oplus \ldots \oplus \mathfrak{H}_r$$

and all the spaces $\mathfrak{H}_k$ are invariant relative to each of the orthoprojectors $P(x)$ $(x \in \mathfrak{M})$, it follows that the system $h_1, \cdots, h_r$, where h_k is the projection of g_k onto $\mathfrak{H}_k$, is reproducing for the function $P(x)$ and satisfies conditions (23.1).

In what follows we shall again need the Hilbert spaces $L_2^{(r)}(0, l)$ introduced in §8. Denote by $P_{(0,l)}^{(r)}(x)$ the orthoprojector operating in $L_2^{(r)}(0, l)$ onto the subspace $L_2^{(r)}(0, x, l)$ consisting of all matrices

$$f(x) = \| f^{(1)}(x) f^{(2)}(x) \ldots f^{(r)}(x) \| \in L_2^{(r)}(0, l)$$

whose elements are equal to zero almost everywhere on the interval (x, l). Obviously $P_{(0,l)}^{(r)}(x)$ $(0 \leq x \leq l)$ is a spectral function. It is absolutely continuous, inasmuch as

$$(P_{(0,l)}^{(r)}(x) f(t), g(t)) = \int_0^x f(t) g^*(t)\, dt.$$

Lemma 23.3. *Suppose that the spectral function $P(x)$ $(0 \leq x \leq l)$ is given in the space $\mathfrak{H}$ and has rank r. If it is absolutely continuous, then there exists an isometric mapping U of the space $\mathfrak{H}$ onto some subspace in $L_2^{(r)}(0, l)$, invariant relative to all orthoprojectors $P_{(0,l)}^{(r)}(x)$, such that*

$$UP(x) = P^{(r)}_{(0,l)}(x)\,U \qquad (0 \leqslant x \leqslant l). \tag{23.2}$$

Proof. Consider some fully orthogonal reproducing system $h_1, \cdots, h_r$ of the function $P(x)$. Put

$$h_k(x) = \frac{d}{dx}(P(x)h_k, h_k) \qquad (k = 1, 2, \ldots, r).$$

We observe that the functions $h_k(x)$ are nonnegative and summable.

We introduce a mapping U_0, assigning to each vector of the form $P(x)h_k$ the vector $P^{(r)}_{(0,l)}(x)g_k(t) \in L^{(r)}_2(0, l)$, where

$$g_k(t) = \|\underbrace{0 \ldots 0\, h_k^{1/2}}_{k}(t)\, 0 \ldots 0\|.$$

In view of the relations

$$(P(x)h_j, h_k) = \begin{cases} \int_0^x h_k(t)\,dt, & j = k, \\ 0 & j \neq k, \end{cases}$$

the mapping U_0 is isometric and may be extended to an isometric mapping U of the entire space $\mathfrak{H}$ onto the closure $\mathfrak{L}$ of the linear envelope of vectors of the form $P^{(r)}_{(0,l)}(x)g_k(t)$ $(0 \leq x \leq l,\ k = 1, 2, \cdots, r)$. The invariance of $\mathfrak{L}$ relative to the orthoprojectors $P^{(r)}_{(0,l)}(x)$ is obvious. From the equation

$$\begin{aligned} UP(x)P(x')h_k &= UP(x'')h_k = P^{(r)}_{(0,l)}(x'')g_k(t) \\ &= P^{(r)}_{(0,l)}(x)P^{(r)}_{(0,l)}(x')g_k(t) = P^{(r)}_{(0,l)}(x)UP(x')h_k \\ &(x, x' \in [0, l],\ x'' = \min\{x, x'\}) \end{aligned}$$

it follows that (23.2) holds. The lemma is proved.

Suppose that $F(x) = \|f_{ij}(x)\|$ is a square or rectangular matrix consisting of summable functions given on some interval $[\alpha, \beta]$. The *general rank* of the matrix $F(x)$ is the largest of the orders of its minors which are distinct from zero on a set of positive measure.

Theorem 23.1. *If $P(x)$ $(0 \leq x \leq l)$ is an absolutely continuous spectral function and $h_1, \cdots, h_s$ $(s \leq \infty)$ a reproducing system for it, then the rank r of the function $P(x)$ is equal to the general rank of the square matrix of sth order $(d/dx)\|P(x)h_j, h_k)\|$. In addition*

$$\| (P(x) h_j, h_k) \| = \int_0^x \Pi(t) \Pi^*(t)\, dt,$$

where $\Pi(t)$ *is a rectangular matrix of general rank whose rows are elements of the space* $L_2^{(r)}(0, l)$.

Proof. By Lemma 23.3 there exists an isometric mapping U of the space $\mathfrak{H}$ into the subspace $\mathfrak{L} \subset L_2^{(r)}(0, l)$ satisfying the condition $UP(x) = P_{(0,l)}^{(r)}(x)U$ $(0 \le x \le l)$. Putting

$$Uh_k = f_k \quad (f_k = f_k(x) = \| f_k^{(1)}(x) f_k^{(2)}(x) \dots f_k^{(r)}(x) \|),$$

we obtain

$$(P(x) h_j, h_k) = (P_{(0,l)}^{(r)}(x) f_j, f_k) = \int_0^x f_j(t) f_k^*(t)\, dt,$$

$$\frac{d}{dx} \| (P(x) h_j, h_k) \| = \| f_j(x) f_k^*(x) \| = \Pi(x) \Pi^*(x),$$

$$\Pi(x) = \left\| \begin{matrix} f_1^{(1)}(x) & f_1^{(2)}(x) & \dots & f_1^{(r)}(x) \\ f_2^{(1)}(x) & f_2^{(2)}(x) & \dots & f_2^{(r)}(x) \\ \cdot & \cdot & \cdot & \cdot \\ f_s^{(1)}(x) & f_s^{(2)}(x) & \dots & f_s^{(r)}(x) \end{matrix} \right\|.$$

The general rank of the matrix $\Pi(x)$ is not higher than r, and, as is easily seen, it coincides with the general rank of the matrix $\Pi(x)\Pi^*(x)$. Suppose that it is equal to r_1, where $r_1 < r$. Then there exist measurable sets $\mathfrak{d}_\alpha \in [0, l]$ $(\alpha = 1, 2, \cdots, m;\ m \le \infty)$ of positive measure, having the following properties:

1) $$\mathfrak{d}_\alpha \cap \mathfrak{d}_\beta = 0 \quad (\alpha \ne \beta), \qquad \bigcup_{\alpha=1}^{m} \mathfrak{d}_\alpha = [0, l];$$

2) for almost every fixed $x \in \mathfrak{d}_\alpha$ all the rows of the matrix $\Pi(x)$ are linear combinations of certain rows $f_{\alpha_1}(x), f_{\alpha_2}(x), \cdots, f_{\alpha_{r_1}}(x)$.

Inasmuch as $\mathfrak{L}$ is invariant relative to all orthoprojectors $P_{(0,l)}^{(r)}(x)$, it follows that the vectors

$$f_{\alpha_j}^{(0)}(x) = \begin{cases} f_{\alpha_j}(x), & x \in \mathfrak{d}_\alpha, \\ 0, & x \notin \mathfrak{d}_\alpha \end{cases}$$
$$(\alpha = 1, 2, \dots, m;\ j = 1, 2, \dots, r_1)$$

lie in $\mathfrak{L}$, and therefore

$$g_j(x) = \sum_{\alpha=1}^{m} \frac{f_{\alpha_j}^{(0)}(x)}{2^{\alpha}\left\| f_{\alpha_j}^{(0)}(x)\right\|} \in \mathfrak{L} \qquad (j = 1, 2, \ldots, r_1).$$

Suppose that $h(x) \in \mathfrak{L}$ and $(P^{(r)}_{(0,l)}(t)g_j(x), h(x)) = 0$ $(0 \le t \le l,\ j = 1, 2, \cdots, r_1)$. Then almost everywhere $g_j(x)h^*(x) = 0$, which means that almost everywhere $f_k(x)h^*(x) = 0$ $(k = 1, 2, \cdots, s)$, i.e. $(P^{(r)}_{(0,l)}(t)f_k(x), h(x)) = 0$. Since the collection of vectors of the form $P^{(r)}_{(0,l)}(t)f_k(x)$ $(0 \le t \le l,\ k = 1, 2, \cdots, s)$ is dense in $\mathfrak{L}$, $h(x) = 0$. We have proved that the set of vectors of the form $P^{(r)}_{(0,l)}(t)g_j(x)$ $(0 \le t \le l,\ j = 1, 2, \cdots, r_1)$ is dense in $\mathfrak{L}$. Accordingly the system $U^{-1}g_j(x)$ $(j = 1, 2, \cdots, r_1)$ is reproducing for the spectral function $P(x)$, which is impossible.

The theorem is proved.

We note that *the rank of the spectral function* $P^{(r)}_{(0,l)}(x)$ *is equal to* r. This assertion follows directly from Theorem 23.1, since the vectors

$$h_k = \|\underbrace{0 \ldots 0\ 1}_{k}\ 0 \ldots 0\| \qquad (k = 1, 2, \ldots, r)$$

form a reproducing system of the function $P^{(r)}_{(0,l)}(x)$ and

$$\frac{d}{dx}\left\|(P^{(r)}_{(0,l)}(x)h_l,\ h_k)\right\| = I.$$

2. Canonical spectral functions. The spectral functions $P_1(x)$ $(x \in \mathfrak{M})$ and $P_2(x)$ $(x \in \mathfrak{M})$, given in the spaces $\mathfrak{H}_1$ and $\mathfrak{H}_2$ respectively, are said to be *unitarily equivalent* if there exists an isometric mapping U of $\mathfrak{H}_1$ onto $\mathfrak{H}_2$ such that $UP_1(x) = P_2(x)U$ $(x \in \mathfrak{M})$. Obviously, unitarily equivalent spectral functions have the same rank.

The spectral function $P(x)$ $(0 \le x \le l)$ of rank r is said to be *canonical* if it is unitarily equivalent to the function $P^{(r)}_{(0,l)}(x)$.

Theorem 2.32. *If π is a maximal chain of rank r, given in the space $\mathfrak{H}$, and $h_1, \cdots, h_s$ $(r \le s \le \infty)$ is a reproducing system for it, normalized by the condition $l = \sum_{k=1}^{s}\|h_k\|^2 < \infty$, then there exists a spectral function $P(x)$ $(x \in \mathfrak{M}$, $\min \mathfrak{M} = 0$, $\max \mathfrak{M} = l)$ whose range coincides with π, and a canonical spectral function $\tilde{P}(x)$ $(0 \le x \le l)$ of rank r, given in some space $\tilde{\mathfrak{H}} \supseteq \mathfrak{H}$, such that*

$$\tilde{P}(x)h = P(x)h \qquad (x \in \mathfrak{M},\ h \in \mathfrak{H}) \tag{23.3}$$

and

$$\sum_{k=1}^{s}(\tilde{P}(x)h_k,\ h_k)=x \qquad (0\leqslant x\leqslant l). \tag{23.4}$$

Proof. *By* Theorem 18.1 there exists a spectral function $P(x)$ ($x \in \mathfrak{M}$, $\min \mathfrak{M} = 0$, $\max \mathfrak{M} = l$), such that its range coincides with π and

$$\sum_{k=1}^{s}(P(x)h_k,\ h_k)=x \qquad (x\in\mathfrak{M}).$$

Applying Theorem 18.3, we find in some space $\mathfrak{H}_1 \supseteq \mathfrak{H}$ a spectral function $P_1(x)$ $(0 \le x \le l)$ having the following properties:

1) $$P_1(x)h=P(x)h \qquad (x\in\mathfrak{M},\ h\in\mathfrak{H}); \tag{23.5}$$

2) the rank of $P_1(x)$ is equal to r;

3) the system $h_1, \cdots, h_s$ is reproducing for $P_1(x)$;

4) $$\sum_{k=1}^{s}(P_1(x)h_k,\ h_k)=x \qquad (0\leqslant x\leqslant l). \tag{23.6}$$

It follows from 3), 4) and Lemma 23.1 that the function $P_1(x)$ is absolutely continuous. In view of Lemma 23.3 there exists an isometric mapping V of the space $\mathfrak{H}_1$ onto some subspace $\mathfrak{L} \subseteq L_2^{(r)}(0,\ l)$ such that $VP_1(x) = P^{(r)}_{(0,l)}(x)V$ $(0 \le x \le l)$. Let $\tilde{\mathfrak{H}} = \mathfrak{H}_1 \oplus \mathfrak{H}_2$, where $\mathfrak{H}_2$ is any Hilbert space whose dimension coincides with the dimension of the space $L_2^{(r)}(0,\ l) \ominus \mathfrak{L}$. We extend V to an isometric mapping U of the space $\tilde{\mathfrak{H}}$ onto $L_2^{(r)}(0,\ l)$, and put $\tilde{P}(x) = U^{-1}P^{(r)}_{(0,l)}(x)U$. Since $\mathfrak{L}$ is invariant relative to the orthoprojectors $P^{(r)}_{(0,l)}(x)$, we have

$$\tilde{P}(x)h=V^{-1}P^{(r)}_{(0,\ l)}(x)Vh=P_1(x)h \qquad (h\in\mathfrak{H}_1). \tag{23.7}$$

Comparing formulas (23.5)–(23.7), we obtain (23.3) and (23.4). The Theorem is proved.

In the case of a chain of rank 1 Theorem 23.2 may be sharpened as follows.

Theorem 23.3. *If π is a continuous maximal chain of rank* 1, *given in a space $\mathfrak{H}$, and $h_1, \cdots, h_s$ $(1 \le s \le \infty)$ is a reproducing system for it, normalized by the condition $l = \sum_{k=1}^{s}\|h_k\|^2 < \infty$, then there exists in $\mathfrak{H}$ a canonical spectral function $P(x)$ $(0 \le x \le l)$ such that its range coincides with π and*

$$\sum_{k=1}^{s}(P(x)h_k,\ h_k)=x \qquad (0\leqslant x\leqslant l).$$

Proof. By the preceding theorem there exists a spectral function $P(x)$ $(0 \leq x \leq l)$ whose range coincides with π, and a canonical spectral function $\tilde{P}(x)$ $(0 \leq x \leq l)$, of the first rank, given in some space $\tilde{\mathfrak{H}} \supseteq \mathfrak{H}$, such that $\tilde{P}(x)h = P(x)h$ $(0 \leq x \leq l,\ h \in \mathfrak{H})$ and $\sum_{k=1}^{s}(\tilde{P}(x)h_k, h_k) = x$ $(0 \leq x \leq l)$. It suffices to show that $\tilde{\mathfrak{H}} = \mathfrak{H}$.

We consider an isometric mapping U of the space $\tilde{\mathfrak{H}}$ onto $L_2(0, l)$ for which

$$U\tilde{P}(x) = P^{(1)}_{(0,\,l)}(x)\,U \qquad (0 \leqslant x \leqslant l).$$

Put $Uh_k = h_k(t)$. Then

$$x = \sum_{k=1}^{s}(\tilde{P}(x)\,h_k,\ h_k) = \sum_{k=1}^{s}(P^{(1)}_{(0,\,l)}(x)\,h_k(t),\ h_k(t))$$
$$= \sum_{k=1}^{s}\int_0^x |h_k(t)|^2\,dt = \int_0^x \sum_{k=1}^{s}|h_k(t)|^2\,dt,$$

so that

$$\sum_{k=1}^{s}|h_k(t)|^2 = 1 \tag{23.8}$$

almost everywhere. If $h \in \tilde{\mathfrak{H}} \ominus \mathfrak{H}$ and $Uh = h(t)$, then

$$0 = (h,\ P(x)\,h_k) = (h,\ \tilde{P}(x)\,h_k)$$
$$= (h(t),\ P^{(1)}_{(0,l)}(x)h_k(t)) = \int_0^x h(t)\overline{h_k(t)}\,dt.$$

Thus each of the functions $h(t)\overline{h_k(t)}$ is equal to zero almost everywhere, which along with (23.8) leads to the equation $h = 0$.

Theorem 23.4. *Suppose given in the space* $\mathfrak{H}$ *a Volterra operator* A *having a maximal chain* π *of rank* r, *and suppose that* $h_1, \cdots, h_s$ $(r \leq s \leq \infty)$ *is a reproducing system of the chain* π, *normalized by the condition* $l = \sum_{k=1}^{s}\|h_k\|^2 < \infty$. *Then some inessential extension* $\tilde{A}$ *of* A *has a canonical spectral function* $\tilde{P}(x)$ $(0 \leq x \leq l)$ *of rank* r *for which*

$$\sum_{k=1}^{s}(\tilde{P}(x)\,h_k,\ h_k) = x \qquad (0 \leqslant x \leqslant l).$$

Proof. In view of Theorem 23.2 there exists a spectral function $P(x)$ $(x \in \mathfrak{M},\ \min \mathfrak{M} = 0,\ \max \mathfrak{M} = l)$ whose range coincides with π, and a canonical spectral function $\tilde{P}(x)$ $(0 \leq x \leq l)$ of rank r, given in some space $\tilde{\mathfrak{H}} \supseteq \mathfrak{H}$, such that $\widetilde{P}(x)h = P(x)h$ $(x \in \mathfrak{M},\ h \in \mathfrak{H})$ and $\sum_{k=1}^{s}(\widetilde{P}(x)h_k, h_k) = x.$ $(0 \le x \le l)$. By Theorem 19.7 the spectral function $\widetilde{P}(x)$ lies in the inessential extension $\widetilde{A}$ of A in $\widetilde{\mathfrak{H}}$.

§24. Volterra operators in the spaces $L_2^{(r)}$

Suppose we are given matrices

$$\varphi_\alpha(x) = \| \varphi_\alpha^{(1)}(x)\, \varphi_\alpha^{(2)}(x) \ldots \varphi_\alpha^{(r)}(x) \| \qquad (\alpha = 1,\ 2,\ \ldots,\ n),$$

lying in the space $L_2^{(r)}(0, l)$ $(r < \infty,\ l < \infty)$, and a square Hermitian matrix $j = \|j_{\alpha\beta}\|_1^n$ with constant elements. The operator

$$\vec{H}f(x) = \sum_{\alpha,\beta=1}^{n} \int_0^l f(t)\varphi_\alpha^*(t)\,dt\, j_{\alpha\beta}\varphi_\beta(x) = \sum_{\alpha,\beta=1}^{n} (f, \varphi_\alpha)\, j_{\alpha\beta}\varphi_\beta(x) \tag{24.1}$$

$$(f(x) = \| f^{(1)}(x)\, f^{(2)}(x) \ldots f^{(r)}(x) \| \in L_2^{(r)}(0, l)),$$

operating in $L_2^{(r)}(0, l)$, is not more than n-dimensional, while

$$(\vec{H}f, g) = \sum_{\alpha,\beta=1}^{n} (f, \varphi_\alpha)\, j_{\alpha\beta}(\varphi_\beta, g) = \left(f, \sum_{\alpha,\beta=1}^{n} (g, \varphi_\alpha)\, j_{\alpha\beta}\varphi_\beta\right) = (f, \vec{H}g).$$

According to Theorem 17.5, there exists in $L_2^{(r)}(0, l)$ a Volterra operator $\vec{A}$ having a spectral function $P(x) = P_{(0,l)}^{(r)}(x)$ and imaginary component $\vec{H}$. By formulas (19.8) and (19.3)

$$\vec{A} = 2i\int_0^l P(x)\vec{H}\,dP(x) = 2i\vec{H} - 2i\int_0^l dP(x)\vec{H}P(x)$$
$$= 2i\int_0^l dP(x)\vec{H}(E - P(x)).$$

For any subdivision $0 = x_0 < x_1 < \cdots < x_m = l$ we will have

$$\left(\sum_{k=1}^{m} (P(x_k) - P(x_{k-1}))\vec{H}(E - P(x_k))f,\ g\right)$$
$$= \sum_{k=1}^{m}\sum_{\alpha,\beta=1}^{n} ((E - P(x_k))f, \varphi_\alpha)\, j_{\alpha\beta}((P(x_k) - P(x_{k-1}))\varphi_\beta, g) =$$

$$= \sum_{\alpha,\beta=1}^{n} \sum_{k=1}^{m} \int_{x_k}^{l} f(t)\varphi_\alpha^*(t)\,dt\, j_{\alpha\beta} \int_{x_{k-1}}^{x_k} \varphi_\beta(t) g^*(t)\,dt.$$

Passing to the limit as $\max(x_k - x_{k-1}) \to 0$, we get

$$(\vec{A}f, g) = 2i \sum_{\alpha,\beta=1}^{n} \int_0^l \int_x^l f(t)\varphi_\alpha^*(t)\,dt\, j_{\alpha\beta} d\int_0^x \varphi_\beta(t) g^*(t)\,dt$$

$$= 2i \sum_{\alpha,\beta=1}^{n} \int_0^l \int_x f(t)\varphi_\alpha(t)\,dt\, j_{\alpha\beta}\varphi_\beta(x) g^*(x)\,dx.$$

Thus

$$\vec{A}f(x) = 2i \int_x^l f(t)\Pi^*(t)\,dt\, j\Pi(x), \tag{24.2}$$

where

$$\Pi(x) = \left\| \begin{matrix} \varphi_1^{(1)}(x) & \varphi_1^{(2)}(x) & \dots & \varphi_1^{(r)}(x) \\ \varphi_2^{(1)}(x) & \varphi_2^{(2)}(x) & \dots & \varphi_2^{(r)}(x) \\ \cdot & \cdot & \cdot & \cdot \\ \varphi_n^{(1)}(x) & \varphi_n^{(2)}(x) & \dots & \varphi_n^{(r)}(x) \end{matrix} \right\|. \tag{24.3}$$

From our considerations it follows at the same time that *each operator of the form* (24.2), *where* j *is a Hermitian matrix with constant elements and* $\Pi(x)$ *is a rectangular matrix whose rows lie in* $L_2^{(r)}(0, l)$, *is a Volterra operator and has the imaginary component*

$$\vec{A}_I f(x) = \int_0^l f(t)\Pi^*(t)\,dt\, j\Pi(x). \tag{24.4}$$

Theorem 24.1. *Suppose that the completely nonselfadjoint Volterra operator* A, *operating in a space* $\mathfrak{H}$, *has a maximal chain of rank* r *and a finite-dimensional imaginary component*

$$A_I f = \sum_{\alpha=1}^{n} (f, e_\alpha)\omega_\alpha e_\alpha \quad \left(\omega_\alpha \neq 0, \quad (e_\alpha, e_\beta) = \begin{cases} 1, & \alpha=\beta \\ 0, & \alpha\neq\beta \end{cases}\right). \tag{24.5}$$

Then $r \leq n$ *and some inessential extension* $\tilde{A}$ *of* A *is unitarily equivalent to an operator of the type* (24.2) *operating in* $L_2^{(r)}(0, l)$ $(l = \Sigma_{\alpha=1}^{n}|\omega_\alpha|)$, *where*

$$j = \| j_{\alpha\beta} \|_1^n, \quad j_{\alpha\beta} = \begin{cases} \operatorname{sign}\omega_\alpha, & \alpha = \beta, \\ 0, & \alpha \neq \beta \end{cases} \tag{24.6}$$

and $\Pi(x)$ *is a matrix of the type* (24.3) *such that*

$$\int_0^l \Pi(x)\Pi^*(x)\,dx = \begin{Vmatrix} |\omega_1| & & & & \\ & |\omega_2| & & & \\ & & \cdot & & \\ & & & \cdot & \\ & & & & |\omega_n| \end{Vmatrix} \tag{24.7}$$

and almost everywhere $\operatorname{tr}(\Pi(x)\Pi^*(x)) = 1$.

Proof. In view of Theorem 16.5 the system $h_\alpha = |\omega_\alpha|^{1/2} e_\alpha$ $(\alpha = 1, 2, \cdots, n)$ is reproducing for any maximal chain of the operator A. Thus $r \le n$.

Suppose that $\tilde{A}$ is an inessential extension of the operator A, operating in some space $\tilde{\mathfrak{H}} \supseteq \mathfrak{H}$. By Theorem 23.4 $\tilde{\mathfrak{H}}$ may be chosen so that the operator $\tilde{A}$ has a canonical spectral function

$$\tilde{P}(x) \quad (0 \leqslant x \leqslant l, \ l = \sum_{\alpha=1}^n \|h_\alpha\|^2 = \sum_{\alpha=1}^n |\omega_\alpha|)$$

of rank r, satisfying the condition $\sum_{\alpha=1}^n (\tilde{P}(x)h_\alpha, h_\alpha) = x$ $(0 \le x \le l)$. By the definition of a canonical spectral function there exists an isometric mapping U of the space $\tilde{\mathfrak{H}}$ onto $L_2^{(r)}(0, l)$ for which $U\tilde{P}(x) = P_{(0,l)}^{(r)}(x)U$ $(0 \le x \le l)$.

Consider the operator $\vec{A} = UAU^{-1}$. Since the spectral function $P_{(0,l)}^{(r)}(x)$ lies in $\vec{A}$ and

$$\vec{A}_I f(x) = U\tilde{A}_I U^{-1} f(x) = \sum_{\alpha=1}^n (U^{-1}f, e_\alpha)\,\omega_\alpha U e_\alpha = \sum_{\alpha=1}^n (f, \varphi_\alpha)\, j_\alpha \varphi_\alpha \ .$$

$$(j_\alpha = \operatorname{sign}\omega_\alpha, \ \varphi_\alpha(x) = \|\varphi_\alpha^{(1)}(x)\,\varphi_\alpha^{(2)}(x) \ldots \varphi_\alpha^{(r)}(x)\| = Uh_\alpha),$$

we have $\vec{A}f(x) = 2i\int_x^l f(t)\Pi^*(t)\,dt\, j\Pi(x)$, where $\Pi(x)$ and j are given by formulas (24.3) and (24.6). In addition

$$\int_0^l \Pi(x)\Pi^*(x)\,dx = \left\| \int_0^l \varphi_\alpha(x)\varphi_\beta^*(x)\,dx \right\|_1^n$$

$$= \|(h_\alpha, h_\beta)\|_1^n = \begin{Vmatrix} |\omega_1| & & & & \\ & |\omega_2| & & & \\ & & \cdot & & \\ & & & \cdot & \\ & & & & |\omega_n| \end{Vmatrix}$$

and

$$\operatorname{tr}\int_0^x \Pi(t)\Pi^*(t)\,dt = \sum_{\alpha=1}^n \int_0^l P_{(0,l)}^{(r)}(x)\,\varphi_\alpha(t)\,\varphi_\alpha^*(t)\,dt$$

$$= \sum_{\alpha=1}^n (\tilde{P}(x)h_\alpha, h_\alpha) = x,$$

since almost everywhere $\operatorname{tr}(\Pi(x)\Pi^*(x)) = 1$.

Theorem 24.2. *If the completely nonselfadjoint Volterra operator A, operating in the space $\mathfrak{H}$, has a continuous maximal chain of the first rank and a finite-dimensional imaginary component* (24.5), *then it is unitarily equivalent to an operator of the type*

$$\vec{A}f(x) = 2i\int_x^l f(t)\Pi^*(t)\,dt\,j\Pi(x),$$

operating in $L_2(0, l)$ $(l = \Sigma_{\alpha=1}^n |\omega_\alpha|)$, where

$$j = \| j_{\alpha\beta} \|_1^n, \quad j_{\alpha\beta} = \begin{cases} \operatorname{sign}\omega_\alpha, & \alpha=\beta; \\ 0 & \alpha\neq\beta, \end{cases}$$

$$\Pi(x) = \begin{Vmatrix} \varphi_1(x) \\ \varphi_2(x) \\ \cdot \\ \cdot \\ \cdot \\ \varphi_n(x) \end{Vmatrix} \qquad (\varphi_\alpha(x) \in L_2(0, l)),$$

$$\int_0^l \Pi(x)\Pi^*(x)\,dx = \begin{Vmatrix} |\omega_1| & & & & \\ & |\omega_2| & & & \\ & & \cdot & & \\ & & & \cdot & \\ & & & & \cdot \\ & & & & & |\omega_n| \end{Vmatrix}$$

and almost everywhere $\Pi^*(x)\Pi(x) = 1$.

Proof. It follows from Theorems 16.5 and 23.3 that the operator A has a canonical spectral function $P(x)$ $(0 \leq x \leq l)$, of the first rank, such that

$$\sum_{\alpha=1}^n (P(x)h_\alpha, h_\alpha) = x \qquad (0 \leqslant x \leqslant l, \quad h_\alpha = |\omega_\alpha|^{1/2} e_\alpha).$$

Put $\vec{A} = UAU^{-1}$, where U is an isometric mapping of $\mathfrak{H}$ onto $L_2(0, l)$ satisfying the condition $UP(x) = P_{(0,l)}^{(1)}(x)U$ $(0 \leq x \leq l)$.

The remainder of the argument does not differ in any respect from that applied in the proof of the preceding theorem.

Theorem 24.3. *If the completely nonselfadjoint dissipative operator A has a maximal chain of the first rank and a finite-dimensional imaginary component*

$$A_I f = \sum_{\alpha=1}^{n} (f, e_\alpha)\, \omega_\alpha e_\alpha \quad \left(\omega_\alpha > 0, \ (e_\alpha, e_\beta) = \begin{cases} 1, & \alpha = \beta, \\ 0, & \alpha \neq \beta \end{cases} \right),$$

then it is unitarily equivalent to an operator of the type

$$\vec{A} f(x) = 2i \int_x^l f(t)\, \Pi^*(t)\, dt \Pi(x),$$

operating in $L_2(0, l)$ $(l = \Sigma_{\alpha=1}^n \omega_\alpha)$, *where*

$$\Pi(x) = \begin{Vmatrix} \varphi_1(x) \\ \varphi_2(x) \\ \vdots \\ \varphi_n(x) \end{Vmatrix} \qquad (\varphi_\alpha(x) \in L_2(0, l)),$$

$$\int_0^l \Pi(x)\, \Pi^*(x)\, dx = \begin{Vmatrix} \omega_1 & & & \\ & \omega_2 & & \\ & & \ddots & \\ & & & \omega_n \end{Vmatrix}$$

and almost everywhere $\Pi^*(x)\Pi(x) = 1$.

The proof follows from Theorems 15.5 and 24.2.

Applying Theorem 24.3 in particular to a completely nonselfadjoint Volterra operator A with a one-dimensional imaginary component

$$A_I f = (f, e)\, l e \qquad (\| e \| = 1, \ l > 0),$$

we obtain a new proof of Theorem 20.2. Indeed, the operator A is unitarily equivalent to the operator

$$\vec{A} f(x) = 2i \int_x^l f(t)\, \overline{\varphi(t)}\, dt \varphi(x),$$

operating in $L_2(0, l)$, where $\phi(x)$ is a function whose modulus is almost everywhere equal to 1. Consider a unitary operator V in $L_2(0, l)$ which assigns the function $f(x)\phi(x)$ to the function $f(x)$. Since $V^{-1}f(x) = f(x)\overline{\phi(x)}$, we have

$$\vec{B}f(x) = V^{-1}\vec{A}Vf(x) = 2iV^{-1}\int_x^l [Vf(t)]\,\overline{\varphi(t)}\,dt\varphi(x) = 2i\int_x^l f(t)\,dt.$$

If the operators A and B are unitarily equivalent, then each of them is said to be a *model* of the other. Above we considered the problem of models of Volterra operators with finite-dimensional imaginary components, operating in $L_2^{(r)}(0, l)$. We shall not consider the general problem of models with infinite-dimensional imaginary components in its full extent. In this connection see the book of Gohberg and Kreĭn [5]. What we shall do is to consider the special case of dissipative Volterra operators with nuclear imaginary components.

Suppose that we are given a dissipative Volterra operator $\vec{A}$ in the space $L_2(0, l)$ $(l < \infty)$. Suppose that $\vec{A}$ has a spectral function $P^{(1)}_{(0,l)}(x)$ and a nuclear imaginary component

$$\vec{A}_I f(x) = \sum_{\alpha=1}^{\infty} (f, \varphi_\alpha)\, \omega_\alpha \varphi_\alpha(x), \tag{24.9}$$

$$\left(\omega_\alpha > 0,\ \sum_{\alpha=1}^{\infty} \omega_\alpha < \infty,\ (\varphi_\alpha, \varphi_\beta) = \int_0^l \varphi_\alpha(x)\,\overline{\varphi_\beta(x)}\,dx = \begin{cases} 1, & \alpha = \beta, \\ 0, & \alpha \neq \beta \end{cases}\right).$$

Denote by $\overrightarrow{A^{(n)}}$ the Volterra operator with the same spectral function and the n-dimensional imaginary component

$$\overrightarrow{A_I^{(n)}} f(x) = \sum_{\alpha=1}^{n} (f, \varphi_\alpha)\, \omega_\alpha \varphi_\alpha(x).$$

According to the estimate (22.4),

$$\begin{aligned} \|\vec{A} - \overrightarrow{A^{(n)}}\| &\leqslant \|(\vec{A} - \overrightarrow{A^{(n)}})_R\| + \|(\vec{A} - \overrightarrow{A^{(n)}})_I\| \\ &\leqslant \left(\frac{4}{\pi} + 1\right)\|(\vec{A} - \overrightarrow{A^{(n)}})_I\|_\omega \leqslant \left(\frac{4}{\pi} + 1\right)(\omega_{n+1} + \omega_{n+2} + \ldots), \end{aligned}$$

so that $\lim_{n\to\infty}\|\vec{A} - \overrightarrow{A^{(n)}}\| = 0$. Accordingly,

$$\left\| \vec{A}f(x) - 2i\int_x^l K_n(x, y) f(y)\,dy \right\| \to 0 \quad (f(x) \in L_2(0, l)), \tag{24.10}$$

where

$$K_n(x, y) = \sum_{\alpha=1}^{n} \omega_\alpha \varphi_\alpha(x) \overline{\varphi_\alpha(y)}. \tag{24.11}$$

Consider the series

$$K(x) = \omega_1 |\varphi_1(x)|^2 + \omega_2 |\varphi_2(x)|^2 + \dots . \tag{24.12}$$

The function $K(x)$ is measurable, and in view of the estimate

$$\int_0^l K(x)\,dx = \sum_{\alpha=1}^{\infty} \omega_\alpha \int_0^l |\varphi_\alpha(x)|^2\,dx = \sum_{\alpha=1}^{\infty} \omega_\alpha < \infty,$$

summable. Therefore $K(x)$ is finite almost everywhere, which means that the series (24.12) converges at all points of some set $O \subset [0, l]$ of measure l. Since

$$\left| \sum_{\alpha=m}^{n} \omega_\alpha \varphi_\alpha(x) \overline{\varphi_\alpha(y)} \right|^2 \leqslant \sum_{\alpha=m}^{n} \omega_\alpha |\varphi_\alpha(x)|^2 \sum_{\alpha=m}^{n} \omega_\alpha |\varphi_\alpha(y)|^2,$$

the series

$$K(x, y) = \omega_1 \varphi_1(x) \overline{\varphi_1(y)} + \omega_2 \varphi_2(x) \overline{\varphi_2(y)} + \dots \tag{24.13}$$

converges for all $x, y \in O$. Moreover,

$$|K_n(x, y)|^2 \leqslant K(x) K(y) \qquad (n = 1, 2, \dots;\ x, y \in O)$$

and

$$K_n(x, y) \to K(x, y).$$

By the Lebesgue theorem on the passage to the limit under the integral sign,

$$\lim_{n\to\infty} \int_x^l K_n(x, y) f(y)\,dy = \int_x^l K(x, y) f(y)\,dy \tag{24.14}$$

for each fixed $x \in O$. Comparing (24.10) and (24.14), we get

$$\vec{A} f(x) = 2i \int_x^l K(x, y) f(y)\,dy. \tag{24.15}$$

It is easy to see that

$$\vec{A}_I f(x) = \int_0^l K(x, y) f(y)\,dy. \tag{24.16}$$

The operator $\vec{A}$ *is completely nonselfadjoint if and only if the function* $K(x)$

is almost everywhere nonzero. Indeed, a necessary and sufficient condition for the operator $\vec{A}$ to be completely nonselfadjoint is that $\vec{A}f(x) = 0$ should imply $f(x) = 0$. If the function $K(x)$ is equal to zero at all the points of some set $\mathfrak{d}$ of positive measure, then for a function $f(x)$ identically zero outside $\mathfrak{d}$ we will have

$$\vec{A}f(x) = 2i \int_x^l \sum_{\alpha=1}^{\infty} \omega_\alpha \varphi_\alpha(x) \overline{\varphi_\alpha(y)} f(y)\, dy = 0.$$

Conversely, if $\vec{A}g(x) = 0$, then

$$\sum_{\alpha=1}^{\infty} \omega_\alpha \left| \int_x^l \varphi_\alpha(t) \overline{g(t)}\, dt \right|^2 = - \lim_{n\to\infty} \int_x^l \frac{d}{dy} \sum_{\alpha=1}^{n} \omega_\alpha \left| \int_y^l \varphi_\alpha(t) \overline{g(t)}\, dt \right|^2 dy$$

$$= \lim_{n\to\infty} \int_x^l \int_y^l [K_n(y, t)\, g(t) \overline{g(y)} + \overline{K_n(y, t)}\, \overline{g(t)}\, g(y)]\, dt\, dy$$

$$= \lim_{n\to\infty} \operatorname{Im} \int_x^l (\overrightarrow{A^{(n)}} g)(y) \overline{g(y)}\, dy$$

$$= \lim_{n\to\infty} \operatorname{Im} \left((E - P^{(1)}_{(0,l)}(x)) \overrightarrow{A^{(n)}} g, g \right)$$

$$= \operatorname{Im} \left((E - P^{(1)}_{(0,l)}(x)) \vec{A} g, g \right) = 0.$$

Accordingly, almost everywhere $\phi_\alpha(x)\overline{g(x)} = 0$ $(\alpha = 1, 2, \cdots)$, and therefore $K(x) = 0$ for almost all points x for which $g(x) \neq 0$.

Theorem 24.4. *Suppose that we are given in the space $\mathfrak{H}$ a completely nonselfadjoint dissipative Volterra operator A with a nuclear imaginary component*

$$A_I f = \sum_{\alpha=1}^{\infty} (f, e_\alpha)\, \omega_\alpha e_\alpha$$

$$\left(\omega_\alpha > 0,\ l = \sum_{\alpha=1}^{\infty} \omega_\alpha < \infty,\ (e_\alpha, e_\beta) = \begin{cases} 1, & \alpha = \beta, \\ 0, & \alpha \neq \beta \end{cases} \right).$$

If A has a maximal chain π of the first rank,[1] *then it is unitarily equivalent to an operator of the form* (24,15) *operating in $L_2(0, l)$, where the function $K(x, y)$ is representable in the form*

1) It will be proved in §34 that this property of the operator A follows from the preceding ones.

$$K(x, y) = \sum_{\alpha=1}^{\infty} \omega_\alpha \varphi_\alpha(x) \overline{\varphi_\alpha(y)} \tag{24.17}$$

$$\left((\varphi_\alpha, \varphi_\beta) = \int_0^l \varphi_\alpha(x) \overline{\varphi_\beta(x)}\, dx = \begin{cases} 1, & \alpha=\beta, \\ 0, & \alpha \neq \beta \end{cases} \right)$$

and satisfies almost everywhere on $[0, l]$ *the condition*

$$K(x) = \sum_{\alpha=1}^{\infty} \omega_\alpha |\varphi_\alpha(x)|^2 = 1. \tag{24.18}$$

Proof. It follows from Theorems 15.5 and 16.5 that the chain π is continuous, and the vectors $h_\alpha = \omega_\alpha^{1/2} e_\alpha$ form a reproducing system for it. By Theorem 23.3 there exists a canonical spectral function $P(x)$ $(0 \le x \le l)$ such that its range coincides with π and

$$\sum_{\alpha=1}^{\infty} (P(x) h_\alpha, h_\alpha) = x \qquad (0 \leqslant x \leqslant l).$$

Suppose that U is an isometric mapping of $\mathfrak{H}$ onto $L_2(0, l)$ for which $UP(x) = P^{(1)}_{(0,l)}(x)U$ $(0 \le x \le l)$. We put $\vec{A} = UAU^{-1}$ and $\phi_\alpha(x) = Ue_\alpha$. Since $\vec{A}$ is a Volterra operator with imaginary component $\vec{A}_I f(x) = \Sigma_{\alpha=1}^{\infty} (f, \phi_\alpha)\omega_\alpha \phi_\alpha(x)$ and spectral function $P^{(1)}_{(0,l)}(x)$, equation (24.15) holds, in which $K(x, y)$ is defined by formula (24.17). Since

$$\int_0^x K(t)\, dt = \sum_{\alpha=1}^{\infty} \omega_\alpha \int_0^l |P^{(1)}_{(0,l)}(x) \varphi_\alpha(t)|^2\, dt$$

$$= \sum_{\alpha=1}^{\infty} \omega_\alpha \| P(x) e_\alpha \|^2 = \sum_{\alpha=1}^{\infty} (P(x) h_\alpha, h_\alpha) = x,$$

$K(x) = 1$ almost everywhere.

§25. Multiplicative representations of characteristic functions of Volterra nodes

1. **Definition of multiplicative integral.** Suppose that $\mathfrak{M}$ is a bounded closed set of real numbers and $F(x)$ $(x \in \mathfrak{M})$ some function whose values are bounded linear operators operating in a Hilbert space $\mathfrak{G}$. We assign to each subdivision

$$\alpha = x_0 < x_1 < \ldots < x_n = \beta \qquad (\alpha = \min \mathfrak{M},\ \beta = \max \mathfrak{M})$$

of the set $\mathfrak{M}$ the integral product

$$\overset{n}{\overrightarrow{\prod_{j=1}}} (E + \Delta F_j) = (E + \Delta F_1)(E + \Delta F_2) \ldots (E + \Delta F_n) \tag{25.1}$$
$$(\Delta F_j = F(x_j) - F(x_{j-1})).$$

The operator $\mathcal{J}$ is called a *right multiplicative Stieltjes integral* relative to the function $F(x)$, and is denoted by the symbol $\overrightarrow{\int}_{\mathfrak{M}} (E + dF(x))$, if for each $\epsilon > 0$ there exists a $\delta > 0$ such that any integral product (25.1) constructed relative to a subdivision of the set $\mathfrak{M}$, whose weight [1] is less than δ, satisfies the inequality

$$\left\| \mathcal{J} - \overset{n}{\overrightarrow{\prod_{j=1}}} (E + \Delta F_j) \right\| < \varepsilon.$$

Analogously, using the integral products

$$\overset{n}{\overleftarrow{\prod_{j=1}}} (E + \Delta F_j) = (E + \Delta F_n) \ldots (E + \Delta F_2)(E + \Delta F_1) \tag{25.2}$$

one defines the *left multiplicative integral* $\overleftarrow{\int}_{\mathfrak{M}} (E + dF(x))$.

We shall also need right and left multiplicative Stieltjes integrals *of exponential type*, which are defined respectively as uniform limits of the integral products $\overrightarrow{\prod}_{j=1}^{n} e^{\Delta F_j}$ and $\overleftarrow{\prod}_{j=1}^{n} e^{\Delta F_j}$, and denoted by the symbols $\overrightarrow{\int}_{\mathfrak{M}} e^{dF(x)}$ and $\overleftarrow{\int}_{\mathfrak{M}} e^{dF(x)}$.

In the case when $\mathfrak{M} = [\alpha, \beta]$ we shall write $\overrightarrow{\int}_{\alpha}^{\beta}$ and $\overleftarrow{\int}_{\alpha}^{\beta}$ instead of $\overrightarrow{\int}_{\mathfrak{M}}$ and $\overleftarrow{\int}_{\mathfrak{M}}$.

The operator $\mathcal{J}$ is called a *right multiplicative Riemann integral* of the operator-function $G(x)$ $(\alpha \le x \le \beta)$, and denoted by the symbol $\overrightarrow{\int}_{\alpha}^{\beta} e^{G(x)dx}$, if for each $\epsilon > 0$ there exists a $\delta > 0$ such that any integral product

$$\overset{n}{\overrightarrow{\prod_{j=1}}} e^{G(\xi_j)\Delta x_j} = e^{G(\xi_1)\Delta x_1} e^{G(\xi_2)\Delta x_2} \ldots e^{G(\xi_n)\Delta x_n} \tag{25.3}$$
$$(\Delta x_j = x_j - x_{j-1}),$$

constructed relative to the subdivision

$$\alpha = x_0 \leqslant \xi_1 \leqslant x_1 \leqslant \xi_2 \leqslant x_2 \leqslant \ldots \leqslant x_{n-1} \leqslant \xi_n \leqslant x_n = \beta,$$

---

1) Concerning the concept of "weight of a subdivision", see §19.

in which $\max \Delta x_j < \delta$, satisfies the inequality

$$\left\| \mathcal{J} - \overset{\curvearrowright}{\prod_{j=1}^{n}} e^{G(\xi_j)\Delta x_j} \right\| < \varepsilon.$$

Analogously one defines the left multiplicative Riemann integral $\overset{\curvearrowleft}{\int_\alpha^\beta} e^{G(x)\,dx}$.

2. Theorems connecting multiplicative integrals of various types. We again suppose that $F(x)$ is an operator-function given on a set $\mathfrak{M}$. To each subdivision $x_0 < x_1 < \cdots < x_n$ of this set there corresponds a number $\sum_{j=1}^{n} \|F(x_j) - F(x_{j-1})\|$, which is called the *variation sum*. If the set of all variation sums is bounded, then we say that $F(x)$ has *bounded variation* on $\mathfrak{M}$. The least upper bound of all variation sums is called the *variation* of $F(x)$.

Theorem 25.1. *Suppose that $F(x)$ $(x \in \mathfrak{M})$ is continuous in the sense of the norm and has bounded variation on $\mathfrak{M}$. If for each interval (a, b) adjacent to $\mathfrak{M}$ the equation*

$$(F(b) - F(a))^2 = 0 \tag{25.4}$$

holds, and if one of the integrals $\mathfrak{I}_1 = \overset{\curvearrowright}{\int_{\mathfrak{M}}} (E + dF(x))$, $\mathfrak{I}_2 = \overset{\curvearrowright}{\int_{\mathfrak{M}}} e^{dF(x)}$ exists, then the other exists as well and they are equal.

Proof. We construct a subdivision $x_0 < x_1 < \cdots < x_n$ of the set $\mathfrak{M}$ and introduce the notation

$$R_j = E + \Delta F_j, \quad \tilde{R}_j = e^{\Delta F_j} \qquad (\Delta F_j = F(x_j) - F(x_{j-1})).$$

If the interval (x_{j-1}, x_j) is adjacent to $\mathfrak{M}$, then in view of (25.4) $R_j = \tilde{R}_j$. For the remaining integrals we get

$$\|\tilde{R}_j - R_j\| = \left\| \frac{1}{2!}\Delta F_j^2 + \frac{1}{3!}\Delta F_j^3 + \ldots \right\|$$
$$\leqslant \frac{1}{2}\|\Delta F_j\|^2 \left(1 + \|\Delta F_j\| + \frac{1}{2!}\|\Delta F_j\|^2 + \ldots\right) \leqslant \frac{1}{2}\|\Delta F_j\|^2 e^v,$$

where v is the variation of the function $F(x)$ on $\mathfrak{M}$.

Using the obvious equation

$$\overset{\curvearrowright}{\prod_{j=1}^{n}} \tilde{R}_j - \overset{\curvearrowright}{\prod_{j=1}^{n}} R_j = (\tilde{R}_1 - R_1) \overset{\curvearrowright}{\prod_{j=2}^{n}} R_j + \tilde{R}_1 (\tilde{R}_2 - R_2) \overset{\curvearrowright}{\prod_{j=3}^{n}} R_j$$

$$+ \overset{\curvearrowright}{\prod_{j=1}^{2}} \tilde{R}_j (\tilde{R}_3 - R_3) \overset{\curvearrowright}{\prod_{j=4}^{n}} R_j + \ \dots \ + \overset{\curvearrowright}{\prod_{j=1}^{n-2}} \tilde{R}_j (\tilde{R}_{n-1} - R_{n-1}) R_n + \overset{\curvearrowright}{\prod_{j=1}^{n-1}} \tilde{R}_j (\tilde{R}_n - R_n) \tag{25.5}$$

and noting that

$$\left\| \overset{\curvearrowright}{\prod_{j=k}^{n}} R_j \right\| \leqslant \prod_{j=k}^{n} e^{\|\Delta F_j\|} \leqslant e^{v} \quad (k = 2, 3, \dots, n),$$

$$\left\| \overset{\curvearrowright}{\prod_{j=1}^{k}} \tilde{R}_j \right\| \leqslant \prod_{j=1}^{k} e^{\|\Delta F_j\|} \leqslant e^{v} \quad (k = 1, 2, \dots, n-1),$$

we arrive at the estimate

$$\left\| \overset{\curvearrowright}{\prod_{j=1}^{n}} \tilde{R}_j - \overset{\curvearrowright}{\prod_{j=1}^{n}} R_j \right\| \leqslant \frac{1}{2} e^{3v} \sum \|\Delta F_j\|^2 \leqslant \frac{v}{2} e^{3v} \max \|\Delta F_j\|,$$

where the summation and maximum are extended over only those intervals not adjacent to $\mathfrak{M}$.

For a given $\epsilon > 0$ there exists a $\delta > 0$ such that for any $x', x'' \in \mathfrak{M}$ satisfying the condition $|x' - x''| < \delta$ the inequality $\|F(x') - F(x'')\| < 2\epsilon e^{-3v}/v$ will be satisfied. Thus for a subdivision $x_0 < x_1 < \cdots < x_n$ of weight less than δ the estimate

$$\left\| \overset{\curvearrowright}{\prod_{j=1}^{n}} \tilde{R}_j - \overset{\curvearrowright}{\prod_{j=1}^{n}} R_j \right\| < \varepsilon$$

holds.

The theorem is proved.

We will say that $G(x)$ $(\alpha \leq x \leq \beta)$ is a *function of Riemann type* if for each $\epsilon > 0$ there exists a $\delta > 0$ such that for any subdivision $\alpha = x_0 < x_1 < \cdots < x_n = \beta$ satisfying the condition $\max(x_j - x_{j-1}) < \delta$ the inequality

$$\sum_{j=1}^{n} \sup_{x_{j-1} \leqslant \xi, \eta \leqslant x_j} \| G(\xi) - G(\eta) \| \Delta x_j < \varepsilon \qquad (\Delta x_j = x_j - x_{j-1})$$

holds.

$$\alpha = x_0 = y_0^{(1)} < y_1^{(1)} < \ldots < y_{k_1}^{(1)} = x_1 = y_0^{(2)} < y_1^{(2)} < \ldots < y_{k_2}^{(2)}$$
$$= x_2 = y_0^{(3)} < \ldots < x_{n-1} = y_0^{(n)} < y_1^{(n)} < \ldots < y_{k_n}^{(n)} = x_n = \beta,$$
$$x_{j-1} \leqslant \xi_j \leqslant x_j, \quad y_{i-1}^{(j)} \leqslant \eta_i^{(j)} \leqslant y_i^{(j)},$$

then

$$\left\| \sum_{j=1}^{n} G(\xi_j)\Delta x_j - \sum_{j=1}^{n}\sum_{i=1}^{k_j} G(\eta_i^{(j)})\Delta y_i^{(j)} \right\| \leqslant \sum_{j=1}^{n}\sum_{i=1}^{k_j} \| G(\xi_j) - G(\eta_i^{(j)}) \| \Delta y_i^{(j)}$$
$$\leqslant \sum_{j=1}^{n} \sup_{x_{j-1}\leqslant \xi, \eta \leqslant x_j} \| G(\xi) - G(\eta) \| \Delta x_j \quad (\Delta y_i^{(j)} = y_i^{(j)} - y_{i-1}^{(j)}),$$

from which it follows that for a function $G(x)$ $(\alpha \le x \le \beta)$ of Riemann type the Riemann integral $\int_\alpha^\beta G(x)dx$ exists, the convergence being in the sense of the norm

Theorem 25.2. *Suppose that* $F(x) = F(\alpha) + \int_\alpha^x G(t)dt$ $(\alpha \le x \le \beta)$, *where* $G(x)$ *is a function of Riemann type. If one of the integrals* $\mathfrak{I}_1 = \overset{\curvearrowright}{\int_\alpha^\beta} (E + dF(x))$, $\mathfrak{I}_2 = \overset{\curvearrowright}{\int_\alpha^\beta} e^{dF(x)}$, $\mathfrak{I}_3 = \overset{\curvearrowright}{\int_\alpha^\beta} e^{G(x)\,dx}$ *exists, then the others exist as well and they are all equal.*

Proof. Obviously the norm of the function $G(x)$ is bounded: $\|G(x)\| \le M$, $\alpha \le x \le \beta$. Accordingly $\|F(x') - F(x'')\| \le M|x' - x''|$, so that the function $F(x)$ is continuous in the sense of the norm and all its variation sums are bounded by the number $M(\beta - \alpha)$. By the preceding theorem the existence of one of the integrals $\mathfrak{I}_1$ or $\mathfrak{I}_2$ implies the existence of the other and the equation $\mathfrak{I}_1 = \mathfrak{I}_2$. We will show that the analogous assertion is valid for the integrals $\mathfrak{I}_1$ and $\mathfrak{I}_3$.

We consider a subdivision $\alpha = x_0 < x_1 < \cdots < x_n = \beta$ and introduce the notation

$$R_j = E + \Delta F_j, \quad \tilde{\tilde{R}}_j = e^{G(\xi_j)\,\Delta x_j},$$
$$= F(x_j) - F(x_{j-1}), \quad x_{j-1} \leqslant \xi_j \leqslant x_j, \quad \Delta x_j = x_j - x_{j-1}).$$

In view of the estimates

$$\| \tilde{\tilde{R}}_j - R_j \| \leqslant \| G(\xi_j)\Delta x_j - \Delta F_j \| + \sum_{p=2}^{\infty} \frac{1}{p!} \| G(\xi_j)\Delta x_j \|^p,$$

$$\| G(\xi_j)\Delta x_j - \Delta F_j \| = \left\| \int_{x_{j-1}}^{x_j} (G(\xi_j) - G(t))\,dt \right\|$$
$$\leqslant \sup_{x_{j-1}\leqslant \xi, \eta \leqslant x_j} \| G(\xi) - G(\eta) \| \Delta x_j,$$

$$\sum_{p=2}^{\infty}\frac{1}{p!}\|G(\xi_j)\Delta x_j\|^p\leqslant\sum_{p=2}^{\infty}\frac{1}{p!}(M\Delta x_j)^p\leqslant\frac{M^2}{2}e^{M(\beta-\alpha)}\Delta x_j^2,$$

$$\left\|\overset{\curvearrowright}{\prod_{j=k}^{n}}R_j\right\|\leqslant\prod_{j=k}^{n}e^{\|\Delta F_j\|}\leqslant e^{M(\beta-\alpha)},\quad\left\|\overset{\curvearrowright}{\prod_{j=1}^{k}}\tilde{\tilde{R}}_j\right\|\leqslant\prod_{j=1}^{k}e^{\|G(\xi_j)\|\Delta x_j}\leqslant e^{M(\beta-\alpha)}$$

and the formulas obtained from (25.5) by replacing $\tilde{R}_j$ by $\tilde{\tilde{R}}_j$, we will have

$$\left\|\overset{\curvearrowright}{\prod_{j=1}^{n}}\tilde{\tilde{R}}_j-\overset{\curvearrowright}{\prod_{j=1}^{n}}R_j\right\|$$

$$\leqslant e^{2M(\beta-\alpha)}\left[\sum_{j=1}^{n}\sup_{x_{j-1}\leqslant\xi,\,\eta\leqslant x_j}\|G(\xi)-G(\eta)\|\Delta x_j+\frac{M^2(\beta-\alpha)}{2}e^{M(\beta-\alpha)}\max\Delta x_j\right]. \tag{25.6}$$

Fix an $\epsilon>0$. In view of (25.6) there exists a $\delta>0$ such that for each subdivision $\alpha=x_0<x_1<\cdots<x_n=\beta$ satisfying the condition $\max(x_j-x_{j-1})<\delta$ the inequality

$$\left\|\overset{\curvearrowright}{\prod_{j=1}^{n}}\tilde{\tilde{R}}_j-\overset{\curvearrowright}{\prod_{j=1}^{n}}R_j\right\|<\varepsilon$$

will hold. The theorem is proved.

It is not hard to see that in Theorems 25.1 and 25.2 the right multiplicative integrals may be replaced by left multiplicative integrals.

3. Multiplicative representations of characteristic functions of Volterra nodes.

Theorem 25.3. *If*

$$\Theta=\begin{pmatrix}A & K & J\\ \mathfrak{H} & & \mathfrak{G}\end{pmatrix}$$

is a Volterra node and $P(x)$ $(x\in\mathfrak{M})$ *is a spectral function of the operator* A, *then*

$$W_\Theta(\lambda)=\overset{\curvearrowright}{\int_{\mathfrak{M}}}\left(E+\frac{2i}{\lambda}K^*dP(x)KJ\right)\quad(\lambda\neq0). \tag{25.7}$$

Proof. Denote by $\mathfrak{z}$ some subdivision $\alpha=x_0<x_1<\cdots<x_n=\beta$ of the set $\mathfrak{M}$ and consider the integral sum

$$A_{\mathfrak{z}} = 2i \sum_{s<t} \Delta P_s A_I \Delta P_t \qquad (\Delta P_s = P(x_s) - P(x_{s-1}))$$

of the integral $A = 2i \int_{\mathfrak{M}} P(x) \dot{A}_I \, dP(x)$. Since $A_{\mathfrak{z}}^n = 0$ and $A_I = KJK^*$, for $\lambda \neq 0$ we have

$$(A_{\mathfrak{z}} - \lambda E)^{-1} = -\frac{E}{\lambda} - \frac{A_{\mathfrak{z}}}{\lambda^2} - \frac{A_{\mathfrak{z}}^2}{\lambda^3} - \dots - \frac{A_{\mathfrak{z}}^{n-1}}{\lambda^n}$$

$$= -\frac{1}{\lambda}\Big(\sum_{s=1}^{n} \Delta P_s + \frac{2i}{\lambda} \sum_{s<t} \Delta P_s A_I \Delta P_t$$

$$+ \left(\frac{2i}{\lambda}\right)^2 \sum_{s<t<r} \Delta P_s A_I \Delta P_t A_I \Delta P_r + \dots$$

$$\dots + \left(\frac{2i}{\lambda}\right)^{n-1} \Delta P_1 A_I \Delta P_2 \dots A_I \Delta P_n\Big),$$

$$E - 2iK^* (A_{\mathfrak{z}} - \lambda E)^{-1} KJ = E + \frac{2i}{\lambda} \sum_{s=1}^{n} K^* \Delta P_s KJ$$

$$+ \left(\frac{2i}{\lambda}\right)^2 \sum_{s<t} K^* \Delta P_s KJK^* \Delta P_t KJ$$

$$+ \left(\frac{2i}{\lambda}\right)^3 \sum_{s<t<r} K^* \Delta P_s KJK^* \Delta P_t KJK^* \Delta P_r KJ + \dots$$

$$\dots + \left(\frac{2i}{\lambda}\right)^n K^* \Delta P_1 KJK^* \Delta P_2 KJ \dots K^* \Delta P_n KJ$$

$$= \prod_{s=1}^{\overset{n}{\curvearrowright}} \left(E + \frac{2i}{\lambda} K^* \Delta P_s KJ\right),$$

so that

$$\left\| W_\Theta(\lambda) - \prod_{s=1}^{\overset{n}{\curvearrowright}} \left(E + \frac{2i}{\lambda} K^* \Delta P_s KJ\right) \right\|$$

$$\leqslant 2 \| K \|^2 \| (A - \lambda E)^{-1} \| \| A - A_{\mathfrak{z}} \| \| (A_{\mathfrak{z}} - \lambda E)^{-1} \|.$$

It remains to be noted that for a given $\epsilon > 0$ and $\lambda \neq 0$ there exists a $\delta > 0$ such that the right side of the last inequality will be less than ϵ for any subdivision $\mathfrak{z}$ of weight less than δ.

Theorem 25.4. *Suppose that*

$$\Theta = \begin{pmatrix} A & K & J \\ \mathfrak{H} & & \mathfrak{G} \end{pmatrix}$$

is a Volterra node and $P(x)$ $(x \in \mathfrak{M})$ *is a spectral function of the operator* A. *If the function* $K^* P(x)K$ *is of bounded variation on* $\mathfrak{M}$, *then*

$$W_\Theta(\lambda) = \overset{\curvearrowright}{\int_{\mathfrak{M}}} e^{\frac{2i}{\lambda} K^* dP(x) KJ}. \tag{25.8}$$

Proof. By the preceding theorem the characteristic operator-function $W_\Theta(\lambda)$ is representable in the form (25.7). For each fixed $\lambda \neq 0$ the function $F(x) = (2i/\lambda)K^* P(x)KJ$ is of bounded variation on $\mathfrak{M}$, and, in view of Lemma 19.1, continuous on $\mathfrak{M}$ in the sense of the norm. Moreover, if the integral (a, b) is adjacent to $\mathfrak{M}$, then in view of (15.3)

$$(F(b) - F(a))^2 = \left(\frac{2i}{\lambda}\right)^2 K^*(P(b) - P(a))\, A_I\, (p(b) - P(a))\, KJ = 0.$$

In view of Theorem 25.1 the integral (25.7) may be replaced by the integral (25.8).

Lemma 25.1. *If* H *is a completely continuous positive operator operating in the space* $\mathfrak{H}$, *and* $\{h_\alpha\}$ *is an arbitrary orthonormalized basis in* $\mathfrak{H}$, *then*

$$\sum_{\alpha=1}^{\infty} s_\alpha(H) = \sum_{\alpha=1}^{\infty} (Hh_\alpha, h_\alpha). \tag{25.9}$$

Proof. Using an orthonormalized basis $\{e_\alpha\}$ consisting of eigenvectors of the operator H, we obtain

$$\sum_{\alpha=1}^{\infty} s_\alpha(H) = \sum_{\alpha=1}^{\infty} (He_\alpha, e_\alpha) = \sum_{\alpha=1}^{\infty} \| H^{1/2} e_\alpha \|^2$$

$$= \sum_{\alpha=1}^{\infty} \sum_{\beta=1}^{\infty} |(H^{1/2} e_\alpha, h_\beta)|^2 = \sum_{\beta=1}^{\infty} \sum_{\alpha=1}^{\infty} |(H^{1/2} h_\beta, e_\alpha)|^2$$

$$= \sum_{\beta=1}^{\infty} \| H^{1/2} h_\beta \|^2 = \sum_{\alpha=1}^{\infty} (Hh_\alpha, h_\alpha).$$

Theorem 25.5. *Suppose that* Θ *is a Volterra node and* $P(x)$ $(x \in \mathfrak{M})$ *is a spectral function of the operator* A. *If* KK^* *is a nuclear operator, then*

$$W_\Theta(\lambda)=\int_{\mathfrak{M}}^{\curvearrowright} e^{\frac{2i}{\lambda}K^*\,dP(x)\,KJ}. \qquad (25.10)$$

Proof. It suffices to show that the function $K^*P(x)K$ is of bounded variation on $\mathfrak{M}$.

Suppose that $\{g_\alpha\}$ is an orthonormalized basis in $\mathfrak{G}$ and $x_0<x_1<\cdots<x_n$ is some subdivision of the set $\mathfrak{M}$. In view of Lemma 25.1

$$\begin{aligned}\sum_{j=1}^{n}\|K^*\Delta P_jK\|&\leqslant\sum_{j=1}^{n}\sum_{\alpha=1}^{\infty}s_\alpha(K^*\Delta P_jK)\\ &=\sum_{j=1}^{n}\sum_{\alpha=1}^{\infty}(K^*\Delta P_jKg_\alpha,g_\alpha)=\sum_{\alpha=1}^{\infty}(K^*Kg_\alpha,g_\alpha)\\ &=\sum_{\alpha=1}^{\infty}s_\alpha(K^*K)\qquad(\Delta P_j=P(x_j)-P(x_{j-1})).\end{aligned}$$

Theorem 25.6. *Suppose that Θ is a Volterra node and $P(x)$ $(\alpha\le x\le\beta)$ is a spectral function of the operator A. If*

$$K^*P(x)K=\int_0^x H(t)\,dt\qquad(\alpha\leqslant x\leqslant\beta),$$

where $H(t)$ is a function of Riemann type, then

$$W_\Theta(\lambda)=\int_\alpha^{\beta\,\curvearrowright} e^{\frac{2i}{\lambda}H(t)\,J\,dt}. \qquad (25.11)$$

The proof follows from Theorems 25.3 and 25.2.

Suppose that $F(x)=\|f_{\alpha\beta}(x)\|_1^r$ $(r<\infty)$ is a matrix-function given on the set $\mathfrak{M}$. The multiplicative integral $\int_{\mathfrak{M}} e^{dF(x)}$ is defined as the limit of the integral products

$$\prod_{s=1}^{\overset{\curvearrowleft}{n}} e^{\Delta F_s}=e^{\Delta F_n}\dots e^{\Delta F_2}e^{\Delta F_1}$$

$$(\min\mathfrak{M}=x_0<x_1<\dots<x_n=\max\mathfrak{M},\ \Delta F_s=F(x_s)-F(x_{s-1}))$$

as the weight of the subdivision tends to zero.

Theorem 25.7. *If $\theta=(A,k_1,\cdots,k_r,j)$ is a Volterra matrix node and $P(x)$ $(x\in\mathfrak{M})$ is a spectral function of the operator A, then*

$$w_\Theta(\lambda) = \overset{\curvearrowleft}{\int_{\mathfrak{M}}} e^{\frac{2i}{\lambda} j d \|(P(x) k_\alpha, k_\beta)\|_1^r}. \tag{25.12}$$

Proof. There exists a Volterra operator node

$$\Theta = \begin{pmatrix} A & K & J \\ \mathfrak{H} & & \mathfrak{G} \end{pmatrix}$$

and an orthonormalized basis $g_1, \cdots, g_r$ in the space $\mathfrak{G}$ such that

$$w_\Theta(\lambda) = \|(W_\Theta(\lambda) g_\alpha, g_\beta)\|_1^r, \quad K g_\alpha = k_\alpha, \quad j = \|(J g_\alpha, g_\beta)\|_1^r.$$

For any subdivision $x_0 < x_1 < \cdots < x_n$ of the set $\mathfrak{M}$ we have

$$\left\| \left(\overset{\curvearrowright}{\prod_{s=1}^{n}} e^{\frac{2i}{\lambda} K^* \Delta P_s K J} g_\alpha, g_\beta \right) \right\|_1^r = \overset{\curvearrowleft}{\prod_{s=1}^{n}} \left\| \left(e^{\frac{2i}{\lambda} K^* \Delta P_s K J} g_\alpha, g_\beta \right) \right\|_1^r$$

$$= \overset{\curvearrowleft}{\prod_{s=1}^{n}} e^{\frac{2i}{\lambda} \|(K^* \Delta P_s K J g_\alpha, g_\beta)\|_1^r} = \overset{\curvearrowleft}{\prod_{s=1}^{n}} e^{\frac{2i}{\lambda} j \|(\Delta P_s k_\alpha, k_\beta)\|_1^r} \quad (\Delta P_s = P(x_s) - P(x_{s-1})).$$

Passing to the limit in the last equation as the weight of the subdivision tends to zero and using formula (25.10), we obtain (25.12).

4. **Application of the above results to the problem of multiplicative representations of entire operator-functions.** Suppose that the operators $B_1, \cdots, B_m$ operate in a finite-dimensional space $\mathfrak{G}$. If the polynomial $T(z) = E + \sum_{k=1}^{m} z^k B_k$ satisfies the conditions

$$T^*(z) J T(z) - J \geqslant 0 \quad (\operatorname{Im} z < 0),$$
$$T^*(z) J T(z) - J = 0 \quad (\operatorname{Im} z = 0)$$
$$(J = J^*, \ J^2 = E),$$

then, by Theorem 11.2 it is representable in the form

$$T(z) = \overset{\curvearrowright}{\prod_{s=1}^{n}} (E + 2iz F_s J) = \overset{\curvearrowright}{\prod_{s=1}^{n}} e^{2iz F_s J}, \tag{25.13}$$

where the F_s $(s = 1, \cdots, n)$ are one-dimensional positive operators such that $F_s J F_s = 0$.

This result may be generalized as follows.

Theorem 25.8. *Suppose that $T(z)$ is an entire function whose values are bounded linear operators operating in a Hilbert space $\mathfrak{G}$. Let J be a linear* *operator in* $\mathfrak{G}$ *satisfying the conditions* $J = J^*$ *and* $J^2 = E$. *Let the following conditions be satisfied:*

1) $T(0) = E$,

2) $T^*(z) J T(z) - J \geq 0 \quad (\mathrm{Im}\, z < 0)$,

3) $T^*(z) J T(z) - J = 0 \quad (\mathrm{Im}\, z = 0)$,

4) *all the operators* $T(z) - E$ *are completely continuous.*

Then

$$T(z) = \int_0^{\overset{\curvearrowright}{1}} (E + 2iz\, dF(x) J), \tag{25.14}$$

where $F(x)$ $(\alpha, x \leq 1)$ *is some strictly increasing absolutely continuous function whose values are completely continuous. In the case when*

$$\lim_{z \to 0} \frac{T(z) - E}{z} \in \mathfrak{S}_1,$$

formula (25.14) *may be rewritten in the form*

$$T(z) = \int_0^{\overset{\curvearrowright}{1}} e^{2iz\, dF(x) J}. \tag{25.15}$$

Conversely, suppose that $F(x)$ $(0 \leq x \leq 1)$ *is a strictly increasing and strongly continuous function with completely continuous values, normalized by the condition* $F(0) = 0$. *If*

$$F^{1/2}(1) J F^{1/2}(1) \in \mathfrak{S}_\omega, \tag{25.16}$$

then the integral (25.14) *exists and is an entire function satisfying conditions* 1)–4). *Under the restriction* $F(1) \in \mathfrak{S}_1$ *the integral* (25.15) *exists and coincides with* (25.14) *for all* z.

Proof. If the entire function $T(z)$ has properties 1)–4), then the function $T(1/\lambda)$, by Theorem 10.6, is characteristic for some simple Volterra node Θ. Suppose that π is a maximal chain of the operator A and $h_1, \cdots, h_s$ $(s \leq \infty)$ a reproducing system for it, normalized by the condition $\Sigma_{k=1}^{s} \|h_k\|^2 = 1$. By Theorem 18.1 π is the range of a spectral function $P(x)$ $(x \in \mathfrak{M})$ satisfying the relation

$$\sum_{k=1}^{s}(P(x)h_k,\ h_k)=x \qquad (x\in\mathfrak{M}).$$

It then follows from Theorem 18.3 that in some space $\tilde{\mathfrak{H}}\supset\mathfrak{H}$ there exists a spectral function $\tilde{P}(x)$ $(0\le x\le 1)$ with the same reproducing system, for which

$$\tilde{P}(x)h=P(x)h \qquad (x\in\mathfrak{M},\ h\in\mathfrak{H}),$$

$$\sum_{k=1}^{s}(\tilde{P}(x)h_k,\ h_k)=x \qquad (0\leqslant x\leqslant 1).$$

Consider the function $F(x)=K^*\tilde{P}(x)K$. In view of Lemma 23.1 this function is absolutely continuous. We shall show that it is strictly increasing. We suppose to this end that $F(x_1)=F(x_2)$ with $x_1<x_2$. Then

$$K^*\Delta\tilde{P}K=0 \qquad (\Delta\tilde{P}=\tilde{P}(x_2)-\tilde{P}(x_1)),$$

so that $\Delta\tilde{P}K=0$, whence

$$\Delta\tilde{P}P(x)K=\Delta\tilde{P}\tilde{P}(x)K=\tilde{P}(x)\Delta\tilde{P}K=0.$$

The linear envelope of vectors of the form $P(x)Kg$ $(x\in\mathfrak{M},\ g\in\mathfrak{G})$ is dense in $\mathfrak{H}$, since in the contrary case the node Θ would be simple. Accordingly, $\Delta\tilde{P}h_k=0$ for $k=1,\cdots,s$, which contradicts the density in $\tilde{\mathfrak{H}}$ of the linear envelope of vectors $\tilde{P}(x)h_k$ $(0\le x\le 1,\ k=1,\cdots,s)$.

In view of Theorem 19.7 the spectral function $\tilde{P}(x)$ belongs to an inessential extension $\tilde{A}$ of the operator A in $\tilde{\mathfrak{H}}$. Using Theorem 25.3 and the fact that the characteristic operator-functions of the nodes Θ and

$$\tilde{\Theta}=\begin{pmatrix}\tilde{A} & K & J\\ \tilde{\mathfrak{H}} & & \mathfrak{G}\end{pmatrix}$$

coincide, we obtain formula (25.14).

Obviously

$$\lim_{z\to 0}\frac{T(z)-E}{z}=2iK^*KJ.$$

If the left side of the last equation belongs to $\mathfrak{S}_1$, then $K^*K\in\mathfrak{S}_1$, and by Theorem 25.5 the function $T(z)$ is representable in the form (25.15).

Suppose that a strongly continuous and strictly increasing operator-function $F(x)$ $(0\le x\le 1)$, with completely continuous operators operating in a space $\mathfrak{G}$ as values, satisfies the conditions

$$F(0)=0, \quad F^{1/2}(1)\, J F^{1/2}(1) \in \mathfrak{S}_\omega.$$

By Theorem 1 of Appendix I

$$F(x) = K^* P(x) K \qquad (0 \leqslant x \leqslant 1),$$

where K is a bounded linear mapping of the space $\mathfrak{G}$ into some space $\mathfrak{H}$, and $P(x)$ $(0 \leq x \leq 1)$ is a spectral function in $\mathfrak{H}$. Inasmuch as $F(1) = K^* K$, the operator K is completely continuous and there exists an isometric mapping U of the closure $\overline{\Re(K)}$ of the range of the operator K onto the closure $\overline{\Re(K^*)}$ of the range of the operator K^* such that $UK = F^{1/2}(1)$. Accordingly,

$$KJK^*h = \begin{cases} U^{-1} F^{1/2}(1)\, J F^{1/2}(1)\, Uh, & h \in \overline{\Re(K)}, \\ 0, & h \perp \overline{\Re(K)}, \end{cases}$$

so that $KJK^* \in \mathfrak{S}_\omega$. By Theorem 22.1 there exists a Volterra operator A in $\mathfrak{H}$ having the spectral function $P(x)$ and imaginary component KJK^*. Applying Theorem 25.3 to the node Θ, we get

$$W_\Theta\left(\frac{1}{z}\right) = \overset{\curvearrowright}{\int_0^1} (E + 2iz\, dF(x)\, J).$$

If $F(1) \in \mathfrak{S}_1$, then by Theorem 25.5

$$W_\Theta\left(\frac{1}{z}\right) = \overset{\curvearrowright}{\int_0^1} e^{2iz\, dF(x)\, J}.$$

The remaining assertions of the theorem follow from the properties of the characteristic operator-function of a Volterra node, as enumerated in §10.

§26. Definition and properties of the function $W(x, \lambda)$

Suppose that

$$\Theta = \begin{pmatrix} A & K & J \\ \mathfrak{H} & & \mathfrak{G} \end{pmatrix}$$

is a Volterra node, $P(x)$ $(x \in \mathfrak{M})$ a spectral function of the operator A, and $W(x, \lambda)$ the characteristic operator-function of the projection of Θ onto the subspace $P(x)\mathfrak{H}$. Obviously

$$W(x, \lambda) = E - 2iK^*P(x)(A - \lambda E)^{-1} P(x) KJ$$
$$= E - 2iK^*(A - \lambda E)^{-1} P(x) KJ. \tag{26.1}$$

By Theorem 25.3

$$W(x, \lambda) = \overset{\curvearrowright}{\int_{\mathfrak{M}_x}} \left(E + \frac{2i}{\lambda} K^* dP(t) KJ\right)$$
$$(\mathfrak{M}_x = \mathfrak{M} \cap [\alpha, x], \quad \alpha = \min \mathfrak{M}). \tag{26.2}$$

Since

$$W(x, \lambda) K^* = K^* - 2iK^*(A - \lambda E)^{-1} P(x) A_I,$$

the following integral exists:

$$\int_{\mathfrak{M}_x} W(t, \lambda) K^* dP(t) \cdot$$
$$= K^*\left(E - (A - \lambda E)^{-1} A\right) P(x) = -\lambda K^*(A - \lambda E)^{-1} P(x).$$

Thus

$$K^*(A - \lambda E)^{-1} P(x) = -\frac{1}{\lambda} \int_{\mathfrak{M}_x} W(t, \lambda) K^* dP(t). \tag{26.3}$$

Theorem 26.1. *The function $W(x, \lambda)$ is the unique solution of the integral equation* [1]

$$W(x, \lambda) = E + \frac{2i}{\lambda} \int_{\mathfrak{M}_x} W(t, \lambda) K^* dP(t) KJ. \tag{26.4}$$

Proof. It follows from (26.1) and (26.3) that $W(x, \lambda)$ satisfies equation (26.4). On the other hand, if $V(x, \lambda)$ is some solution of that equation, then

$$V(x, \lambda) = E + \frac{2i}{\lambda} \int_{\mathfrak{M}_x} \left[E + \frac{2i}{\lambda} \int_{\mathfrak{M}_{x_1}} V(x_2, \lambda) K^* dP(x_2) KJ\right] K^* dP(x_1) KJ$$

$$= E + \frac{2i}{\lambda} K^*P(x) KJ + \left(\frac{2i}{\lambda}\right)^2 \int_{\mathfrak{M}_x} \int_{\mathfrak{M}_{x_1}} V(x_2, \lambda) K^* dP(x_2) A_I dP(x_1) KJ =$$

1) The integral equation (26.4) reduces in a natural way to the differential system considered in §13, examples 3 and 5.

$$= E + \frac{2i}{\lambda} K^*P(x)KJ + \frac{2i}{\lambda^2} K^*A\dot{P}(x)KJ$$

$$+ \left(\frac{2i}{\lambda}\right)^3 \int_{\mathfrak{M}_x} \int_{\mathfrak{M}_{x_1}} \int_{\mathfrak{M}_{x_2}} V(x_3, \lambda) K^* dP(x_3) A_I dP(x_2) A_I dP(x_2) KJ = \dots$$

$$\dots = E + \frac{2i}{\lambda} K^*P(x)KJ + \frac{2i}{\lambda^2} K^*AP(x)KJ + \dots + \frac{2i}{\lambda^n} K^* A^{n-1} P(x) KJ$$

$$+ \left(\frac{2i}{\lambda}\right)^{n+1} \int_{\mathfrak{M}_x} \int_{\mathfrak{M}_{x_1}} \dots \int_{\mathfrak{M}_{x_n}} V(x_{n+1}, \lambda) K^* \, dP(x_{n+1}) A_I \dots dP(x_2) A_I dP(x_1) KJ.$$

Inasmuch as

$$\int_{\mathfrak{M}_{x_n}} V(x_{n+1}, \lambda) K^* dP(x_{n+1}) = \int_{\mathfrak{M}} V(t, \lambda) K^* dP(t) P(x_n),$$

we have

$$V(x, \lambda) = E + \frac{2i}{\lambda} K^* \sum_{s=0}^{n-1} \left(\frac{A}{\lambda}\right)^s P(x) KJ$$

$$+ \frac{2i}{\lambda^{n+1}} \int_{\mathfrak{M}} V(t, \lambda) K^* dP(t) A^n P(x) KJ.$$

Passing to the limit in the last equation as $n \to \infty$ and taking account of the fact that

$$\left\| (A - \lambda E)^{-1} + \frac{1}{\lambda} \sum_{s=0}^{n-1} \left(\frac{A}{\lambda}\right)^s \right\| \to 0, \quad \left\| \frac{A^n}{\lambda^{n+1}} \right\| \to 0,$$

we get

$$V(x, \lambda) = E - 2iK^* (A - \lambda E)^{-1} P(x) KJ = W(x, \lambda).$$

Theorem 26.2. *The function $W(x, \lambda)$ satisfies the relation*

$$W(x, \lambda) J W^*(x, \mu) - J$$
$$= 2i \left(\frac{1}{\lambda} - \frac{1}{\bar{\mu}}\right) \int_{\mathfrak{M}_x} W(t, \lambda) K^* dP(t) K W^*(t, \mu). \qquad (26.5)$$

Proof. From the equation

$$
\begin{aligned}
&(A-\lambda E)^{-1} P(x)-P(x)(A^{*}-\bar{\mu} E)^{-1} \\
&=(A-\lambda E)^{-1}[P(x)(A^{*}-\bar{\mu} E)-(A-\lambda E) P(x)](A^{*}-\bar{\mu} E)^{-1} \\
&=-2 i(A-\lambda E)^{-1} P(x) K J K^{*} P(x)(A^{*}-\bar{\mu} E)^{1-} \\
&\qquad +(\lambda-\bar{\mu})(A-\lambda E)^{-1} P(x)(A^{*}-\bar{\mu} E)^{-1}
\end{aligned}
$$

and formula (26.3) it follows that

$$
\begin{aligned}
W(x, \lambda) J W^{*}(x, \mu)-J &= \left[E-2 i K^{*}(A-\lambda E)^{-1} P(x) K J\right] \\
&\quad \times J\left[E+2 i J K^{*} P(x)(A^{*}-\bar{\mu} E)^{-1} K\right]-J \\
&=-2 i K^{*}\left[(A-\lambda E)^{-1} P(x)-P(x)(A^{*}-\bar{\mu} E)^{-1}\right. \\
&\quad \left.+2 i(A-\lambda E)^{-1} P(x) K J K^{*} P(x)(A^{*}-\bar{\mu} E)^{-1}\right] K \\
&=2 i(\bar{\mu}-\lambda) K^{*}(A-\lambda E)^{-1} P(x)(A^{*}-\bar{\mu} E)^{-1} K \\
&=2 i \frac{\bar{\mu}-\lambda}{\bar{\mu} \lambda} \int_{\mathfrak{M}_{x}} W(t, \lambda) K^{*} d P(t) \int_{\mathfrak{M}_{x}} d P(s) K W^{*}(s, \mu) \\
&=2 i\left(\frac{1}{\lambda}-\frac{1}{\bar{\mu}}\right) \int_{\mathfrak{M}_{x}} W(t, \lambda) K^{*} d P(t) K W^{*}(t, \mu).
\end{aligned}
$$

Theorem 26.3. *The resolvent of the Volterra operator A possesses the triangular representation*

$$
(A-\lambda E)^{-1}=-\frac{E}{\lambda}
-\frac{2 i}{\lambda^{2}} \int_{\mathfrak{M}} \int_{\mathfrak{M}_{x}} d P(t) K W^{*}(t, \bar{\lambda}) J W(x, \lambda) K^{*} d P(x). \tag{26.6}
$$

Proof. Since

$$
\begin{aligned}
&(A-\lambda E)^{-1} P(x)-P(x)(A^{*}-\lambda E)^{-1} \\
&=P(x)(A^{*}-\lambda E)^{-1}[(A^{*}-\lambda E)-(A-\lambda E)](A-\lambda E)^{-1} P(x) \\
&\qquad=-2 i P(x)(A^{*}-\lambda E)^{-1} K J K^{*}(A-\lambda E)^{-1} P(x),
\end{aligned}
$$

we have

$$
\begin{aligned}
&(A-\lambda E)^{-1} P(x) K \\
&=P(x)(A^{*}-\lambda E)^{-1} K(E-2 i J K^{*}(A-\lambda E)^{-1} P(x) K) \\
&\qquad=P(x)(A^{*}-\lambda E)^{-1} K J W(x, \lambda) J .
\end{aligned}
$$

In view of formula (26.3)

$$
(A-\lambda E)^{-1} P(x) K=-\frac{1}{\lambda} \int_{\mathfrak{M}_{x}} d P(t) K W^{*}(t, \bar{\lambda}) J W(x, \lambda) J,
$$

so that

$$(A-\lambda E)^{-1} = -\frac{E}{\lambda} + \frac{1}{\lambda}(A-\lambda E)^{-1} A$$
$$= -\frac{E}{\lambda} + \frac{2i}{\lambda} \int_{\mathfrak{M}} (A-\lambda E)^{-1} P(x) KJK^* \, dP(x)$$
$$= -\frac{E}{\lambda} + \frac{2i}{\lambda^2} \int_{\mathfrak{M}} \int_{\mathfrak{M}_x} dP(t) \, KW^*(t, \bar{\lambda}) \, JW(x, \lambda) \, K^* \, dP(x).$$

Theorem 26.4. *Suppose that Θ is a simple Volterra node and that $P_1(x)$ $(x \in \mathfrak{M}_1)$ and $P_2(x)$ $(x \in \mathfrak{M}_2)$ are spectral functions of A. If $\mathfrak{M}_1 = \mathfrak{M}_2 = \mathfrak{M}$, and*

$$K^* P_1(x) K = K^* P_2(x) K \qquad (x \in \mathfrak{M}),$$

then $P_1(x) = P_2(x)$ $(x \in \mathfrak{M})$.

Proof. Put

$$W_j(x, \lambda) = E - 2iK^* (A-\lambda E)^{-1} P_j(x) KJ \qquad (j = 1, 2).$$

By formula (26.2)

$$W_1(x, \lambda) = \overset{\curvearrowright}{\int_{\mathfrak{M}_x}} \left(E + \frac{2i}{\lambda} K^* \, dP_1(x) \, KJ\right)$$
$$= \overset{\curvearrowright}{\int_{\mathfrak{M}_x}} \left(E + \frac{2i}{\lambda} K^* \, dP_2(x) \, KJ\right) = W_2(x, \lambda).$$

The equation $P_1(x) = P_2(x)$ $(x \in \mathfrak{M})$ now follows from Theorem 3.3.

CHAPTER III

UNICELLULAR OPERATORS. JORDAN REPRESENTATIONS OF DISSIPATIVE VOLTERRA OPERATORS WITH NUCLEAR IMAGINARY COMPONENTS

§ 27. Criteria for unicellularity of Volterra operators

We recall that the characteristic operator function $W(\lambda)$ of the Volterra node $\Theta = \begin{pmatrix} A & K & J \\ \mathfrak{H} & & \mathfrak{G} \end{pmatrix}$ has the following properties:

($\mathrm{I}^{(0)}$) $W(\lambda)$ is an entire function of $\mu = 1/\lambda$;

($\mathrm{II}^{(0)}$) $\lim_{\lambda\to\infty} \|W(\lambda) - E\| = 0$;

($\mathrm{III}^{(0)}$) $W^*(\lambda)JW(\lambda) - J \geq 0$ $(\operatorname{Im}\lambda > 0)$;

($\mathrm{IV}^{(0)}$) $W^*(\lambda)JW(\lambda) - J = 0$ $(\operatorname{Im}\lambda = 0,\ \lambda \neq 0)$;

($\mathrm{V}^{(0)}$) all the operators $W(\lambda) - E$ $(\lambda \neq 0)$ are completely continuous.

The class of all functions satisfying the conditions ($\mathrm{I}^{(0)}$)–($\mathrm{V}^{(0)}$) was denoted in § 10 by $\Omega_J^{(0)}$. In the same place it was proved that each function $W(\lambda) \in \Omega_J^{(0)}$ coincides with the characteristic operator-function of some simple Volterra node. Obviously the class $\Omega_J^{(0)}$ contains, along with any two functions, their product.

If $W(\lambda)$ lies in the class $\Omega_J^{(0)}$, then all of its left and right regular divisors also lie in $\Omega_J^{(0)}$. As to nonregular divisors, all that we can assert is the following.

Suppose that in the neighborhood of the point at infinity

$$W(\lambda) = W_1(\lambda)\, W_2(\lambda) \qquad \left(W(\lambda) \in \Omega_J^{(0)},\ \ W_i(\lambda) \in \Omega_J\right).$$

If one of the functions $W_1(\lambda)$, $W_2(\lambda)$ lies in the class $\Omega_J^{(0)}$, then the other has the same property.

We present the proof for the case when $W_1(\lambda) \in \Omega_J^{(0)}$. Consider the simple $\Theta_i = \begin{pmatrix} A & K_i & J \\ \mathfrak{H}_i & & \mathfrak{G} \end{pmatrix}$ with characteristic operator-functions $W_i(\lambda)$ $(i = 1, 2)$ and put

$\Theta = \Theta_1\Theta_2 = \begin{pmatrix} A & K & J \\ \mathfrak{H} & & \mathfrak{G} \end{pmatrix}$. From the equation $W_\Theta(\lambda) = W(\lambda)$ it follows that the operator K is completely continuous. Since $K_2 = P_2K$, where P_2 is an orthoprojector onto $\mathfrak{H}_2$, then K_2 is also completely continuous. Moreover, $W_2(\lambda)$ $(= W_1^{-1}(\lambda)W(\lambda))$ is an entire function of $\mu = 1/\lambda$. By Theorem 9.3 the operator A does not have spectral points other than zero, and in view of Theorem 10.1 it is a Volterra operator. Thus $W_2(\lambda) \in \Omega_J^{(0)}$.

We imbed the Volterra operator A into the simple Volterra node $\Theta = \begin{pmatrix} A & K & J \\ \mathfrak{H} & & \mathfrak{G} \end{pmatrix}$. By Theorem 5.7 the operator A is unicellular if and only if all regular left divisors of the function $W_\Theta(\lambda)$ form an ordered set. It is not always easy to determine whether a given divisor of $W_\Theta(\lambda)$ is regular or not. Therefore the following remark is useful. *The operator A is unicellular if the set of all left divisors of the function $W_\Theta(\lambda)$ lying in the class $\Omega_J^{(0)}$ (both regular and nonregular) is ordered.*

The object of the present section is the proof of the converse theorem. Indeed, we shall prove that if A is a unicellular operator and $W_1(\lambda)$ $(\in \Omega_J^{(0)})$ and $W_2(\lambda)$ $(\in \Omega_J^{(0)})$ are left divisors of $W_\Theta(\lambda)$, then either $W_1(\lambda) \prec W_2(\lambda)$ or $W_2(\lambda) \prec W_1(\lambda)$. We note that this last assertion is trivial when $J = E$, since in this case, by Theorem 10.5, $W_\Theta(\lambda)$ has no nonregular divisors in $\Omega_E^{(0)}$.

1. Criteria for unicellularity. A function of the form

$$W(\lambda) = E + \frac{2i\sigma}{\lambda} PJ, \tag{27.1}$$

where σ is a positive number and P is an orthoprojector onto a one-dimensional subspace, satisfying the condition $PJP = 0$, will be said to be *elementary*. The elementary function (27.1) lies in the class $\Omega_J^{(0)}$, since

$$W^*(\lambda) JW(\lambda) - J = \frac{4\sigma \operatorname{Im}\lambda}{|\lambda|^2} JPJ.$$

Consider the operator node Θ. *If $A = 0$ and* $\dim(K\mathfrak{G}) = 1$, *then $W_\Theta(\lambda)$ is an elementary function.* Indeed, $\dim(K\mathfrak{G}) = \dim(K^*K\mathfrak{G})$. Therefore

$$W_\Theta(\lambda) = E + \frac{2i}{\lambda} K^*KJ = E + \frac{2i\sigma}{\lambda} PJ,$$

where $\sigma > 0$ and P is an orthoprojector onto a one-dimensional subspace.

Moreover,

$$PJP = \frac{1}{\sigma^2} K^*KJK^*K = \frac{1}{\sigma^2} K^*A_IK = 0.$$

Lemma 27.1. *If the characteristic operator function of the node* Θ *is elementary, then* $A = 0$ *and* $\dim(K\mathfrak{G}) = \dim(\mathfrak{H}) = 1$.

Proof. In some neighborhood of the point at infinity

$$E - 2iK^*(A - \lambda E)^{-1}KJ = E + \frac{2i\sigma}{\lambda} PJ.$$

Accordingly,

$$K^*A^nK = 0 \quad (n = 1, 2, \ldots),$$
$$(AA^mKg, A^{*n}Kg') = 0 \quad (g, g' \in \mathfrak{G};\ m, n = 0, 1, \ldots),$$

and since the linear envelopes of the sets of vectors of the form A^mKg ($g \in \mathfrak{G}$, $m = 0, 1, \cdots$) and $A^{*n}Kg'$ ($g' \in \mathfrak{G}$, $n = 0, 1, \cdots$) are dense in $\mathfrak{H}$, we have $A = 0$. Moreover, $KK^* = \sigma P$, and therefore $\dim(K\mathfrak{G}) = 1$. In view of the simplicity of the node in question $\dim \mathfrak{H} = 1$.

Lemma 27.2. *If* $W(\lambda) = E + (2i\sigma/\lambda)PJ$ *is an elementary function, then every function of the form* $E + (2i\sigma_1/\lambda)PJ$ $(0 \le \sigma_1 \le \sigma)$ *is both a left and a right divisor of the function* $W(\lambda)$. *There are no other divisors of* $W(\lambda)$.

Proof. The first assertion of the lemma follows from the relation

$$E + \frac{2i\sigma}{\lambda} PJ = \left(E + \frac{2i\sigma_1}{\lambda} PJ\right)\left(E + \frac{2i(\sigma - \sigma_1)}{\lambda} PJ\right)$$
$$= \left(E + \frac{2i(\sigma - \sigma_1)}{\lambda} PJ\right)\left(E + \frac{2i\sigma_1}{\lambda} PJ\right).$$

Suppose that in the neighborhood of the point at infinity the equation $W(\lambda) = W_1(\lambda)W_2(\lambda)$, is satisfied, where $W_i(\lambda)$ $(i = 1, 2)$ are certain functions of the class Ω_J. We construct the simple nodes

$$\Theta_i = \begin{pmatrix} A_i & K_i & J \\ \mathfrak{H}_i & & \mathfrak{G} \end{pmatrix},$$

for which $W_{\Theta_i}(\lambda) = W_i(\lambda)$, and put

$$\Theta = \Theta_1\Theta_2 = \begin{pmatrix} A & K & J \\ \mathfrak{H} & & \mathfrak{G} \end{pmatrix}.$$

The elementary function $W(\lambda)$ is characteristic for the principal part

$$\Theta_{\mathfrak{E}} = \begin{pmatrix} A_\Theta & K & J \\ \mathfrak{E}_\Theta & & \mathfrak{G} \end{pmatrix}$$

of the node Θ. By Lemma 27.1 $A_\Theta = 0$, i.e. $A\mathfrak{E}_\Theta = 0$.

Applying Lemma 2.1, we obtain

$$\mathfrak{E}_\Theta^{(0)} \cap \mathfrak{H}_1 = \mathfrak{E}_\Theta^{(0)} \cap \mathfrak{H}_2 = 0.$$

Since every vector of the form Ah $(h \in \mathfrak{H}_1)$ lies in both the subspaces $\mathfrak{E}_\Theta^{(0)}$ and $\mathfrak{H}_1$, we have $A\mathfrak{H}_1 = 0$. At the same time $\mathfrak{E}_\Theta \cup \mathfrak{H}_1 = \mathfrak{H}$, since in the contrary case the subspace $\mathfrak{E}_\Theta^{(0)} \cap \mathfrak{H}_2$ would be different from zero. Thus $A\mathfrak{H} = 0$, which means that $A_1 = A_2 = 0$. We arrive at the equation

$$E + \frac{2i\sigma}{\lambda} PJ = \left(E + \frac{2i}{\lambda} K_1^* K_1 J\right)\left(E + \frac{2i}{\lambda} K_2^* K_2 J\right),$$

from which it easily follows that

$$W_1(\lambda) = E + \frac{2i\sigma_1}{\lambda} PJ, \quad W_2(\lambda) = E + \frac{2i\sigma_2}{\lambda} PJ$$
$$(\sigma_1 \geqslant 0, \quad \sigma_2 \geqslant 0, \quad \sigma_1 + \sigma_2 = \sigma).$$

Lemma 27.3. *Suppose that the functions $W_j(\lambda)$ $(j = 1, 2, 3)$ lie in the class Ω_J. If $W_1(\lambda) \ll W_2(\lambda) \ll W_3(\lambda)$ and $W_1^{-1}(\lambda)W_3(\lambda)$ is an elementary function, then either $W_1(\lambda) = W_2(\lambda)$ or $W_2(\lambda) = W_3(\lambda)$.*

Proof. We construct the simple node

$$\Theta_3 = \begin{pmatrix} A_3 & K_3 & J \\ \mathfrak{H}_3 & & \mathfrak{G} \end{pmatrix},$$

satisfying the condition $W_{\Theta_3}(\lambda) = W_3(\lambda)$. By Theorem 5.6 there exist subspaces $\mathfrak{H}_1$ and $\mathfrak{H}_2$ with $\mathfrak{H}_1 \subseteq \mathfrak{H}_2 \subseteq \mathfrak{H}_3$ such that

$$W_{\Theta_1}(\lambda) = W_1(\lambda), \quad W_{\Theta_2}(\lambda) = W_2(\lambda) \quad (\Theta_1 = \mathrm{pr}_{\mathfrak{H}_1}\Theta_3, \ \Theta_2 = \mathrm{pr}_{\mathfrak{H}_2}\Theta_3).$$

Since the characteristic operator-function of the node $\mathrm{pr}_{\mathfrak{H}_3 \ominus \mathfrak{H}_1}\Theta_3$ is elementary, $\dim(\mathfrak{H}_3 \ominus \mathfrak{H}_1) = 1$. Therefore either $\mathfrak{H}_1 = \mathfrak{H}_2$ or $\mathfrak{H}_2 = \mathfrak{H}_3$, which is equivalent to the assertion of the lemma in question.

Lemma 27.4. *If the set of all regular left divisors of the function* $W(\lambda) \in \Omega_J^{(0)}$ *is ordered, then for each of its nonregular left divisors* $W_1(\lambda)$ $(\in \Omega_J^{(0)})$ *there exist regular left divisors* $W'(\lambda)$ *and* $W''(\lambda)$ *such that:*

1) $W'(\lambda) \prec W_1(\lambda) \prec W''(\lambda)$,

2) $(W'(\lambda))^{-1}W''(\lambda)$ *is an elementary function.*

Proof. There exist simple Volterra nodes

$$\Theta_1 = \begin{pmatrix} A_1 & K_1 & J \\ \mathfrak{H}_1 & & \mathfrak{G} \end{pmatrix} \text{ and } \Theta_2 = \begin{pmatrix} A_2 & K_2 & J \\ \mathfrak{H}_2 & & \mathfrak{G} \end{pmatrix}$$

such that $W_{\Theta_1}(\lambda) = W_1(\lambda)$, $W_{\Theta_2}(\lambda) = W_1^{-1}(\lambda)W(\lambda)$. By the hypotheses of the lemma the node

$$\Theta = \Theta_1\Theta_2 = \begin{pmatrix} A & K & J \\ \mathfrak{H} & & \mathfrak{G} \end{pmatrix}$$

is excess. Suppose that

$$\Theta_{\mathfrak{E}} = \begin{pmatrix} A_\Theta & K & J \\ \mathfrak{E}_\Theta & & \mathfrak{G} \end{pmatrix} \text{ and } \Theta_{\mathfrak{E}(0)} = \begin{pmatrix} 0 & 0 & J \\ \mathfrak{E}_\Theta^{(0)} & & \mathfrak{G} \end{pmatrix}$$

are the principal and the excess parts of the node Θ, respectively. Since

$$\mathfrak{H}' \subset \mathfrak{H}_1 \subset \mathfrak{H}'' \oplus \mathfrak{E}_\Theta^{(0)},$$

where $\mathfrak{H}' = \mathfrak{H}_1 \cap \mathfrak{E}_\Theta$ and $\mathfrak{H}''$ is the closure of the projection of the subspace $\mathfrak{H}_1$ onto $\mathfrak{E}_\Theta$, we have

$$W'(\lambda) \prec W_1(\lambda) \prec W''(\lambda)$$

$$\left(W'(\lambda) = W_{\Theta'}(\lambda),\ W''(\lambda) = W_{\Theta''}(\lambda),\ \Theta' = \operatorname{pr}_{\mathfrak{H}'}\Theta,\quad \Theta'' = \operatorname{pr}_{\mathfrak{H}'' \oplus \mathfrak{E}_\Theta^{(0)}}\Theta\right).$$

At the same time

$$\Theta' = \operatorname{pr}_{\mathfrak{H}'}\Theta_{\mathfrak{E}},\ \Theta'' = \Theta_{\mathfrak{E}(0)}\operatorname{pr}_{\mathfrak{H}''}\Theta = \Theta_{\mathfrak{E}(0)}\operatorname{pr}_{\mathfrak{H}''}\Theta_{\mathfrak{E}},$$

from which it follows, in view of the simplicity of the node $\Theta_{\mathfrak{E}}$ and the equation $W_{\Theta_{\mathfrak{E}}} = W(\lambda)$, that $W'(\lambda)$ and $W''(\lambda)$ are regular left divisors of the function $W(\lambda)$.

According to Theorem 5.7 the operator A_Θ is unicellular. Since $\mathfrak{H}' \subset \mathfrak{H}''$ $(\mathfrak{H}' \neq \mathfrak{H}'')$ and $A_\Theta\mathfrak{H}'' \subset \mathfrak{H}'$, we have $\dim(\mathfrak{H}'' \ominus \mathfrak{H}') = 1$. Therefore the function $(W'(\lambda))^{-1}W''(\lambda)$, coinciding with the characteristic operator-function of the projection of the node $\Theta_{\mathfrak{G}}$ onto the subspace $\mathfrak{H}'' \ominus \mathfrak{H}'$, is elementary.

Theorem 27.1. *Imbed the Volterra operator A in the simple Volterra node Θ. For A to be unicellular, it is necessary and sufficient that the set of all left divisors of the function $W_{\Theta}(\lambda)$ lying in the class $\Omega_J^{(0)}$ be ordered.*

Proof. The sufficiency follows directly from Theorem 5.7.

Now suppose that A is unicellular and that $W_1(\lambda)$ $(\in \Omega_J^{(0)})$ and $W_2(\lambda)$ $(\in \Omega_J^{(0)})$ are left divisors of $W_{\Theta}(\lambda)$. If they are regular, then, as was shown in Theorem 5.6, either $W_1(\lambda) \prec W_2(\lambda)$ or $W_2(\lambda) \prec W_1(\lambda)$.

If $W_1(\lambda)$ and $W_2(\lambda)$ are regular divisors, then by Lemma 27.4 there exist regular left divisors $W_j'(\lambda)$, $W_j''(\lambda)$ $(j = 1, 2)$ such that

$$W_j'(\lambda) \prec W_j(\lambda) \prec W_j''(\lambda),$$

and the functions $(W_j'(\lambda))^{-1} W_j''(\lambda)$ are elementary. In view of Lemma 27.3 and the corollary of Theorem 5.3, the only possible relations for the functions $W_j'(\lambda)$ and $W_j''(\lambda)$ are the following:

1) $W_1''(\lambda) \preccurlyeq W_2'(\lambda)$,

2) $W_1'(\lambda) = W_2'(\lambda)$, $W_1''(\lambda) = W_2''(\lambda)$,

3) $W_2''(\lambda) \preccurlyeq W_1'(\lambda)$.

In cases 1) and 3) we obtain respectively $W_1(\lambda) \prec W_2(\lambda)$ and $W_2(\lambda) \prec W_1(\lambda)$. Now if case 2) holds, then according to Lemma 27.2

$$W_1(\lambda) = W_1'(\lambda)\left(E + \frac{2i\sigma_1}{\lambda} PJ\right), \quad W_2(\lambda) = W_1'(\lambda)\left(E + \frac{2i\sigma_2}{\lambda} PJ\right)$$
$$(\sigma_1 > 0, \ \sigma_2 > 0, \quad PJP = 0),$$

and therefore again either $W_1(\lambda) \prec W_2(\lambda)$ or $W_2(\lambda) \prec W_1(\lambda)$.

Analogously one treats the situation when only one of the divisors $W_1(\lambda)$ or $W_2(\lambda)$ is nonregular.

2. Criteria for the existence of nonregular divisors. The elementary functions $E + (2i\sigma_1/\lambda)P_1J$ and $E + (2i\sigma_2/\lambda)P_2J$ are said to be similar if $P_1 = P_2$.

The left elementary divisor $E + (2i\sigma/\lambda)PJ$ of the function $W(\lambda) \in \Omega_J^{(0)}$ is said to be *maximal* if $E + (2i(\sigma + \epsilon)/\lambda)PJ$ is not a left divisor of the function $W(\lambda)$ for any positive ϵ. Analogously one introduces the concept of right maximal elementary divisor.

Lemma 27.5. *If the function* $W(\lambda) \in \Omega_J^{(0)}$ *has a left (right) elementary divisor* $W_0(\lambda)$, *then it has a maximal left (right) elementary divisor similar to the divisor* $W_0(\lambda)$.

Proof. Suppose that $W_0(\lambda) = E + (2i\sigma_0/\lambda)PJ$ is a left elementary divisor of the function $W(\lambda)$. The least upper bound of the set of all numbers σ for which $E + (2i\sigma/\lambda)PJ$ is a left divisor of $W(\lambda)$ is denoted by $\bar{\sigma}$. It is not hard to see that $\bar{\sigma} < \infty$.

Put $\overline{W(\lambda)} = E + (2i\bar{\sigma}/\lambda)PJ$. It suffices to show that the function

$$\overline{W(\lambda)}^{-1}\, W(\lambda) = \left(E - \frac{2i\bar{\sigma}}{\lambda} PJ\right) W(\lambda) \tag{27.2}$$

lies in the class $\Omega_J^{(0)}$. One verifies directly that it satisfies points ($\mathrm{I}^{(0)}$), ($\mathrm{II}^{(0)}$) and ($\mathrm{V}^{(0)}$) of the definition of that class. Consider a strictly increasing sequence of positive numbers σ_n converging to $\bar{\sigma}$. All the functions

$$\left(E - \frac{2i\sigma_n}{\lambda} PJ\right) W(\lambda)$$

lie in the class $\Omega_j^{(0)}$, which means that they satisfy the relations ($\mathrm{III}^{(0)}$) and ($\mathrm{IV}^{(0)}$). Passing now to the limit as $n \to \infty$, we find that the function (27.2) satisfies these relations as well.

The proof of the assertion of the lemma on a right maximal elementary divisor is carried out analogously.

The lemma is proved.

Suppose that $W(\lambda) = W_1(\lambda)W_2(\lambda)$ ($W_k(\lambda) \in \Omega_J^{(0)}$). The function $W_1(\lambda)$ is said to be a *left*, and $W_2(\lambda)$ a *right, admissible divisor of the function* $W(\lambda)$, if $W_1(\lambda)$ and $W_2(\lambda)$ cannot be represented in the form

$$W_1(\lambda) = W_{11}(\lambda)\, W_{12}(\lambda), \quad W_2(\lambda) = W_{21}(\lambda)\, W_{22}(\lambda) \qquad (W_{ij}(\lambda) \in \Omega_J^{(0)}), \tag{27.3}$$

where $W_{12}(\lambda)$ and $W_{21}(\lambda)$ are similar elementary functions.

Lemma 27.6. *For a divisor of the function* $W(\lambda) \in \Omega_J^{(0)}$ *to be admissible it is necessary and sufficient that it be regular.*

Proof. If $W_1(\lambda)$ ($\in \Omega_J^{(0)}$) is a nonregular left divisor of $W(\lambda)$, then the node $\Theta = \Theta_1\Theta_2$, where

$$\Theta_1=\begin{pmatrix} A_1 & K_1 & J \\ \mathfrak{H}_1 & & \mathfrak{G}\end{pmatrix} \quad\text{and}\quad \Theta_2=\begin{pmatrix} A_2 & K_2 & J \\ \mathfrak{H}_2 & & \mathfrak{G}\end{pmatrix}$$

are simple Volterra nodes for which $W_{\Theta_1}(\lambda)=W_1(\lambda)$ and $W_{\Theta_2}(\lambda)=W_2(\lambda)=W_1^{-1}(\lambda)W(\lambda)$, is excess. By Theorem 10.4 there exist one-dimensional subspaces $\mathfrak{H}_{12}\subset\mathfrak{H}_1$ and $\mathfrak{H}_{21}\subset\mathfrak{H}_2$ such that

$$A_1^*\mathfrak{H}_{12}=A_2\mathfrak{H}_{21}=0 \tag{27.4}$$

and

$$K_1^*\mathfrak{H}_{12}=K_2^*\mathfrak{H}_{21}. \tag{27.5}$$

In view of (27.4)

$$\Theta_1=\Theta_{11}\Theta_{12},\ \Theta_2=\Theta_{21}\Theta_{22}\ (\Theta_{ij}=\mathrm{pr}_{\mathfrak{H}_{ij}}\Theta_i,\ \mathfrak{H}_{11}=\mathfrak{H}_1\ominus\mathfrak{H}_{12},\ \mathfrak{H}_{22}=\ \mathfrak{H}_2\ominus\mathfrak{H}_{21})$$

$$W_1(\lambda)=W_{11}(\lambda)\,W_{12}(\lambda),\ \ W_2(\lambda)=W_{21}(\lambda)\,W_{22}(\lambda)\ \ (W_{ij}(\lambda)=W_{\Theta_{ij}}(\lambda)).$$

From Theorem 2.2 and equation (27.5) it follows that $W_{12}(\lambda)$ and $W_{21}(\lambda)$ are similar elementary functions.

Conversely, suppose that equation (27.3) holds, and suppose that $W_{12}(\lambda)$ and $W_{21}(\lambda)$ are similar elementary functions. Without loss of generality one can suppose that $W_{12}(\lambda)$ and $W_{21}(\lambda)$ are maximal, so that they are admissible divisors of the functions $W_1(\lambda)$ and $W_2(\lambda)$ respectively. By the part of the lemma already proved $W_{12}(\lambda)$ is a right regular divisor of the function $W_1(\lambda)$, and $W_{21}(\lambda)$ a left regular divisor of the function $W_2(\lambda)$.

Introduce the simple Volterra nodes

$$\Theta_{ij}=\begin{pmatrix} A_{ij} & K_{ij} & J \\ \mathfrak{H}_{ij} & & \mathfrak{G}\end{pmatrix} \qquad (i,\ j=1,\ 2)$$

for which $W_{\Theta_{ij}}(\lambda)=W_{ij}(\lambda)$. Then

$$\Theta_1=\Theta_{11}\Theta_{12}=\begin{pmatrix} A_1 & K_1 & J \\ \mathfrak{H}_1 & & \mathfrak{G}\end{pmatrix} \quad\text{and}\quad \Theta_2=\Theta_{21}\Theta_{22}=\begin{pmatrix} A_2 & K_2 & J \\ \mathfrak{H}_2 & & \mathfrak{G}\end{pmatrix}$$

are simple nodes satisfying the relations $W_{\Theta_1}(\lambda)=W_1(\lambda)$ and $W_{\Theta_2}(\lambda)=W_2(\lambda)$. Obviously $\mathfrak{H}_{12}$

and $\mathfrak{H}_{21}$ are one-dimensional subspaces, annihilated respectively by the operators A_1^* and A_2^*. Moreover $K_{12}^*\mathfrak{H}_{12} = K_{21}^*\mathfrak{H}_{21}$, which is equivalent to the equation $K_1^*\mathfrak{H}_{12} = K_2^*\mathfrak{H}_{21}$. In view of Theorem 10.4 the node $\Theta = \Theta_1\Theta_2$ is excess.

Theorem 27.2. *Suppose that $W(\lambda)$ lies in the class $\Omega_J^{(0)}$ and that Θ is a simple Volterra node for which the function $W(\lambda)$ is characteristic. For the class $\Omega_J^{(0)}$ not to contain nonregular divisors of $W(\lambda)$ it is necessary and sufficient that for each subspace $\mathfrak{H}_0$ invariant relative to A the equation $\overline{A\mathfrak{H}_0} = \mathfrak{H}_0$ should hold.*

Proof. Suppose that there exists a subspace $\mathfrak{H}_0$ invariant relative to A such that $\mathfrak{H}_1 = \overline{A\mathfrak{H}_0} \neq \mathfrak{H}_0$. Represent $\mathfrak{H}$ in the form $\mathfrak{H}_1 \oplus \mathfrak{H}_2 \oplus \mathfrak{H}_3$, where $\mathfrak{H}_2$ is a one-dimensional subspace lying in the difference $\mathfrak{H}_0 \ominus \mathfrak{H}_1$, and note that $\mathfrak{H}_1$ and $\mathfrak{H}_1 \oplus \mathfrak{H}_2$ are invariant relative to A. Putting

$$\Theta_j = \begin{pmatrix} A_j & K_j & J \\ \mathfrak{H}_j & & \mathfrak{G} \end{pmatrix} = \mathrm{pr}_{\mathfrak{H}_j}\Theta \qquad (j = 1, 2, 3),$$

we obtain

$$W(\lambda) = W_1(\lambda)\,W_2(\lambda)\,W_3(\lambda) \qquad (W_j(\lambda) = W_{\Theta_j}(\lambda)).$$

Since Θ_3 is a simple node and $A_2 = 0$, $W_2(\lambda)$ is an elementary function:

$$W_2(\lambda) = E + \frac{2i\sigma}{\lambda}PJ \qquad (\sigma > 0,\ PJP = 0).$$

In view of Lemma 27.6 the function $W_1(\lambda)\,(E + (2i\sigma_1/\lambda)PJ)$ $(0 < \sigma_1 < \sigma)$ is a nonregular left divisor of $W(\lambda)$.

Conversely, if $W(\lambda)$ has a nonregular left divisor $W_1(\lambda)$ $(\in \Omega_J^{(0)})$, then the node $\Theta = \Theta_1\Theta_2$, where

$$\Theta_1 = \begin{pmatrix} A_1 & K_1 & J \\ \mathfrak{H}_1 & & \mathfrak{G} \end{pmatrix} \text{ and } \Theta_2 = \begin{pmatrix} A_2 & K_2 & J \\ \mathfrak{H}_2 & & \mathfrak{G} \end{pmatrix}$$

are simple Volterra nodes whose characteristic operator-functions are equal respectively to $W_1(\lambda)$ and $W_1^{-1}(\lambda)W(\lambda)$, is excess. Suppose that $\mathfrak{H}' = \mathfrak{H}_1 \cap \mathfrak{G}_\Theta$ and $\mathfrak{H}''$ is the closure of the projection of $\mathfrak{H}_1$ onto $\mathfrak{G}_\Theta$. It is not hard to see that $\mathfrak{H}' \neq \mathfrak{H}''$. At the same time $\mathfrak{H}''$ is invariant relative to A, and $A\mathfrak{H}'' \subset \mathfrak{H}'$.

§ 28. Certain information on nuclear operators

1. **Trace of an operator.** We recall that a completely continuous operator A is said to be *nuclear* if the series $\Sigma_{j=1}^{\infty} s_j(A)$ consisting of its singular numbers converges. The class of all nuclear operators is denoted by $\mathfrak{S}_1$. The Banach space $\mathfrak{S}_\pi$ introduced in §21 coincides with $\mathfrak{S}_1$ for $\pi_j = 1$ $(j = 1, 2, \cdots)$. Therefore:

1) if $A \in \mathfrak{S}_1$ and $B \in \mathfrak{S}_1$, then $\alpha A + \beta B \in \mathfrak{S}_1$, where α and β are arbitrary complex numbers;

2) if $A \in \mathfrak{S}_1$, then $A^* \in \mathfrak{S}_1$;

3) if $A \in \mathfrak{S}_1$ and B is any bounded operator, then $AB \in \mathfrak{S}_1$ and $BA \in \mathfrak{S}_1$.

We shall say that the bounded linear operator A operating in a separable Hilbert space $\mathfrak{H}$ has a *matrix trace* if for any orthonormalized basis $\{e_j\}_1^\infty$ of the space $\mathfrak{H}$ the series $\Sigma_{j=1}^{\infty} (Ae_j, e_j)$ converges.

Lemma 28.1. *Suppose that H is a positive bounded operator. If for some orthonormalized basis $\{g_j\}_1^\infty$ the series $\Sigma_{j=1}^{\infty} (Hg_j, g_j)$ converges, then $H \in \mathfrak{S}_1$. Conversely, if $H \in \mathfrak{S}_1$, then H has a matrix trace, and for any orthonormalized basis $\{f_j\}_1^\infty$ the equation*

$$\sum_{j=1}^{\infty}(Hf_j, f_j) = \sum_{j=1}^{\infty} s_j(H) \tag{28.1}$$

holds.

Proof. The second assertion of the lemma follows directly from Lemma 25.1. Taking account of the fact that the series $\Sigma_{j=1}^{\infty} (Hg_j, g_j)$ converges, we consider the sequence

$$K_n h = \sum_{j=1}^{n} (h, g_j) H^{1/2} g_j$$

of finite-dimensional operators. Since

$$\|H^{1/2}h - K_n h\|^2 \leqslant \left(\sum_{j=n+1}^{\infty} |(h, g_j)| \, \|H^{1/2} g_j\|\right)^2$$
$$\leqslant \sum_{j=n+1}^{\infty} |(h, g_j)|^2 \sum_{j=n+1}^{\infty} \|H^{1/2} g_j\|^2 \leqslant \|h\|^2 \sum_{j=n+1}^{\infty} (Hg_j, g_j),$$

it follows that $\|H^{1/2} - K_n\| \to 0$, so that H is a completely continuous operator. By Lemma 25.1

$$\sum_{j=1}^{\infty} s_j(H) = \sum_{j=1}^{\infty} (Hg_j, g_j) < \infty,$$

so that $H \in \mathfrak{S}_1$. The lemma is proved.

Suppose that H is a selfadjoint bounded operator and $E(\lambda)$ an orthogonal resolution of unity corresponding to it. We introduce the notations

$$\mathfrak{H}^+ = (E - E(0))\,\mathfrak{H}, \quad \mathfrak{H}^- = E(0)\,\mathfrak{H},$$

$$H^+ = \int_0^\infty \lambda\, dE(\lambda), \quad H^- = -\int_{-\infty}^0 \lambda\, dE(\lambda).$$

Obviously

$$\mathfrak{H} = \mathfrak{H}^+ \oplus \mathfrak{H}^-, \ H^+ \geqslant 0, \ H^- \geqslant 0, \ H = H^+ - H^-, \ H^+\mathfrak{H}^- = H^-\mathfrak{H}^+ = 0.$$

Lemma 28.2. *If the selfadjoint bounded operator H has a matrix trace, then the operators H^+ and H^- have the same property.*

Proof. Denote by $\{e_j^+\}$ and $\{e_j^-\}$ orthonormalized bases in the spaces $\mathfrak{H}^+$ and $\mathfrak{H}^-$, and by $\{e_j\}_1^\infty$ the orthonormal basis of $\mathfrak{H}$ gotten by joining them. Inasmuch as a permutation of terms in the basis $\{e_j\}_1^\infty$ leads again to an orthonormalized basis, the series $\sum_{j=1}^\infty (He_j, e_j)$ converges absolutely. Therefore

$$\sum_{j=1}^\infty (H^+e_j, e_j) = \sum_j (H^+e_j^+, e_j^+) = \sum_j (He_j^+, e_j^+) \leqslant \sum_{j=1}^\infty |(He_j, e_j)| < \infty.$$

By Lemma 28.1 the operator H^+ has a matrix trace. From the equation $H^- = H^+ - H$ it follows that H^- has a matrix trace as well.

Theorem 28.1. *For the bounded linear operator H to have a matrix trace, it is necessary and sufficient that it lie in the class $\mathfrak{S}_1$.*

Proof. If A has a matrix trace, then obviously the operators A^*, A_R and A_I have matrix traces as well, and so, by Lemma 28.2, do the operators A_R^+, A_R^-, A_I^+, and A_I^-. Since

$$A = A_R^+ - A_R^- + iA_I^+ - iA_I^- \tag{28.2}$$

and all the terms, by Lemma 28.1, lie in $\mathfrak{S}_1$, we see that $A \in \mathfrak{S}_1$.

Conversely, suppose that $A \in \mathfrak{S}_1$. Then $A_R \in \mathfrak{S}_1$ and $A_I \in \mathfrak{S}_1$, and in view of the relations $(A_R^\pm)^2 \leq (A_R)^2$ and $(A_I^\pm)^2 \leq (A_I)^2$ we also have $A_R^\pm \in \mathfrak{S}_1$, $A_I^\pm \in \mathfrak{S}_1$. From (28.2) and Lemma 28.1 it follows that A has a matrix trace.

The theorem is proved.

The results just obtained make it possible to assert that the sum of the series $\sum_{j=1}^{\infty}(Ag_j, g_j)$, where $A \in \mathfrak{S}_1$ and $\{g_j\}_1^{\infty}$ is an orthonormalized basis, is finite and does not depend on the choice of the basis. This sum is called the *trace* of the operator A and is denoted by the symbol $\operatorname{tr} A$. If $A > 0$ and $A \notin \mathfrak{S}_1$, then for any choice of orthonormalized basis $\{g_j\}_1^{\infty}$ the sum $\sum_{j=1}^{\infty}(Ag_j, g_j)$ is infinite. In accordance with this we will say that the operator A has an infinite trace, and write $\operatorname{tr} A = \infty$.

Theorem 28.2. *Suppose that A is a completely continuous operator. If B is a bounded operator and $AB \in \mathfrak{S}_1$ and $BA \in \mathfrak{S}_1$, then*

$$\operatorname{tr}(AB) = \operatorname{tr}(BA). \tag{28.3}$$

Proof. By Lemma 21.1 there exist orthonormalized systems $\{g_j\}_1^{\infty}$ and $\{h_j\}_1^{\infty}$ such that

$$Af = \sum_{j=1}^{\infty} s_j(A)(f, g_j)h_j, \quad A^*f = \sum_{j=1}^{\infty} s_j(A)(f, h_j)g_j.$$

Since A and A^* annihilate respectively the orthogonal complements and the linear envelopes of the vectors g_j $(j = 1, 2, \cdots)$ and h_j $(j = 1, 2, \cdots)$, we have

$$\operatorname{tr}(AB) = \sum_{j=1}^{\infty}(ABh_j, h_j), \quad \operatorname{tr}(BA) = \sum_{j=1}^{\infty}(BAg_j, g_j).$$

It remains to be noted that

$$(ABh_j, h_j) = (Bh_j, A^*h_j) = s_j(A)(Bh_j, g_j) = (BAg_j, g_j).$$

Theorem 28.3. *If $A \in \mathfrak{S}_1$ and $P_1, P_2, \cdots$ is a sequence of orthoprojectors weakly converging to an orthoprojector P_0, then*

$$\operatorname{tr}(P_0AP_0) = \lim_{m\to\infty} \operatorname{tr}(P_mAP_m). \tag{28.4}$$

Proof. Suppose that $\{e_k\}_1^{\infty}$ is an arbitrary orthonormalized basis and that P is some orthoprojector. Using Lemma 21.1, we obtain the equation

$$\operatorname{tr}(PAP) = \sum_{k=1}^{\infty}(PAPe_k, e_k) = \sum_{k=1}^{\infty}\sum_{j=1}^{\infty} s_j(A)(Pe_k, g_j)(Ph_j, e_k)$$
$$= \sum_{j=1}^{\infty} s_j(A)\sum_{k=1}^{\infty}(Ph_j, e_k)\overline{(Pg_j, e_k)} = \sum_{j=1}^{\infty} s_j(A)(Ph_j, g_j).$$

Accordingly,

$$| \operatorname{tr}(P_0AP_0) - \operatorname{tr}(P_mAP_m) | \leqslant \sum_{j=1}^{\infty} s_j(A) | ((P_0 - P_m) h_j, g_j) |$$

$$\leqslant \sum_{j=1}^{N} s_j(A) | ((P_0 - P_m) h_j, g_j) | + \sum_{j=N+1}^{\infty} s_j(A).$$

We may now make the second term arbitrarily small by taking N sufficiently large, and then the first by an appropriate choice of m.

2. Trace formula. Suppose that $\mathfrak{H}_1$ and $\mathfrak{H}_2$ are subspaces of the same space $\mathfrak{H}$. We will denote by the symbol $\mathfrak{H}_1 \tilde{\cup} \mathfrak{H}_2$ the smallest subspace of $\mathfrak{H}$ containing $\mathfrak{H}_1$ and $\mathfrak{H}_2$.

Lemma 28.3. *Suppose that the bounded linear operator A operating in the space $\mathfrak{H}$ has invariant subspaces $\mathfrak{H}^{(1)}$ and $\mathfrak{H}^{(2)}$, and induces in them respectively, operators $A^{(1)}$ and $A^{(2)}$. In the subspace $\mathfrak{H}^{(3)} = \mathfrak{H} \ominus \mathfrak{H}^{(2)}$ we suppose given an operator $A^{(3)}h = P^{(3)}Ah$ $(h \in \mathfrak{H}^{(3)})$, where $P^{(3)}$ is an orthoprojector onto $\mathfrak{H}^{(3)}$, and denote by T the mapping of $\mathfrak{H}^{(1)}$ into $\mathfrak{H}^{(3)}$ defined by the formula $Th = P^{(3)}h$ $(h \in \mathfrak{H}^{(1)})$. Then the following assertions are valid.*

1) $TA^{(1)} = A^{(3)}T$.

2) $\mathfrak{H}^{(1)} = \mathfrak{G}_0 \oplus \mathfrak{G}_1 \oplus \mathfrak{G}_2$, *where* $\mathfrak{G}_0 = \mathfrak{H}^{(1)} \cap \mathfrak{H}^{(2)}$, $\mathfrak{G}_0 \oplus \mathfrak{G}_1$ *is the collection of all vectors* $h \in \mathfrak{H}^{(1)}$ *for which* $A^{(1)}h \in \mathfrak{H}^{(2)}$, *and* $\mathfrak{G}_2$ *is the closure of the range of the operator adjoint to the operator* $B = TA^{(1)}$.

3) *If the operator B is completely continuous, then the subspaces $\mathfrak{G}_2$ and $\mathfrak{H}^{(4)} = A_3[(\mathfrak{H}^{(1)} \tilde{\cup} \mathfrak{H}^{(2)}) \ominus \mathfrak{H}^{(2)}]$ have the same dimension, and for certain orthonormalized bases $\{g_j\}_1^{\omega}$ and $\{h_j\}_1^{\omega}$ $(\omega \leq \infty)$ of these spaces the equations*

$$(A^{(1)}g_j, g_j) = (A^{(3)}h_j, h_j) \qquad (j = 1, 2, \ldots, \omega)$$

hold.

Proof. 1) Since $\mathfrak{H}^{(2)}$ is invariant relative to A, $P^{(3)}A = P^{(3)}AP^{(3)}$. Using the invariance of the space $\mathfrak{H}^{(1)}$, we get

$$TA^{(1)}h = P^{(3)}Ah = P^{(3)}AP^{(3)}h = A^{(3)}Th \qquad (h \in \mathfrak{H}^{(1)}).$$

2) Denote by $\mathfrak{G}^{(1)}$ the subspace consisting of all vectors $h \in \mathfrak{H}^{(1)}$ annihilated by the operator $B = TA^{(1)}$. It consists of those and only those vectors of the space $\mathfrak{H}^{(1)}$ which are carried by the operator $A^{(1)}$ into $\mathfrak{H}^{(2)}$, and its orthogonal

complement $\mathfrak{G}_2 = \mathfrak{H}^{(1)} \ominus \mathfrak{G}^{(1)}$ is obviously the closure of the range of the operator B^*. The subspace $\mathfrak{G}^{(1)}$ is representable in the form $\mathfrak{G}^{(1)} = \mathfrak{G}_0 \oplus \mathfrak{G}_1$, where $\mathfrak{G}_0 = \mathfrak{H}^{(1)} \cap \mathfrak{H}^{(2)}$.

3) The subspaces $\mathfrak{G}_2$ and $\mathfrak{H}^{(4)}$ are the closures of the ranges of the operators B^*B and BB^*. Suppose that B is a completely continuous operator. Then in $\mathfrak{G}_2$ there exists an orthonormalized basis $\{g_j\}_1^\omega$ $(\omega \leq \infty)$, consisting of eigenvectors of the operator B^*B:

$$B^*Bg_j = s_j^2 g_j \quad (s_j > 0,\ j = 1,\ 2,\ \ldots,\ \omega).$$

It is easy to see that the system $h_j = Bg_j/s_j$ $(j = 1, 2, \cdots, \omega)$ is also orthonormalized and forms a basis in $\mathfrak{H}^{(4)}$. Inasmuch as

$$Bg_j = s_j h_j, \quad B^*h_j = s_j g_j$$

and $BA^{(1)} = A^{(3)}TA^{(1)} = A^{(3)}B$, we have

$$(A^{(1)}g_j,\ g_j) = \frac{1}{s_j}(A^{(1)}g_j,\ B^*h_j) = \frac{1}{s_j}(A^{(3)}Bg_j,\ h_j) = (A^{(3)}h_j,\ h_j).$$

Theorem 28.4. *If A is a nuclear operator and $\mathfrak{H}^{(1)}$ and $\mathfrak{H}^{(2)}$ are invariant subspaces of it, then*

$$\operatorname{tr} A^{(1)} + \operatorname{tr} A^{(2)} = \operatorname{tr} A^{\cup} + \operatorname{tr} A^{\cap}, \tag{28.5}$$

where $A^{(1)}$, $A^{(2)}$, $A^{\cup}$ and $A^{\cap}$ are induced by A in $\mathfrak{H}^{(1)}$, $\mathfrak{H}^{(2)}$, $\mathfrak{H}^{\cup} = \mathfrak{H}_1 \tilde{\cup} \mathfrak{H}_2$, $\mathfrak{H}^{\cap} = \mathfrak{H}_1 \cap \mathfrak{H}_2$ respectively.

Proof. Suppose that $\mathfrak{H}^{(3)} = \mathfrak{H}^{\cup} \ominus \mathfrak{H}^{(2)}$, $P^{(3)}$ is an orthoprojector operating in $\mathfrak{H}^{\cup}$ onto $\mathfrak{H}^{(3)}$ and $A^{(3)}h = P^{(3)}A^{\cup}h$ $(h \in \mathfrak{H}^{(3)})$. Put $\mathfrak{H}^{(1)} = \mathfrak{G}_0 \oplus \mathfrak{G}_1 \oplus \mathfrak{G}_2$, where $\mathfrak{G}_0 = \mathfrak{H}^{(1)} \cap \mathfrak{H}^{(2)}$ and $\mathfrak{G}_0 \oplus \mathfrak{G}_1$ is the collection of all vectors $h \in \mathfrak{H}^{(1)}$ such that $A^{(1)}h \in \mathfrak{H}^{(2)}$. By Lemma 28.3 there exist orthonormalized bases $\{g_j\}_1^\omega$ and $\{h_j\}_1^\omega$ $(\omega \leq \infty)$ of the subspaces $\mathfrak{G}_2$ and $\overline{A^{(3)}\mathfrak{H}^{(3)}}$, for which

$$(A^{(1)}g_j,\ g_j) = (A^{(3)}h_j,\ h_j) \quad (j = 1,\ 2,\ \ldots,\ \omega).$$

Since $(A^{(1)}g, g) = 0$ $(g \in \mathfrak{G}_1)$ and $(A^{(3)}h, h) = 0$ $(h \perp \overline{A^{(3)}\mathfrak{H}^{(3)}})$, we have

$$\operatorname{tr} A^{(3)} = \sum_{j=1}^{\omega} (A^{(3)}h_j,\ h_j) = \sum_{j=1}^{\omega} (A^{(1)}g_j,\ g_j) = \operatorname{tr} A^{(1)} - \operatorname{tr} A^{\cap}. \tag{28.6}$$

Moreover, obviously

$$\operatorname{tr} A^{\cup} = \operatorname{tr} A^{(2)} + \operatorname{tr} A^{(3)}. \tag{28.7}$$

(28.5) now follows from (28.6) and (28.7).

Theorem 28.5. *Suppose that A is a bounded linear operator with nuclear imaginary component. If the spaces $\mathfrak{H}^{(1)}$ and $\mathfrak{H}^{(2)}$ are invariant relative to A and if at least one of the operators induced in them is completely continuous, then, in the notation of the preceding theorem,*

$$\operatorname{tr}(A^{(1)})_I + \operatorname{tr}(A^{(2)})_I = \operatorname{tr}(A^{\cup})_I + \operatorname{tr}(A^{\cap})_I. \tag{28.8}$$

Proof. For definiteness we suppose that $A^{(1)}$ is completely continuous. Choosing orthonormalized bases $\{g_j\}_1^{\omega}$ and $\{h_j\}_1^{\omega}$ as was done in Theorem 28.4, we get

$$\operatorname{tr}(A^{\cup})_I - \operatorname{tr}(A^{(2)})_I = \operatorname{tr}(A^{(3)})_I = \sum_{j=1}^{\omega}((A^{(3)})_I h_j, h_j)$$
$$= \sum_{j=1}^{\omega}((A^{(1)})_I g_j, g_j) = \operatorname{tr}(A^{(1)})_I - \operatorname{tr}(A^{\cap})_I.$$

Theorem 28.6. *Suppose that the bounded linear dissipative operator A, operating in the space $\mathfrak{H}$, has invariant subspaces $\mathfrak{H}^{(1)}$ and $\mathfrak{H}^{(2)}$ and induces in them completely continuous operators $A^{(1)}$ and $A^{(2)}$ respectively. If the operators $(A^{(1)})_I$ and $(A^{(2)})_I$ are nuclear and $\mathfrak{H} = \mathfrak{H}_1 \tilde{\cup} \mathfrak{H}_2$, then the operator A_I is also nuclear.*

Proof. Suppose given in the space $\mathfrak{H}^{(3)} = \mathfrak{H} \ominus \mathfrak{H}^{(2)}$ an operator $A^{(3)}h = P^{(3)}Ah$ $(h \in \mathfrak{H}^{(3)})$, where $P^{(3)}$ is the orthoprojector onto $\mathfrak{H}^{(3)}$. By Lemma 28.3 there exists an orthonormalized sequence $\{g_j\}_1^{\omega}$ $(\omega \leq \infty)$ in $\mathfrak{H}^{(1)}$ and an orthonormalized basis $\{h_j\}_1^{\omega}$ in $\overline{A^{(3)}\mathfrak{H}^{(3)}}$, such that

$$(A^{(1)}g_j, g_j) = (A^{(3)}h_j, h_j) \quad (j = 1, 2, \ldots, \omega).$$

Suppose that $\{f_j\}_1^{\tau}$ $(\tau \leq \infty)$ is an orthonormalized basis in $\mathfrak{H}^{(2)}$. Since

$$\sum_{j=1}^{\tau}(A_I f_j, f_j) = \sum_{j=1}^{\tau}((A^{(2)})_I f_j, f_j) = \operatorname{tr}(A^{(2)})_I,$$

$$\sum_{j=1}^{\omega}(A_I h_j,\ h_j)=\sum_{j=1}^{\omega}((A^{(3)})_I h_j,\ h_j)=\sum_{j=1}^{\omega}((A^{(1)})_I g_j,\ g_j)\leqslant \operatorname{tr}(A^{(1)})_I,$$

we have

$$\sum_{j=1}^{\tau}(A_I f_j,\ f_j)+\sum_{j=1}^{\omega}(A_I h_j,\ h_j)<\infty,$$

which is equivalent to the assertion of the theorem.

§ 29. Least common multiple and greatest common divisor of functions of the class $\Omega_{\mathfrak{G}}^{(\exp)}$

1. Divisors of functions of the class $\Omega_{\mathfrak{G}}^{(\exp)}$. We continue the study of functions of the class $\Omega_{\mathfrak{G}}^{(\exp)}$ begun in § 8.

Lemma 29.1. *If* $W(\lambda)\in\Omega_{\mathfrak{G}}^{(\exp)}$ *and* $W_1(\lambda)\ll W(\lambda)$, *then* $W_1(\lambda)\in\Omega_{\mathfrak{G}}^{(\exp)}$ *and* $\sigma[W_1]\leq\sigma[W]$.

Proof. Construct the simple node

$$\Theta=\begin{pmatrix} A & K & E\\ \mathfrak{H} & & \mathfrak{G}\end{pmatrix}$$

with characteristic operator-function $W(\lambda)$. By Theorem 8.3, $A\in\Lambda^{(\exp)}$ and $\sigma[A]=\sigma[W]$. In view of Theorem 5.4 there exists a subspace $\mathfrak{H}_1$ invariant relative to A such that

$$W_1(\lambda)=W_{\Theta_1}(\lambda)\qquad\left(\Theta_1=\operatorname{pr}_{\mathfrak{H}_1}\Theta=\begin{pmatrix} A_1 & K_1 & J\\ \mathfrak{H}_1 & & \mathfrak{G}\end{pmatrix}\right).$$

The operator A_1 does not have spectral points other than zero. Moreover,

$$\|(A_1-\lambda E)^{-1}\|\leqslant\|(A-\lambda E)^{-1}\|,$$

so that $A_1\in\Lambda^{(\exp)}$. Thus $W_1(\lambda)\in\Omega_{\mathfrak{G}}^{(\exp)}$. Since $\sigma[W_1]=\sigma[A_1]$ and $\sigma[A_1]\leq\sigma[A]$, we have $\sigma[W_1]\leq\sigma[W]$.

The lemma is proved.

Analogously one may show that *if* $W_1(\lambda)$ *is a right regular divisor of the function* $W(\lambda)\in\Omega_{\mathfrak{G}}^{(\exp)}$, *then* $W_1(\lambda)\in\Omega_{\mathfrak{G}}^{(\exp)}$ *and* $\sigma[W_1]\leq\sigma[W]$.

Lemma 29.2. *If* $W_1(\lambda)\prec W(\lambda)$ $(W_1(\lambda),\ W(\lambda)\in\Omega_{\mathfrak{G}}^{(\exp)})$, *then* $W_1(\lambda)\ll W(\lambda)$.

Proof. There exist simple nodes

$$\Theta_j = \begin{pmatrix} A_j & K_j & J \\ \mathfrak{H}_j & & \mathfrak{G} \end{pmatrix} \quad (j = 1, 2)$$

such that $W_{\Theta_1}(\lambda) = W_1(\lambda)$ and $W_{\Theta_2}(\lambda) = W_1^{-1}(\lambda)W(\lambda)$. By Theorem 7.3 the product $\Theta_1\Theta_2$ is a simple node.

2. Weight of a function of the class $\Omega_{\mathfrak{G}}^{(\exp)}$. Each function $W(\lambda) \in \Omega_{\mathfrak{G}}^{(\exp)}$ decomposes into a series in negative powers of λ and converging in norm:

$$W(\lambda) = E + \frac{2i}{\lambda} H + \dots \qquad (\lambda \neq 0,\ H \geqslant 0).$$

The trace of the operator $2H$, which can be either finite or infinite, will be called the *weight* of the function $W(\lambda)$ and denoted by $\tau[W]$.

We note the following properties of the weight, following directly from its definition and Corollary 2 to Theorem 5.1:

1. *If* $W_j(\lambda) \in \Omega_{\mathfrak{G}}^{(\exp)}$ $(j = 1, 2)$, *then*

$$\tau[W_1W_2] = \tau[W_1] + \tau[W_2]. \tag{29.1}$$

2. *Let* $W_j(\lambda) \in \Omega_{\mathfrak{G}}^{(\exp)}$ $(j = 1, 2)$ *and* $W_1(\lambda) \prec W_2(\lambda)$. *If* $\tau[W_1] = \tau[W_2] < \infty$, *then* $W_1(\lambda) = W_2(\lambda)$.

3. *If* $W_j(\lambda) \in \Omega_{\mathfrak{G}}^{(\exp)}$ $(j = 1, 2)$ *and* $W_1(\lambda) \prec W_2(\lambda)$, *then* $\tau[W_1] \leq \tau[W_2]$.

Theorem 29.1. *If* $W(\lambda) \in \Omega_{\mathfrak{G}}^{(\exp)}$, *then*

$$\sigma[W] \leqslant \tau[W]. \tag{29.2}$$

Proof. We need only consider the case when $\tau[W] < \infty$. The function $W(\lambda)$ is characteristic for some node

$$\Theta = \begin{pmatrix} A & K & E \\ \mathfrak{H} & & \mathfrak{G} \end{pmatrix},$$

whose basic operator does not have nonzero spectral points. Suppose that $\{h_j\}$ and $\{g_j\}$ are orthonormalized bases in $\mathfrak{H}$ and $\mathfrak{G}$ respectively. Since

$$\sum_j (KK^*h_j, h_j) = \sum_j \| K^*h_j \|^2 = \sum_j \sum_i |(K^*h_j, g_i)|^2$$
$$= \sum_i \sum_j |(Kg_i, h_j)|^2 = \sum_i \| Kg_i \|^2 = \sum_i (K^*Kg_i, g_i),$$

we have

$$\operatorname{tr} A_I = \operatorname{tr}(KK^*) = \operatorname{tr}(K^*K) = \frac{1}{2}\tau[W] < \infty.$$

Accordingly, A_I is a nuclear operator. By Theorem 10.1 it is a Volterra operator.

Suppose that $P(x)$ $(x \in \mathfrak{M})$ is a spectral function of the operator A. By Theorem 25.5

$$W\left(\frac{1}{\mu}\right) = \overset{\curvearrowright}{\int\limits_{\mathfrak{M}}} e^{2i\mu K^* dP(x) K}.$$

For any subdivision $x_0 < x_1 < \cdots < x_n$ of the set $\mathfrak{M}$ we get

$$\left\| \overset{\curvearrowright}{\prod_{j=1}^{n}} e^{2i\mu K^* \Delta P_j K} \right\| \leqslant \prod_{j=1}^{n} e^{2|\mu| \| K^* \Delta P_j K \|}$$

$$\leqslant \prod_{j=1}^{n} e^{2|\mu| \operatorname{tr}(K^* \Delta P_j K)} = e^{2|\mu| \operatorname{tr}(K^*K)}$$

$$(\Delta P_j = P(x_j) - P(x_{j-1})).$$

Thus

$$\left\| W\left(\frac{1}{\mu}\right) \right\| \leqslant e^{2|\mu| \operatorname{tr}(K^*K)},$$

so that $\sigma[W] \leq \tau[W]$.

The theorem is proved. In the process we have obtained the proofs of the following two assertions.

Theorem 29.2. *For the function $W(\lambda)$ to lie in the class $\Omega_{\mathfrak{G}}^{(\exp)}$ and to have a finite weight, it is necessary and sufficient that it should be characteristic for some simple dissipative node*

$$\Theta = \begin{pmatrix} A & K & E \\ \mathfrak{H} & & \mathfrak{G} \end{pmatrix},$$

where A is a Volterra operator with nuclear imaginary component. In addition, $\tau[W] = 2 \operatorname{tr} A_I$.

Theorem 29.3. *If A is a dissipative Volterra operator with nuclear imaginary* *component, then* $(A - (1/\mu)E)^{-1}$ *is a function of exponential type not larger than* $2\,\mathrm{tr}\,A_I$.

We note a case when $\sigma[W]$ admits an estimate from below.

Theorem 29.4. *If* $W(\lambda) \in \Omega_{\mathfrak{G}}^{\exp}$ *and* $\mathfrak{G}$ *is of finite dimension* n, *then*

$$\sigma[W] \geqslant \frac{1}{n}\tau[W]. \tag{29.3}$$

Proof. By Theorem 8.2, $W(\lambda)$ is a divisor of the function $e^{(i\sigma/\lambda)E}$ $(\sigma = \sigma[W])$. Since $\tau[e^{(i\sigma/\lambda)E}] = n\sigma$, we have

$$\tau[W] \leqslant \tau\left[e^{\frac{i\sigma}{\lambda}E}\right] = n\sigma[W].$$

Corollary. *If* A *is a dissipative Volterra operator with an n-dimensional imaginary component, then*

$$\sigma[A] \geqslant \frac{2}{n}\ \mathrm{tr}\ A_I. \tag{29.4}$$

3. **Least common multiple and greatest common divisor.** Suppose that the functions $W_\gamma(\lambda)$, where the index γ runs through some set Γ, lie in the class $\Omega_{\mathfrak{G}}^{(\exp)}$. A function $W(\lambda) \in \Omega_{\mathfrak{G}}^{(\exp)}$ is said to be a *common multiple* of the functions $W_\gamma(\lambda)$ if $W_\gamma(\lambda) \prec W(\lambda)$ for all $\gamma \in \Gamma$. A common multiple of the functions $W_\gamma(\lambda)$ is said to be a *least common multiple* if it is a left divisor of any common multiple of these functions.

Theorem 29.5. *Any collection of functions* $W_\gamma(\lambda) \in \Omega_{\mathfrak{G}}^{(\exp)}$ $(\gamma \in \Gamma)$ *satisfying the condition* $\sup_{\gamma\in\Gamma}\sigma[W_\gamma] < \infty$ *has one and only one least common multiple. If* $W^{\cup}(\lambda)$ *is the least common multiple of the functions* $W_\gamma(\lambda)$ $(\gamma \in \Gamma)$, *then*

$$\sigma\left[W^{\cup}\right] = \sup_{\gamma\in\Gamma}\sigma[W_\gamma]. \tag{29.5}$$

In the case when the functions $W_\gamma(\lambda)$ $(\gamma \in \Gamma)$ *constitute an ordered set,*

$$\tau\left[W^{\cup}\right] = \sup_{\gamma\in\Gamma}\tau[W_\gamma]. \tag{29.6}$$

Proof. The functions $W_\gamma(\lambda)$ $(\gamma \in \Gamma)$ have a common multiple, since, by Theorem 8.2,

$$W_\gamma(\lambda) \prec e^{\frac{i\bar{\sigma}}{\lambda}E} \qquad (\gamma \in \Gamma,\ \bar{\sigma} = \sup_{\gamma \in \Gamma} \sigma[W_\gamma]).$$

Suppose that $W(\lambda)$ is some common multiple of the functions $W_\gamma(\lambda)$ $(\gamma \in \Gamma)$, and

$$\Theta = \begin{pmatrix} A & K & E \\ \mathfrak{H} & & \mathfrak{G} \end{pmatrix}$$

is a simple node whose characteristic operator-function coincides with $W(\lambda)$. The following considerations are based on the properties of the mapping Φ indicated in Theorem 5.6. There exist subspaces $\mathfrak{H}_\gamma$, invariant relative to A, such that

$$W_\gamma(\lambda) = W_{\Theta_\gamma}(\lambda) \qquad (\Theta_\gamma = \mathrm{pr}_{\mathfrak{H}_\gamma}\Theta).$$

Put $W^\cup(\lambda) = W_{\Theta^\cup}(\lambda)$ $(\Theta^\cup = \mathrm{pr}_{\mathfrak{H}^\cup}\Theta)$, where $\mathfrak{H}^\cup$ is the closure of the linear envelope of the subspaces $\mathfrak{H}_\gamma$. Obviously $W^\cup(\lambda)$ is a common multiple of the functions $W_\gamma(\lambda)$, satisfying the following condition: if $W_0(\lambda) \in \Omega_{\mathfrak{G}}^{(\exp)}$ and

$$W_\gamma(\lambda) \prec W_0(\lambda) \prec W^\cup(\lambda) \qquad (\gamma \in \Gamma),$$

then $W_0(\lambda) = W^\cup(\lambda)$.

Suppose that $W_1(\lambda)$ is any common multiple of the functions $W_\gamma(\lambda)$ $(\gamma \in \Gamma)$. Consider some common multiple $\widetilde{W}(\lambda)$ of the functions $W^\cup(\lambda)$ and $W_1(\lambda)$, the simple node

$$\Theta = \begin{pmatrix} \widetilde{A} & \widetilde{K} & E \\ \widetilde{\mathfrak{H}} & & \mathfrak{G} \end{pmatrix}$$

with characteristic function $\widetilde{W}(\lambda)$ and subspace $\widetilde{\mathfrak{H}}_\gamma$ invariant relative to $\widetilde{A}$, for which

$$W_\gamma(\lambda) = W_{\widetilde{\Theta}_\gamma}(\lambda) \qquad (\widetilde{\Theta}_\gamma = \mathrm{pr}_{\widetilde{\mathfrak{H}}_\gamma}\widetilde{\Theta}).$$

Denoting by $\widetilde{\mathfrak{H}}^\cup$ the closure of the linear envelope of the subspaces $\widetilde{\mathfrak{H}}_\gamma$, we get

$$W_\gamma(\lambda) \prec \widetilde{W}^\cup(\lambda) \prec W^\cup(\lambda),$$
$$\widetilde{W}^\cup(\lambda) \prec W_1(\lambda) \qquad \left(\widetilde{W}^\cup(\lambda) = W_{\widetilde{\Theta}^\cup}(\lambda),\quad \widetilde{\Theta}^\cup = \mathrm{pr}_{\widetilde{\mathfrak{H}}^\cup}\widetilde{\Theta}\right).$$

Accordingly $\widetilde{W}^{\cup}(\lambda) = W^{\cup}(\lambda)$ and $W^{\cup}(\lambda) \prec W_1(\lambda)$. Thus we have proved that $W^{\cup}(\lambda)$ is a common multiple of the functions $W_\gamma(\lambda)$ $(\gamma \in \Gamma)$.

If each of the functions $W'(\lambda)$ and $W''(\lambda)$ is a least common multiple of the functions $W_\gamma(\lambda)$, then $W'(\lambda) \prec W''(\lambda)$, $W''(\lambda) \prec W'(\lambda)$, and therefore $W'(\lambda) = W''(\lambda)$.

It follows from the relations $W_\gamma(\lambda) \prec W^{\cup}(\lambda) \prec e^{(i\widetilde{\sigma}/\lambda)E}$ that

$$\sigma[W_\gamma] \leqslant \sigma[W^{\cup}] \leqslant \sigma\left[e^{\frac{i\widetilde{\sigma}}{\lambda}E}\right].$$

Since the right side of this inequality is equal to $\widetilde{\sigma}$, we have

$$\sigma[W^{\cup}] = \sup_{\gamma \in \Gamma} \sigma[W_\gamma].$$

Passing to the proof of formula (29.6), we need to emphasize that the discussion is only required in the case when $\widetilde{\tau} = \sup_{\gamma \in \Gamma} \tau[W_\gamma] < \infty$ and there is no function $W_{\gamma_0}(\lambda)$ $(\gamma_0 \in \Gamma)$ such that $\tau[W_{\gamma_0}] = \widetilde{\tau}$. We select from the set $W_\gamma(\lambda)$ $(\gamma \in \Gamma)$ a sequence $W_{\gamma_j}(\lambda)$ $(j = 1, 2, \cdots)$ such that the relations $W_{\gamma_j}(\lambda) \prec W_{\gamma_{j+1}}(\lambda)$ $(j = 1, 2, \cdots)$ are satisfied and so that the equation $\lim_{j\to\infty} \tau[W_{\gamma_j}] = \widetilde{\tau}$ holds. It is easy to see that $W^{\cup}(\lambda)$ is the least common multiple of the functions $W_{\gamma_j}(\lambda)$.

Suppose that P_j and $P^{\cup}$ are orthoprojectors onto $\mathfrak{H}_{\gamma_j}$ and $\mathfrak{H}^{\cup}$ respectively. From the relations

$$P_j \to P^{\cup}, \quad W_{\gamma_j}(\lambda) = E + \frac{2i}{\lambda} K^* P_j K + \ldots,$$
$$2 \operatorname{tr} (P_j K K^* P_j) = 2 \operatorname{tr} (K^* P_j K) = \tau[W_{\gamma_j}] < \widetilde{\tau}$$

it follows that

$$\sup_{\gamma \in \Gamma} \tau[W_\gamma] = \lim_{j\to\infty} \tau[W_{\gamma_j}] = 2 \lim_{j\to\infty} \operatorname{tr} (P_j K K^* P_j)$$
$$= 2 \operatorname{tr} (P^{\cup} K K^* P^{\cup}) = \tau[W^{\cup}].$$

The theorem is proved.

The function $W(\lambda) \in \Omega_{\mathfrak{G}}^{(\exp)}$ is said to be a common divisor of the functions $W_\gamma(\lambda) \in \Omega_{\mathfrak{G}}^{(\exp)}$ $(\gamma \in \Gamma)$, if $W(\lambda) \prec W_\gamma(\lambda)$ for all $\gamma \in \Gamma$. A common divisor $W(\lambda)$ of the functions $W_\gamma(\lambda)$ $(\gamma \in \Gamma)$ is said to be the *greatest common divisor* if for any common divisor $W_1(\lambda)$ of these functions the relation $W_1(\lambda) \prec W(\lambda)$ is satisfied.

Theorem 29.6. *Every collection of functions* $W_\gamma(\lambda) \in \Omega_{\mathfrak{G}}^{(\exp)}$ $(\gamma \in \Gamma)$ *has one and only one greatest common divisor. If* $W^\frown(\lambda)$ *is the greatest common divisor of the ordered collection of functions* $W_\gamma(\lambda)$ $(\gamma \in \Gamma)$ *and* $\inf_{\gamma\in\Gamma} \tau[W_\gamma] < \infty$, *then*

$$\sigma\left[W^\cap\right] = \inf_{\gamma\in\Gamma} \sigma[W_\gamma], \tag{29.7}$$

$$\tau\left[W^\cap\right] = \inf_{\gamma\in\Gamma} \tau[W_\gamma]. \tag{29.8}$$

Proof. We suppose first that the functions $W_\gamma(\lambda)$ $(\gamma \in \Gamma)$ have a common multiple. Then there exists a simple exponential node

$$\Theta = \begin{pmatrix} A & K & E \\ \mathfrak{H} & & \mathfrak{G} \end{pmatrix}$$

and subspaces $\mathfrak{H}_\gamma$ invariant relative to A such that $W_\gamma(\lambda) = W_{\Theta_\gamma}(\lambda)$ $(\Theta_\gamma = \mathrm{pr}_{\mathfrak{H}_\gamma}\Theta)$. It is not hard to see that the characteristic operator-function $W^\cap(\lambda)$ of the projection of the node Θ onto the subspace $\mathfrak{H}^\cap = \bigcap_{\gamma\in\Gamma}\mathfrak{H}_\gamma$ is the greatest common divisor of the functions $W_\gamma(\lambda)$ $(\gamma \in \Gamma)$. In the case when the functions $W_\gamma(\lambda)$ $(\gamma \in \Gamma)$ do not have a common multiple, we fix on some function $W_{\gamma_0}(\lambda)$ $(\gamma_0 \in \Gamma)$ and denote by $W'_\gamma(\lambda)$ the greatest common divisor of the function $W_{\gamma_0}(\lambda)$ and $W_\gamma(\lambda)$. By what was proved above the functions $W'_\gamma(\lambda)$ $(\gamma \in \Gamma)$ have a greatest common divisor, which at the same time is the greatest common divisor of the functions $W_\gamma(\lambda)$ $(\gamma \in \Gamma)$. The uniqueness of the greatest common divisor is obvious.

Now we proceed to the derivation of formulas (29.7) and (29.8). In view of the fact that the set of functions $W_\gamma(\lambda)$ $(\gamma \in \Gamma)$ is ordered, and in view of the inequality $\inf_{\gamma\in\Gamma}\tau[W_\gamma] < \infty$, we need consider only the case when the functions $W_\gamma(\lambda)$ $(\gamma \in \Gamma)$ have a common multiple of finite weight. If $\mathfrak{H}^\cap$ coincides with one of the spaces $\mathfrak{H}_\gamma$, then the validity of formulas (29.7) and (29.8) is obvious. In the contrary case it is possible to select from the collection of subspaces $\mathfrak{H}_\gamma$ a sequence $\mathfrak{H}_{\gamma_j}$ $(\mathfrak{H}_{\gamma_{j+1}} \subset \mathfrak{H}_{\gamma_j},\ j = 1, 2, \cdots)$ such that the orthoprojectors P_j onto the $\mathfrak{H}_{\gamma_j}$ will converge to an orthoprojector $P^\cap$ onto $\mathfrak{H}^\cap$. In addition the equations

$$\lim_{j\to\infty} \sigma\left[W_{\gamma_j}\right] = \inf_{\gamma\in\Gamma} \sigma[W_\gamma], \quad \lim_{j\to\infty} \tau\left[W_{\gamma_j}\right] = \inf_{\gamma\in\Gamma} \tau[W_\gamma]$$

will be satisfied. Inasmuch as

$$\tau\left[W_{\gamma_j}\right] = 2 \operatorname{tr}(P_j KK^* P_j), \quad \tau\left[W^{\cap}\right] = 2 \operatorname{tr}\left(P^{\cap} KK^* P^{\cap}\right)$$

and KK^* is a nuclear operator, in view of Theorem 28.3 we have $\lim_{j\to\infty} \tau[W_{\gamma_j}] = \tau[W^{\cap}]$. Thus

$$\tau\left[W^{\cap}\right] = \inf_{\gamma \in \Gamma} \tau\left[W_{\gamma}\right].$$

Put $W_{\gamma_j}(\lambda) = W^{\cap}(\lambda) W_{\gamma_j}^{(1)}(\lambda)$. Since

$$\tau\left[W_{\gamma_j}\right] = \tau\left[W^{\cap}\right] + \tau\left[W_{\gamma_j}^{(1)}\right], \quad \sigma\left[W_{\gamma_j}^{(1)}\right] \leqslant \tau\left[W_{\gamma_j}^{(1)}\right],$$

we have $\sigma[W_{\gamma_j}^{(1)}] \to 0$. It follows from the inequality $\sigma[W_{\gamma_j}] \le \sigma[W^{\cap}] + \sigma[W_{\gamma_j}^{(1)}]$ that $\lim_{j\to\infty} \sigma[W_{\gamma_j}] \le \sigma[W^{\cap}]$. At the same time $W^{\cap}(\lambda) \prec W_{\gamma_j}(\lambda)$, and therefore $\sigma[W^{\cap}] \le \lim_{j\to\infty} \sigma[W_{\gamma_j}]$.

The theorem is proved.

The following example shows that if the restriction $\inf_{\gamma\in\Gamma} \tau[W_\gamma] < \infty$ is lacking then formulas (29.7) and (29.8) become generally speaking false.

Suppose that $\mathfrak{G}$ is infinite dimensional, $\{e_j\}$ an orthonormalized basis in $\mathfrak{G}$, and P_n an orthoprojector onto the linear envelope of the vectors $e_1, \cdots, e_n$. The greatest common divisor $W^{\cap}(\lambda)$ of the ordered collection of functions

$$W_n(\lambda) = e^{\frac{i}{\lambda} Q_n} = P_n + e^{\frac{i}{\lambda}} Q_n \qquad (Q_n = E - P_n)$$

is equal to E, so that $\sigma[W^{\cap}] = \tau[W^{\cap}] = 0$. At the same time

$$\sigma\left[W_n\right] = 1, \quad \tau\left[W_n\right] = \infty \qquad (n = 1, 2, \ldots).$$

Theorem 29.7. *If $W^{\cup}(\lambda)$ is the least common multiple and $W^{\cap}(\lambda)$ the greatest common divisor of the functions $W_j(\lambda) \in \Omega_{\mathfrak{G}}^{(\exp)}$ $(j = 1, 2)$, then*

$$\tau\left[W_1\right] + \tau\left[W_2\right] = \tau\left[W^{\cup}\right] + \tau\left[W^{\cap}\right]. \tag{29.9}$$

Proof. We need consider only the case when $\tau[W_1] < \infty$ and $\tau[W_2] < \infty$. Suppose that Θ is a simple node with characteristic operator-function $W^{\cup}(\lambda)$, and $\mathfrak{H}^{(j)}$ $(j = 1, 2)$ are subspaces invariant relative to A, for which

$$W_j(\lambda) = W_{\Theta_j}(\lambda) \qquad (\Theta_j = \mathrm{pr}_{\mathfrak{H}^{(j)}}\Theta).$$

In view of Theorems 29.5 and 29.6,

$$\mathfrak{H} = \mathfrak{H}^{(1)} \tilde{\cup} \mathfrak{H}^{(2)}, \quad W^{\cap}(\lambda) = W_{\Theta^{\cap}}(\lambda) \qquad (\Theta^{\cap} = \mathrm{pr}_{\mathfrak{H}^{\cap}}\Theta, \ \mathfrak{H}^{\cap} = \mathfrak{H}^{(1)} \cap \mathfrak{H}^{(2)}).$$

Moreover,

$$\tau[W_j] = 2 \ \mathrm{tr}\ (A^{(j)})_I, \ \tau\left[W^{\cup}\right] = 2 \ \mathrm{tr}\ A_I, \ \tau\left[W^{\cap}\right] = 2 \ \mathrm{tr}\ (A^{\cap})_I,$$

where $A^{(j)}$ and $A^{\cap}$ are induced in the subspaces $\mathfrak{H}^{(j)}$ and $\mathfrak{H}^{\cap}$ respectively. It follows from Theorem 28.6 that A_j is a nuclear operator. Applying formula (28.8), we obtain (29.9).

Lemma 29.3. *If $W(\lambda) \in \Omega_{\mathfrak{G}}^{(\exp)}$ is an orthoprojector onto some n-dimensional subspace in $\mathfrak{G}$, and $W(\lambda) = W^{\cap}(\lambda) W_0(\lambda)$, where $W^{\cap}(\lambda)$ is the greatest common divisor of the functions $W(\lambda)$ and $e^{i\sigma P^{\perp}/\lambda}$ ($\sigma = [W]$, $P^{\perp} = E - P$), then*

$$\tau[W_0] \leqslant n\sigma. \tag{29.10}$$

Proof. Consider the simple node

$$\tilde{\Theta} = \begin{pmatrix} \tilde{A} & \tilde{K} & E \\ \tilde{\mathfrak{H}} & & \mathfrak{G} \end{pmatrix}$$

with characteristic operator-function $e^{i\sigma E/\lambda}$ and subspaces $\mathfrak{H}_P, \mathfrak{H}_{P\perp}, \mathfrak{H}$, invariant relative to $\tilde{A}$, for which

$$W_{\Theta_P}(\lambda) = e^{\frac{i\sigma P}{\lambda}} \qquad (\Theta_P = \mathrm{pr}_{\mathfrak{H}_P}\tilde{\Theta}),$$

$$W_{\Theta_{P\perp}}(\lambda) = e^{\frac{i\sigma P\perp}{\lambda}} \qquad (\Theta_{P\perp} = \mathrm{pr}_{\mathfrak{H}_{P\perp}}\tilde{\Theta}),$$

$$W_{\Theta}(\lambda) = W(\lambda) \qquad (\Theta = \mathrm{pr}_{\mathfrak{H}}\tilde{\Theta}).$$

In view of the equation

$$\left(e^{\frac{i\sigma P}{\lambda}} - E\right)\left(e^{\frac{i\sigma P\perp}{\lambda}} - E\right) = 0$$

and Theorem 5.8, $\tilde{\mathfrak{H}} = \mathfrak{H}_P \oplus \mathfrak{H}_{P\perp}$. Denote by $A^{(P)}$ and A the operators induced in $\mathfrak{H}_P$ and $\mathfrak{H}$ respectively. From Theorem 29.2 and the formula $\tau[e^{i\sigma P/\lambda}] = n\sigma$ it follows that $A^{(P)}$ is a Volterra operator satisfying the condition $2\,\mathrm{tr}\,(A^{(P)})_I = n\sigma$.

Represent $\mathfrak{H}$ in the form

$$\mathfrak{H} = \mathfrak{G}_0 \oplus \mathfrak{G}_1 \oplus \mathfrak{G}_2,$$

where $\mathfrak{G}_0 = \mathfrak{H}_{P\perp} \cap \mathfrak{H}$ and $\mathfrak{G}_0 \oplus \mathfrak{G}_1$ is the collection of all vectors $h \in \mathfrak{H}$ for which $Ah \in \mathfrak{G}_0$. By Lemma 28.3 there exists an orthonormalized basis $\{g_j\}_1^{\omega}$ ($\omega \le \infty$) in $\mathfrak{G}_2$ and an orthonormalized sequence $\{h_j\}_1^{\omega}$ in $\mathfrak{H}_P$ such that

$$(Ag_j, g_j) = (A^{(P)} h_j, h_j) \qquad (j = 1, 2, \ldots, \omega).$$

Since $\mathfrak{G}_1 = 0$ (in the contrary case the operator $\tilde{A}$ would not be completely nonselfadjoint) and $W^{\cap}(\lambda) = W_{\Theta^{\cap}}(\lambda)$ $(\Theta^{\cap} = \mathrm{pr}_{\mathfrak{G}_0} \tilde{\Theta})$, we have

$$W_0(\lambda) = W_{\Theta_2}(\lambda) \qquad (\Theta_2 = \mathrm{pr}_{\mathfrak{G}_2} \tilde{\Theta}).$$

For the basic operator $A^{(2)}$ of the node Θ_2 we have

$$\sum_{j=1}^{\omega} ((A^{(2)})_I g_j, g_j) = \sum_{j=1}^{\omega} (A_I g_j, g_j) = \sum_{j=1}^{\omega} ((A^{(P)})_I h_j, h_j) \leqslant \frac{n\sigma}{2}.$$

Accordingly,

$$\tau[W_0] = 2 \operatorname{tr} (A^{(2)})_I \leqslant n\sigma.$$

4. Limit theorems on operators of the class $\Lambda^{(\exp)}$. Theorems 8.3 and 29.2 make it possible to give the following formulation for a portion of the assertions of Theorems 29.5 and 29.6.

Suppose that A is a completely nonselfadjoint operator of the class $\Lambda^{(\exp)}$ operating in the space $\mathfrak{H}$. We denote by A_γ the operators induced in certain subspaces $\mathfrak{H}_\gamma$ $(\gamma \in \Gamma)$ which are invariant relative to A.

Theorem 29.5$'$. *Suppose that the smallest subspace containing all the* $\mathfrak{H}_\gamma$ $(\gamma \in \Gamma)$ *coincides with* $\mathfrak{H}$. *Then*

$$\sigma[A] = \sup_{\gamma \in \Gamma} \sigma[A_\gamma]. \tag{29.11}$$

In addition, when the $\mathfrak{H}_\gamma$ *are ordered by inclusion,*

$$\operatorname{tr} A_I = \sup_{\gamma \in \Gamma} \operatorname{tr} (A_\gamma)_I. \tag{29.12}$$

Theorem 29.6′. *Suppose that the subspaces* $\mathfrak{H}_\gamma$ $(\gamma \in \Gamma)$ *are ordered by inclusion. If* $\operatorname{tr} A_I < \infty$, *then*

$$\sigma\left[A^{\cap}\right] = \inf_{\gamma \in \Gamma} \sigma[A_\gamma], \tag{29.13}$$

$$\operatorname{tr}\left(A^{\cap}\right)_I = \inf_{\gamma \in \Gamma} \operatorname{tr}(A_\gamma)_I, \tag{29.14}$$

where $A^{\cap}$ *is the operator induced in the subspace* $\mathfrak{H}^{\cap} = \bigcap_{\gamma \in \Gamma} \mathfrak{H}_\gamma$.

§ 30. Criteria for unicellularity of dissipative operators of exponential type

1. Unicellular dissipative Volterra operators with nuclear imaginary components. The function $W(\lambda) \in \Omega_{\mathfrak{G}}^{(\exp)}$ will be said to be *ordered* if the set of all its regular left divisors is ordered.

Lemma 30.1. *Suppose that the function* $W(\lambda) \in \Omega_{\mathfrak{G}}^{(\exp)}$ *satisfies the condition* $\sigma[W] = \tau[W]$. *If* $W(\lambda) = W_1(\lambda)W_2(\lambda)$ $(W_j(\lambda) \in \Omega_{\mathfrak{G}}^{(\exp)})$, *then* $\sigma[W_j] = \tau[W_j]$ $(j = 1, 2)$.

Proof. In view of relations (8.2), (29.2) and (29.1)

$$\sigma[W] \leqslant \sigma[W_1] + \sigma[W_2] \leqslant \tau[W_1] + \tau[W_2] = \tau[W].$$

Therefore

$$\sigma[W_1] + \sigma[W_2] = \tau[W_1] + \tau[W_2].$$

Inasmuch as each term in the left side of the last equation is not larger than the corresponding term in the right side, we have $\sigma[W_1] = \tau[W_1]$ and $\sigma[W_2] = \tau[W_2]$.

Theorem 30.1. *Suppose that the function* $W(\lambda) \in \Omega_{\mathfrak{G}}^{(\exp)}$ *has a finite weight. For it to be ordered it is necessary and sufficient that*

$$\sigma[W] = \tau[W]. \tag{30.1}$$

Proof. If $W(\lambda)$ satisfies the condition (30.1) and $W'(\lambda)$ and $W''(\lambda)$ are regular left divisors of it, then by Lemma 30.1

$$\sigma[W'] = \tau[W'],\ \sigma[W''] = \tau[W''],$$
$$\sigma\left[W^{\cup}\right] = \tau\left[W^{\cup}\right],\ \sigma\left[W^{\cap}\right] = \tau\left[W^{\cap}\right],$$

where $W^{\cup}(\lambda)$ and $W^{\cap}(\lambda)$ are the least common multiple and greatest common divisor of the functions $W'(\lambda)$ and $W''(\lambda)$ respectively. Applying Theorems 29.5

and 29.7, we obtain

$$\tau\left[W^{\cup}\right] = \max\{\tau[W'],\ \tau[W'']\}, \tag{30.2}$$

$$\tau[W'] + \tau[W''] = \tau\left[W^{\cup}\right] + \tau\left[W^{\cap}\right]. \tag{30.3}$$

Comparing (30.2) and (30.3), we arrive at the conclusion that either $\tau[W'] = \tau[W^{\cap}]$, or $\tau[W''] = \tau[W^{\cap}]$. In the first case $W'(\lambda) = W^{\cap}(\lambda) \prec W''(\lambda)$, and in the second $W''(\lambda) = W^{\cap}(\lambda) \prec W'(\lambda)$.

We turn to the proof of necessity. We suppose that there exists a function $W(\lambda) \in \Omega_{\mathfrak{G}}^{(\exp)}$ such that the set of its left regular divisors is ordered and $\sigma[W] < \tau[W] < \infty$. Suppose that P is an orthoprojector onto some one-dimensional subspace in $\mathfrak{G}$ and $W_P(\lambda)$ is the greatest common divisor of the functions $W(\lambda)$ and $e^{i\sigma P^{\perp}/\lambda}$ ($\sigma = \sigma[W]$, $P^{\perp} = E - P$). It follows from Lemma 29.3 that

$$\tau[W_P] \geqslant \tau[W] - \sigma[W].$$

The functions $W_P(\lambda)$, for all possible P, form an ordered set. Denoting by $W_0(\lambda)$ their greatest common divisor and applying Theorem 29.6, we find that

$$\tau[W_0] = \inf_P \tau[W_P] \geqslant \tau[W] - \sigma[W] > 0. \tag{30.4}$$

On the other hand, taking into account the relation $W_0(\lambda) \prec e^{i\sigma P^{\perp}/\lambda}$ and decomposing $W_0(\lambda)$ into a series $E + (2i/\lambda)H_0 + \cdots$, we find that for any one-dimensional P

$$0 \leqslant 2H_0 \leqslant \sigma(E - P).$$

Thus $H_0 = 0$, so that $\tau[W_0] = 0$, which contradicts inequality (30.4).

The theorem is proved.

We denote by $\Lambda_0^{(\exp)}$ the class of all dissipative Volterra operators with nuclear imaginary components. By Theorem 29.3 $\Lambda_0^{(\exp)} \subset \Lambda^{(\exp)}$. If $A \in \Lambda^{(\exp)}$ and A_I is a nuclear operator, then by Theorem 10.1 $A \in \Lambda_0^{(\exp)}$.

Theorem 30.2. *Imbed the completely nonselfadjoint operator* $A \in \Lambda_0^{(\exp)}$ *in the node* $\Theta = \begin{pmatrix} A & K & E \\ \mathfrak{H} & & \mathfrak{G} \end{pmatrix}$. *For the operator* A *to be unicellular, it is necessary and sufficient that*

$$\sigma[W_\Theta] = \tau[W_\Theta]. \tag{30.5}$$

The proof follows from Theorems 5.7 and 30.1.

Theorem 30.3. *For the completely nonselfadjoint operator $A \in \Lambda_0^{(\exp)}$ to be unicellular, it is necessary and sufficient that*

$$\sigma[A] = 2 \operatorname{tr} A_I. \tag{30.6}$$

Proof. Imbed A in the node

$$\Theta = \begin{pmatrix} A & K & E \\ \mathfrak{H} & & \mathfrak{G} \end{pmatrix}$$

The criterion (30.6) follows from the preceding theorem and the equations

$$\sigma[W_\Theta] = \sigma[A], \quad \tau[W_\Theta] = 2 \operatorname{tr} A_I$$

(see Theorems 8.3 and 29.2).

2. Invariant subspaces of operators of the class $\Lambda^{(\exp)}$. The bounded linear operator A will be assigned to the class $\Lambda_\infty^{(\exp)}$ if $A \in \Lambda^{(\exp)}$ and $A \notin \Lambda_0^{(\exp)}$.

Lemma 30.2. *If $W(\lambda) \in \Omega_{\mathfrak{G}}^{(\exp)}$, then there exist functions*

$$W_n^{(j)}(\lambda) = E + \frac{2i}{\lambda} H_n^{(j)} + \ldots \in \Omega_{\mathfrak{G}}^{(\exp)} \quad (j = 1, 2;\ n = 1, 2, \ldots),$$

satisfying the following conditions: 1) $W(\lambda) = W_n^{(1)}(\lambda) W_n^{(2)}(\lambda)$; 2) $W_{n+1}^{(1)}(\lambda) \prec W_n^{(1)}(\lambda)$; 3) *the sequence $H_n^{(1)}$ converges strongly to zero*; 4) $\tau[W_n^{(2)}] < \infty$.

Proof. Let $\{g_j\}_1^\infty$ be an orthonormalized basis in $\mathfrak{G}$ and P_n an orthoprojector onto the linear envelope of the vectors $g_1, \cdots, g_n$. Consider the equation $W(\lambda) = W_n^{(1)}(\lambda) W_n^{(2)}(\lambda)$, where $W_n^{(1)}(\lambda)$ is the greatest common divisor of the functions $W(\lambda)$ and $e^{i\sigma P_n^\perp/\lambda}$ ($\sigma = \sigma[W]$, $P_n^\perp = E - P_n$). Obviously $W_{n+1}^{(1)}(\lambda) \prec W_n^{(1)}(\lambda)$. Since $H_n^{(1)} \le \sigma P_n^\perp/2$, the sequence $H_n^{(1)}$ converges strongly to zero. By Lemma 29.3

$$\tau\left[W_n^{(2)}\right] \leqslant n\sigma < \infty.$$

Lemma 30.3. *If B is a completely nonselfadjoint operator of the class $\Lambda^{(\exp)}$, operating in a space $\mathfrak{H}$, then there exists a sequence of orthoprojectors Q_n ($n = 1, 2, \cdots$) such that*: 1) *the subspaces $Q_n\mathfrak{H}$ are invariant relative to B*; 2) $Q_{n+1} < Q_n$; 3) *the sequence Q_n converges strongly to zero*; 4) *the operators*

$(E - Q_n)B$, *considered respectively in the spaces* $(E - Q_n)\mathfrak{H}$, *lie in the class* $\Lambda_0^{(\exp)}$.

Proof. Imbed B in the node

$$\Theta = \begin{pmatrix} B & K & E \\ \mathfrak{H} & & \mathfrak{G} \end{pmatrix}.$$

The function $W_\Theta(\lambda)$ lies in the class $\Omega_{\mathfrak{G}}^{(\exp)}$ and is representable in the form $W_\Theta(\lambda) = W_n^{(1)}(\lambda)W_n^{(2)}(\lambda)$, where the function $W_n^{(j)}(\lambda)$ $(j = 1, 2;\ n = 1, 2, \cdots)$ satisfies the conditions of the preceding lemma. There exist subspaces $\mathfrak{H}_n^{(1)}$ invariant relative to B and such that

$$W_n^{(1)}(\lambda) = W_{\Theta_n^{(1)}}(\lambda) \qquad \left(\Theta_n^{(1)} = \operatorname{pr}_{\mathfrak{H}_n^{(1)}}\Theta\right),$$

$$W_n^{(2)}(\lambda) = W_{\Theta_n^{(2)}}(\lambda) \qquad \left(\Theta_n^{(2)} = \operatorname{pr}_{\mathfrak{H}_n^{(2)}}\Theta, \quad \mathfrak{H}_n^{(2)} = \mathfrak{H} \ominus \mathfrak{H}_n^{(1)}\right).$$

The sequence Q_n of orthoprojectors onto to the subspaces $\mathfrak{H}_n^{(1)}$ is not increasing, and accordingly converges strongly to some orthoprojector Q_0. Since the subspace $Q_0\mathfrak{H}$ is invariant relative to B and $KQ_0K = 0$, we have $Q_0 = 0$, since in the contrary case the operator B would not be completely nonselfadjoint. The last assertion of the lemma follows from Theorem 29.2.

Lemma 30.4. *If A is a completely nonselfadjoint operator of the class* $\Lambda^{(\exp)}$ *operating in the space* $\mathfrak{H}$, *then there exists a sequence of orthoprojectors* P_n $(n = 1, 2, \cdots)$ *having the following properties:* 1) *the subspaces* $P_n\mathfrak{H}$ *are invariant relative to* A; 2) $P_n < P_{n+1}$; 3) *the sequence* P_n *converges strongly to* E; 4) *the operators* A_n *induced by* A *in the subspaces* $P_n\mathfrak{H}$ *lie in the class* $\Lambda_0^{(\exp)}$.

Proof. Consider the operator $B = -A^*$ and a sequence of operators Q_n for which all the conditions of Lemma 30.3 are satisfied. It is easy to see that the sequence $P_n = E - Q_n$ satisfies the requirements of the lemma. The lemma is proved.

We shall extend the concept of a spectral function $P(x)$ $(x \in \mathfrak{M})$ by holding to all the points of the definition presented in § 18 except one: we will not require that the set $\mathfrak{M}$ be bounded.

Theorem 30.4. *If A is a completely nonselfadjoint operator of the class* $\Lambda_\infty^{(\exp)}$ *operating in the space* $\mathfrak{H}$, *then there exists in* $\mathfrak{H}$ *a spectral function*

$P(x)$ $(0 \leq x \leq \infty)$ *having the following properties*: 1) *the subspaces* $P(x)\mathfrak{H}$ *are invariant relative to* A; 2) $\operatorname{tr}(P(x)A_I P(x)) = x$ $(0 \leq x \leq \infty)$.

Proof. The chain π of orthoprojectors in $\mathfrak{H}$ will be assigned to the class $\mathfrak{A}$ if: a) $0 \in \pi$ and $E \in \pi$; b) the subspaces $P\mathfrak{H}$ $(P \in \pi)$ are invariant relative to A; c) the operators $PA_I P$ $(P \in \pi,\ P \neq E)$ are nuclear; d) π contains a strictly increasing sequence converging to E. As in Theorem 15.2, we establish in $\mathfrak{A}$ an order relation by putting $\pi_1 \prec \pi_2$ $(\pi_1, \pi_2 \in \mathfrak{A})$ if all the orthoprojectors of the chain π_1 lie in π_2. Since each ordered part of the collection $\mathfrak{A}$ obviously has a least upper bound, according to Zorn's lemma there exists a chain $\pi^* \in \mathfrak{A}$ which is maximal in $\mathfrak{A}$.

We suppose that the chain π^* has a jump (P^-, P^+). If $P^+ \neq E$, then A induces in the subspace $P^+\mathfrak{H}$ an operator A_{P^+} of the class $\Lambda_0^{(\exp)}$. The portion of the chain π^* consisting of the orthoprojectors preceding P^+ belongs to the operator A_{P^+} and may be supplemented, according to Theorems 15.3 and 15.5, to a continuous chain also belonging to A_{P^+}. In this way the chain π^* becomes extended to a certain chain $\pi^{**} \in \mathfrak{A}$, which will contradict the maximality of π^* in $\mathfrak{A}$. In the case $P^+ = E$ we consider the operator $A_0 h = (E - P^-)Ah$, operating on the elements h of the subspace $\mathfrak{H}_0 = (E - P^-)\mathfrak{H}$. It belongs to the class $\Lambda_\infty^{(\exp)}$ and by Lemma 30.4 has a nontrivial invariant subspace $\mathfrak{H}_0' \subset \mathfrak{H}_0$, in which an operator A_0' of the class $\Lambda_0^{(\exp)}$ is induced. Denote by P_0' the orthoprojector onto $\mathfrak{H}_0'$. We again obtain a contradiction, inasmuch as the chain π^* precedes the chain $\pi^{**} \in \mathfrak{A}$ which is obtained as a result of adjunction of the orthoprojector $P^- + P_0'$ to π^*.

The foregoing argument shows that the chain π^* is continuous. Moreover, it is obviously closed. Thus π^* is maximal in the set of all chains, and therefore coincides with the range of some spectral function $\widetilde{P}(t)$ $(0 \leq t \leq 1)$.

In order to complete the proof we note that the function $x = \phi(t) = \operatorname{tr}(\widetilde{P}(t)A_I\widetilde{P}(t))$ $(0 \leq t \leq 1)$ is strictly increasing, since in the contrary case the operator A would not be completely nonselfadjoint. It follows from Theorem 28.3 that it is continuous. Suppose that $t = \psi(x)$ $(0 \leq x \leq \infty)$ is its inverse function. It is easy to see that the function $P(x) = \widetilde{P}[\psi(x)]$ $(0 \leq x \leq \infty)$ satisfies all the requirements of the theorem.

Corollary 1. *The operator* $A \in \Lambda_\infty^{(\exp)}$ *has a spectral function* $P(x)$ $(0 \leq x \leq \infty)$ *such that*

$$A = 2i \lim_{x \to \infty} \int_0^x P(t) A_I \, dP(t). \tag{30.7}$$

In addition the operators $\int_0^x P(t) A_I dP(t)$ $(x < \infty)$ *are Volterra operators, and the passage to the limit is realized in the sense of strong convergence.*

Corollary 2. *If* $W(\lambda) \in \Omega_{\mathfrak{G}}^{(\exp)}$ *and* $0 \le \tau_0 < \tau[W]$, *then there exists a regular left divisor* $W_0(\lambda)$ *of the function* $W(\lambda)$ *such that* $\tau[W_0] = \tau_0$.

Theorem 30.5. *There are no unicellular operators in the class* $\Lambda_\infty^{(\exp)}$.

Proof. According to Theorem 30.4 the operator $A \in \Lambda_\infty^{(\exp)}$ has a maximal chain π such that all the operators PA_IP $(P \in \pi,\ P \neq E)$ are nuclear. On the other hand, as follows from Lemma 30.3, there exists a subspace $\mathfrak{H}'$, distinct from $\mathfrak{H}$ and invariant relative to A, in which there is induced an operator with nonnuclear imaginary component. Since the orthoprojector P' onto $\mathfrak{H}'$ cannot belong to the chain π, A is nonunicellular.

Theorem 30.5 may be given the following formulation.

Theorem 30.5′. *If an operator of the class* $\Lambda^{(\exp)}$ *is unicellular, then it is a Volterra operator and has a nuclear imaginary component.*

Theorem 30.6. *Suppose that the bounded linear operator* A *operating in the space* $\mathfrak{H}$ *does not have spectral points other than zero. For* A *to belong to the class* $\Lambda^{(\exp)}$, *it is necessary and sufficient that there should exist in* $\mathfrak{H}$ *an orthoprojector* P_n *having the following properties:*

1) $P_n \to E$.

2) *The subspaces* $\mathfrak{H}_n = P_n\mathfrak{H}$ *are invariant relative to* A.

3) *The operators* A_n *induced in the subspaces* $\mathfrak{H}_n$ *belong to the class* $\Lambda_0^{(\exp)}$, *and the sequence of types* $\sigma[A_n]$ *is bounded below.*

Proof. The necessity follows easily from Theorem 30.4. If conditions 1), 2) and 3) are satisfied, then the operator A may be imbedded in a dissipative node Θ. In view of Theorems 8.3 and 8.1,

$$\left\| W_{\Theta_n}\left(\frac{1}{\mu}\right) \right\| \leqslant e^{\tilde{\sigma}|\mu|} \quad \left(\Theta_n = \mathrm{pr}_{\mathfrak{H}_n}\Theta,\ \tilde{\sigma} = \sup_n \sigma[A_n]\right).$$

Consequently, $\|W_\Theta(1/\mu)\| \le e^{\tilde{\sigma}|\mu|}$, and, again by Theorem 8.3, $A \in \Lambda^{(\exp)}$.

§ 31. Cyclic dissipative operators of exponential type

1. **Generating vector of a unicellular operator.** Suppose that A is a bounded linear operator operating in a space $\mathfrak{H}$. If there exists a vector $h \in \mathfrak{H}$ such that the sequence $A^n h$, $n = 0, 1, \cdots$, is dense in $\mathfrak{H}$, then the operator A is said to be *cyclic*, and h is called the *generating vector* of the operator A.

Theorem 31.1. *Every unicellular operator is cyclic.*

Proof. Suppose that A is a unicellular operator and that π is the collection of orthoprojectors onto all subspaces invariant relative to A. If π has a jump of the type (P, E), then any vector $h \in (E - P)\mathfrak{H}$ which is distinct from zero is generating. If there is no such discontinuity, then there exists a strictly increasing sequence $P_n \to E$ $(P_n \in \pi)$, and as the generating vector one can choose

$$h = \sum_{n=1}^{\infty} \frac{1}{2^n} f_n \qquad (\| f_n \| = 1,\ f_n \in (P_{n+1} - P_n)\mathfrak{H}).$$

Lemma 31.1. *Suppose that A is a unicellular operator and $\mathfrak{H}_n$ $(n = 1, 2, \cdots)$ is some sequence of subspaces invariant with respect to it. If f_n is a generating vector of the operator A_n induced in $\mathfrak{H}_n$ and $d^2 = \sum_{n=1}^{\infty} \|f_n\|^2 < \infty$, then there exists a sequence of numbers $\{\xi_n\}_1^\infty$ $(\sum_{n=1}^{\infty} |\xi_n|^2 < \infty)$, such that the vector $h = \sum_{n=1}^{\infty} \xi_n f_n$ will be generating for the operator $\tilde{A}$ induced in the closure $\tilde{\mathfrak{H}}$ of the linear envelope of all the subspaces $\mathfrak{H}_n$.*

Proof. Without loss of generality we may suppose that $\mathfrak{H}_n \subset \mathfrak{H}_{n+1}$ $(\mathfrak{H}_n \neq \mathfrak{H}_{n+1})$, $n = 1, 2, \cdots$. Obviously

$$d_n = \| (E - P_n) f_{n+1} \| > 0 \qquad (n = 1,\ 2,\ \ldots),$$

where P_n is an orthoprojector onto $\mathfrak{H}_n$. Construct a sequence $\{\xi_n\}_1^\infty$, choosing ξ_1 and ξ_2 $(\neq 0)$ arbitrarily, and ξ_n $(n = 3, 4, \cdots)$ so that the following equations are satisfied:

$$\left.\begin{aligned} &|\xi_3| = \frac{d_1 |\xi_2|}{\sqrt{4} d}, \\ &|\xi_4| = \min\left\{ \frac{d_1 |\xi_2|}{\sqrt{8} d}, \frac{d_2 |\xi_3|}{\sqrt{4} d} \right\}, \\ &|\xi_5| = \min\left\{ \frac{d_1 |\xi_2|}{\sqrt{16} d}, \frac{d_2 |\xi_3|}{\sqrt{8} d}, \frac{d_3 |\xi_4|}{\sqrt{4} d} \right\}, \\ &\cdots\cdots\cdots\cdots\cdots\cdots \end{aligned}\right\} \tag{31.1}$$

Inasmuch as $\sum_{n=1}^{\infty}|\xi_n|^2 < \infty$, the series $h = \sum_{n=1}^{\infty}\xi_n f_n$ converges strongly.

It suffices to show that h is not contained in any of the subspaces $\mathfrak{H}_n$. We suppose that $h \in \mathfrak{H}_k$. Then

$$\sum_{n=k+2}^{\infty} \xi_n f_n = g - \xi_{k+1} f_{k+1} \qquad (g \in \mathfrak{H}_k),$$

so that

$$\left\| \sum_{n=k+2}^{\infty} \xi_n f_n \right\| \geqslant \| \xi_{k+1} (E - P_k) f_{k+1} \| = | \xi_{k+1} | d_k. \tag{31.2}$$

On the other hand, by formulas (31.1),

$$\left\| \sum_{n=k+2}^{\infty} \xi_n f_n \right\|^2 \leqslant \sum_{n=k+2}^{\infty} | \xi_n |^2 \sum_{n=k+2}^{\infty} \| f_n \|^2 \leqslant \frac{| \xi_{k+1} |^2 d_k^2}{2},$$

which contradicts the estimate (31.2).

Theorem 31.2. *Suppose that A is a unicellular operator operating in the space $\mathfrak{H}$. If $\dim \mathfrak{H} > 1$, then A has a generating vector lying in the range of the operator A_I.*

Proof. We construct a sequence $\{h_n\}_1^{\infty}$ of nonzero elements and denote by $\mathfrak{H}_n$ the closure of the linear envelope of vectors $A^k A_I h_n$ $(k = 0, 1, \cdots)$. Inasmuch as A is unicellular and $\dim \mathfrak{H} > 1$, A is a completely nonselfadjoint operator. Accordingly the closure of the linear envelope of the subspaces $\mathfrak{H}_n$ coincides with $\mathfrak{H}$.

The vector $f_n = A_I h_n / n\|h_n\|$ is generating for the operator A_n induced in $\mathfrak{H}_n$. Applying Lemma 31.1, we find a sequence of numbers $\{\xi_n\}_1^{\infty}$ $(\sum_{n=1}^{\infty}|\xi_n|^2 < \infty)$ such that the vector

$$h = \sum_{n=1}^{\infty} \xi_n f_n = A_I g \qquad \left(g = \sum_{n=1}^{\infty} \frac{\xi_n h_n}{n \| h_n \|} \right)$$

is generating for the operator A.

2. Supplementary information on the universal node. The concepts of universal operator $\widetilde{\mathfrak{I}}_l$ and universal node

$$\tilde{\Theta}_l = \begin{pmatrix} \tilde{\mathfrak{I}}_l & \tilde{K}_l & E \\ \tilde{L}_2(0, l) & l_2 \end{pmatrix}$$

were introduced in § 8.

We denote by $P(t)$ the orthoprojector onto the subspace of the space $\tilde{L}_2(0, l)$ consisting of all functions equal to zero almost everywhere on the segment $[t, l]$:

$$P(t) f(x) = \begin{cases} f(x) & (0 \leqslant x < t), \\ 0 & (t \leqslant x \leqslant l) \end{cases} \quad (f(x) \in \tilde{L}_2(0, l)).$$

We denote by l_f ($f = f(x) \in \tilde{L}_2(0, l)$) the smallest number t for which $f = P(t)f$.

Lemma 31.2. *Suppose that $f(x) \in \tilde{L}_2(0, l)$ and that $\mathfrak{H}$ is the closure of the linear envelope of the sequence $\tilde{\mathfrak{I}}_l^n f(x)$ $(n = 0, 1, \cdots)$. Suppose that A is the operator induced by $\tilde{I}_l$ in $\mathfrak{H}$. Then the type $\sigma[A]$ of the function $(A - (1/\mu)E)^{-1}$ is equal to $2l_f$.*

Proof. Denote by Θ and $\Theta_{l'}$ the projections of the node $\tilde{\Theta}_l$ onto $\mathfrak{H}$ and $P(l')\tilde{L}_2(0, l)$ $(0 \leq l' \leq l)$ respectively. It was shown in § 8 that $W_{\tilde{\Theta}_l}(\lambda) = e^{2il/\lambda}E$. Analogously one can show that $W_{\Theta_{l'}}(\lambda) = e^{2il'/\lambda}E$.

According to the hypotheses of the lemma $\mathfrak{H} \subset P(l_f)\tilde{L}_2(0, l)$. Accordingly $W_\Theta(\lambda) \ll e^{2il_f/\lambda}E$, so that $\sigma[W_\Theta] \leq 2l_f$. Suppose that $\sigma[W_\Theta] < 2l_f$. Then from Theorem 8.2

$$W_\Theta(\lambda) \ll e^{\frac{2il'}{\lambda}} E \qquad (2l' = \sigma[W_\Theta]),$$

so that $\mathfrak{H} \subset P(l')\tilde{L}_2(0, l)$, which contradicts the definition of l_f. Thus $\sigma[W_\Theta] = 2l_f$. It remains to be noted that in view of Theorem 8.2 $\sigma[W_\Theta] = \sigma[A]$.

Lemma 31.3. *If $\mathfrak{H}$ is an invariant subspace of the operator $\tilde{\mathfrak{I}}_l$ and A is the operator induced in $\mathfrak{H}$, then there exists a vector $f(x) \in \mathfrak{H}$ such that $\sigma[A] = 2l_f$.*

Proof. We fix in $\mathfrak{H}$ an orthonormalized basis $\{f_j(x)\}_1^\infty$ and denote by $\mathfrak{H}_j$ the closure of the linear envelope of the sequence $\tilde{\mathfrak{I}}_l^n f_j(x)$ $(n = 0, 1, \cdots)$. By Theorem 29.5′ $\sigma[A] = \sup_j \sigma[A_j]$, where A_j is the operator induced in the subspace $\mathfrak{H}_j$, and in view of Lemma 31.2 $\sigma[A_j] = 2l_{f_j}$.

In the following considerations we need only the case when $2l_{f_j} < \sigma[A]$ for all f. We select from the sequence $\{f_j\}_1^\infty$ a subsequence $\{g_j\}_1^\infty$ such that the sequence of numbers $\{l_{g_j}\}_1^\infty$ is strongly increasing and converges to $\frac{1}{2}\sigma[A]$. We note that

$$g_j = P(l_{g_j})\, g_j, \quad \|g_j - P(l_{g_{j-1}})\, g_j\| = \alpha_j > 0.$$

Suppose that $f(x) = \sum_{j=1}^\infty \gamma_j g_j(x)$, where $\{\gamma_j\}_1^\infty$ is an arbitrary sequence of numbers satisfying the conditions

$$\left.\begin{array}{l} 0 < \gamma_1, \\ 0 < \gamma_2 < \dfrac{\alpha_1\gamma_1}{2}, \\ 0 < \gamma_3 < \min\left\{\dfrac{\alpha_1\gamma_1}{2^2}, \dfrac{\alpha_2\gamma_2}{2}\right\}, \\ 0 < \gamma_4 < \min\left\{\dfrac{\alpha_1\gamma_1}{2^3}, \dfrac{\alpha_2\gamma_2}{2^2}, \dfrac{\alpha_3\gamma_3}{2}\right\}, \\ \cdots\cdots\cdots\cdots\cdots \end{array}\right\} \tag{31.3}$$

In view of the inequalities (31.3)

$$\|f - P(l_{g_n})f\| = \left\|\sum_{j=1}^\infty \gamma_j (E - P(l_{g_n}))\, g_j\right\| = \left\|\sum_{j=n+1}^\infty \gamma_j (E - P(l_{g_n}))\, g_j\right\|$$

$$\geqslant \|\gamma_{n+1}(E - P(l_{g_n}))\, g_{n+1}\| - \left\|\sum_{j=n+2}^\infty \gamma_j (E - P(l_{g_n}))\, g_j\right\| \geqslant \gamma_{n+1}\alpha_{n+1} - \sum_{j=n+2}^\infty \gamma_j > 0.$$

Accordingly $l_f \geq \frac{1}{2}\sigma[A]$, and, since inequality is impossible here $2l_f = \sigma[A]$.

The lemma is proved.

To each function $\alpha(x) \in L_2(0, l)$ we will assign the function $\overleftrightarrow{\alpha(x)} = \alpha(l - x)$.

Lemma 31.4. *Suppose that* $\mathfrak{H}$ *is an invariant subspace of the operator* $\tilde{\mathfrak{J}}_l$ *and that* A *is the operator induced in* $\mathfrak{H}$. *If* A *has the generating vector*

$$f(x) = \| f^{(1)}(x)\, f^{(2)}(x) \ldots \|,$$

then the set of vectors of the form

$$\left\| \overleftrightarrow{\overleftrightarrow{f^{(1)}(x)} * \varphi(x)} \quad \overleftrightarrow{\overleftrightarrow{f^{(2)}(x)} * \varphi(x)} \ldots \right\| \quad (\varphi(x) \in L_2(0, l)),$$

where the $*$ *denotes the convolution operator as in formula* (20.5), *is dense in* $\mathfrak{H}$.

Proof. The linear envelope of the vectors $\widetilde{\mathcal{I}}_l^n f(x)$ $(n = 0, 1, \cdots)$ is dense in $\mathfrak{H}$. The linear envelope $\mathfrak{H}_1$ of the vectors $\widetilde{\mathcal{I}}_l^n f(x)$ $(n = 1, 2, \cdots)$ has the same property, since in the contrary case the operator A would not be completely nonselfadjoint. From the equations

$$\frac{(n-1)!}{(2i)^n}\mathcal{I}_l^n f^{(j)}(x) = \int_x^l (t-x)^{n-1} f^{(j)}(t)\,dt = \int_0^{l-x} f^{(j)}(x+t)\,t^{n-1}\,dt$$

it follows that $\mathfrak{H}_1$ coincides with the set of vectors

$$\left\| \int_0^{l-x} f^{(1)}(x+t)\,p(t)\,dt \quad \int_0^{l-x} f^{(2)}(x+t)\,p(t)\,dt \ \ldots \right\|,$$

where $p(t)$ runs through the collection $\mathcal{P}$ of all polynomials. Denote by $\mathfrak{H}_2$ the set of vectors of the form

$$\left\| \int_0^{l-x} f^{(1)}(x+t)\,\varphi(t)\,dt \quad \int_0^{l-x} f^{(2)}(x+t)\,\varphi(t)\,dt \ \ldots \right\|$$

$$(\varphi(t) \in L_2(0, l)).$$

Since $\mathfrak{H}_1 \subset \mathfrak{H}_2$, $\mathcal{P}$ is dense in $L_2(0, l)$, and $\mathfrak{H}_2 \subset \mathfrak{H}$, we have $\overline{\mathfrak{H}}_2 = \mathfrak{H}$. The assertion of the lemma now follows from the relations

$$\overleftrightarrow{\overleftrightarrow{f^{(j)}}(x) * \varphi(x)} = \overleftrightarrow{\int_0^x f^{(j)}(l-x+t)\,\varphi(t)\,dt} = \int_0^{l-x} f^{(j)}(x+t)\,\varphi(t)\,dt.$$

Lemma 31.5. *Suppose that*

$$\widetilde{\Theta}_l = \begin{pmatrix} \widetilde{\mathcal{I}}_l & \widetilde{K}_l & E \\ \widetilde{L}_2(0, l) & & l_2 \end{pmatrix}$$

is a universal node and that P_ξ *is an orthoprojector onto the one-dimensional subspace in* l_2 *spanned by the unit vector* $g = (\xi_1, \xi_2, \cdots) \in l_2$. *Denote by* L_ξ *and* $L_\xi^\perp$ *the subspaces in* $\widetilde{L}_2(0, l)$ *consisting respectively of vectors of the form*

$$\| \xi_1\varphi(x) \quad \xi_2\varphi(x) \ \ldots \| \qquad (\varphi(x) \in L_2(0, l))$$

and vectors $\|f^{(1)}(x)\ f^{(2)}(x)\ \dots\|$ *satisfying the condition* $\sum_{j=1}^{\infty}\bar{\xi}_j f^{(j)}(x)=0$.

The subspaces L_{ξ} *and* $L_{\xi}^{\perp}$ *are mutually orthogonal and invariant relative to* $\widetilde{\mathcal{J}}_l$, *while* $L_{\xi}\oplus L_{\xi}^{\perp}=L_2(0,l)$. *The characteristic operator-functions of the projections* $(\widetilde{\Theta}_l)_{\xi}$ *and* $(\widetilde{\Theta}_l)_{\xi}^{\perp}$ *of the node* $\widetilde{\Theta}_l$ *onto* L_{ξ} *and* $L_{\xi}^{\perp}$ *are equal respectively to* $e^{2ilP_{\xi}/\lambda}$ *and* $e^{2ilP_{\xi}^{\perp}/\lambda}$ $(P_{\xi}^{\perp}=E-P_{\xi})$.

Proof. The first assertion of the lemma is verified directly.

Inasmuch as the orthoprojector Q_{ξ} onto the subspace L_{ξ} is given by the formula

$$Q_{\xi}\|h^{(1)}(x)\quad h^{(2)}(x)\ \dots\|=\|\xi_1\psi(x)\quad \xi_2\psi(x)\ \dots\|$$
$$\left(\psi(x)=\sum_{j=1}^{\infty}\bar{\xi}_j h^{(j)}(x)\right),$$

and, moreover,

$$\widetilde{K}_l g_j=l^{1/2}h_j$$
$$(g_j=(\underbrace{0,\ \dots,\ 0,\ 1}_{j},\ 0,\ \dots),\qquad h_j=\|\underbrace{0\ \dots\ 0\ l^{-1/2}}_{j}\ 0\ \dots\|),$$

by (8.12) we have

$$\left(W_{(\widetilde{\Theta}_l)_{\xi}}(\lambda)g_i,\ g_j\right)-(g_i,\ g_j)=-2il\left((\widetilde{\mathcal{J}}_l-\lambda E)^{-1}Q_{\xi}h_i,\ Q_{\xi}h_j\right)$$
$$=-2il\xi_i\bar{\xi}_j\sum_{\alpha,\beta=1}^{\infty}\left((\widetilde{\mathcal{J}}_l-\lambda E)^{-1}\bar{\xi}_{\alpha}h_{\alpha},\ \bar{\xi}_{\beta}h_{\beta}\right)=\xi_i\bar{\xi}_j\left(e^{\frac{2il}{\lambda}}-1\right).$$

On the other hand

$$(P_{\xi}g_i,\ g_j)=(g_i,\ g)(g,\ g_j)=\xi_i\bar{\xi}_j,$$

$$\left(e^{\frac{2ilP_{\xi}}{\lambda}}g_i,\ g_j\right)=\left(\left(E+\left(e^{\frac{2il}{\lambda}}-1\right)P_{\xi}\right)g_i,\ g_j\right)=(g_i,\ g_j)+\xi_i\bar{\xi}_j\left(e^{\frac{2il}{\lambda}}-1\right).$$

Thus $W_{(\widetilde{\Theta}_l)_{\xi}}(\lambda)=e^{2ilP_{\xi}/\lambda}$. Since

$$W_{\widetilde{\Theta}_l}(\lambda)=e^{\frac{2il}{\lambda}}E\quad\text{and}\quad W_{\widetilde{\Theta}_l}(\lambda)=W_{(\widetilde{\Theta}_l)_{\xi}}(\lambda)\,W_{(\widetilde{\Theta}_l)_{\xi}^{\perp}}(\lambda),$$

we have

$$W_{(\tilde{\Theta}_l)^{\perp}_{\xi}}(\lambda) = e^{\frac{2ilP^{\perp}_{\xi}}{\lambda}}.$$

3. **Unicellularity of cyclic operators.** The functions $W_1(\lambda) \in \Omega_{\mathfrak{G}}^{(\exp)}$ and $W_2(\lambda) \in \Omega_{\mathfrak{G}}^{(\exp)}$ are said to be *mutually prime* if they do not have a common divisor other than E.

Theorem 31.3. *If $W(\lambda)$ is an unordered function of the class $\Omega_{\mathfrak{G}}^{(\exp)}$ and P is an orthoprojector onto an arbitrary one-dimensional subspace in $\mathfrak{G}$, then the greatest common divisor of the functions $W(\lambda)$ and $e^{i\sigma P^{\perp}/\lambda}$ $(\sigma = \sigma[W],\ P^{\perp} = E - P)$ is distinct from E. If on the other hand $W(\lambda)$ is an ordered function, then there exists a one-dimensional orthoprojector P such that the functions $W(\lambda)$ and $e^{i\sigma P^{\perp}/\lambda}$ are mutually prime for any $\sigma > 0$.*

Proof. Suppose that $W(\lambda)$ is an unordered function of the class $\Omega_{\mathfrak{G}}^{(\exp)}$ and P is an orthoprojector onto some one-dimensional subspace in $\mathfrak{G}$. We suppose that $W(\lambda)$ and $e^{i\sigma P^{\perp}/\lambda}$ $(\sigma = \sigma[W],\ P^{\perp} = E - P)$ are mutually prime. Then, by Lemma 29.3, $\tau[W] \leq \sigma[W]$, which contradicts the inequality $\sigma[W] < \tau[W]$, which follows from Theorem 30.1.

Now suppose that $W(\lambda)$ is an ordered function and that

$$\Theta = \begin{pmatrix} A & K & E \\ \mathfrak{H} & & \mathfrak{G} \end{pmatrix}$$

is a simple node for which it is characteristic. The operators A and $B = -A^*$ are unicellular. Applying Theorem 31.2 and using the equation $B_I = A_I = KK^*$, we choose $g \in \mathfrak{G}$ so that the vector Kg is generating for B. Suppose that P is an orthoprojector onto the one-dimensional subspace spanning g, σ some positive number, and $W_0(\lambda)$ the greatest common divisor of the functions $W(\lambda)$ and $e^{i\sigma P^{\perp}/\lambda}$. If $W_0(\lambda) \neq E$, then there exists a subspace $\mathfrak{H}_0 \neq 0$, invariant relative to A, such that

$$W_0(\lambda) = W_{\Theta_0}(\lambda) \qquad (\Theta_0 = \mathrm{pr}_{\mathfrak{H}_0}\Theta).$$

At the same time $W_0(\lambda) \ll e^{i\sigma P^{\perp}/\lambda}$, and therefore $K^*Q_0K \leq (\sigma/2)P^{\perp}$, where Q_0 is the orthoprojector onto $\mathfrak{H}_0$. Accordingly

$$(K^*Q_0Kg,\ g) \leqslant \frac{\sigma}{2}(P^{\perp}g,\ g) = 0, \qquad Q_0Kg = 0.$$

In view of this last equation the vector Kg belongs to the space $\mathfrak{H} \ominus \mathfrak{H}_0$, which is invariant relative to B, and therefore it cannot be generating for B.

Theorem 31.4. *Suppose that A is a completely nonselfajoint operator of the class $\Lambda^{(\exp)}$. If the operator A is cyclic, then it is unicellular.*

Proof. Consider the universal node

$$\tilde{\Theta}_l = \begin{pmatrix} \tilde{\mathcal{J}}_l & \tilde{K}_l & E \\ \tilde{L}_2(0, l) & l_2 \end{pmatrix},$$

where $l = \frac{1}{2}\sigma[A]$. In view of Theorem 8.4 we may suppose without loss of generality that the operator A is induced by the operator $\tilde{\mathcal{J}}_l$ in some subspace $\mathfrak{H} \subset L_2(0, l)$ invariant for $\tilde{\mathcal{J}}_l$.

Suppose that $h(x) = \|h^{(1)}(x) \;\; h^{(2)}(x), \cdots\|$ is a generating vector for the operator A and L_j is the subspace in $L_2(0, l)$ which is the closure of the linear envelope of the functions

$$\mathcal{J}_l^n h^{(j)}(x) \quad \left(n = 0,\ 1,\ \ldots;\ \mathcal{J}_l h^{(j)}(x) = 2i \int_x^l h^{(j)}(t)\, dt\right).$$

Since $\mathcal{J}_l$ is a unicellular operator, by Lemma 31.1 there exists a unit vector $(\xi_1, \xi_2, \cdots)$ of the space l_2 such that the function $\psi(x) = \sum_{j=1}^{\infty} \bar{\xi}_j h^{(j)}(x)$ will be generating for the operator induced by $\mathcal{J}_l$ in the closure of the linear envelope of all subspaces L_j.

Construct in $\tilde{L}_2(0, l)$ a space $L_\xi^\perp$ invariant relative to $\tilde{\mathcal{J}}_l$, consisting of all vectors $f(x) = \|f^{(1)}(x) \;\; f^{(2)}(x) \cdots\|$ satisfying the condition $\sum_{j=1}^{\infty} \bar{\xi}_j f^{(j)}(x) = 0$. By Lemma 31.5 the characteristic operator-function of the projection of the node $\tilde{\Theta}_l$ onto $L_\xi^\perp$ is equal to $e^{i\sigma P_\xi^\perp/\lambda}$ ($\sigma = \sigma[A]$, $P_\xi^\perp = E - P_\xi$), where P_ξ is the orthoprojector onto the one-dimensional subspace containing the unit vector $(\xi_1, \xi_2, \cdots)$. Suppose that the vector $g(x) = \|g^{(1)}(x) \;\; g^{(2)}(x) \cdots\|$ lies in the intersection of the spaces $\mathfrak{H}$ and $L_\xi^\perp$. In view of Lemma 31.4 there exist functions $\phi_n(x) \in L_2(0, l)$ ($n = 1, 2, \cdots$) such that the sequence

$$\| \overleftrightarrow{h^{(1)}(x) * \varphi_n(x)} \quad \overleftrightarrow{h^{(2)}(x) * \varphi_n(x)} \ldots \|$$

converges strongly to $\overleftrightarrow{g(x)}$. In addition

$$\overleftrightarrow{\psi(x)} * \varphi_n(x) = \sum_{j=1}^{\infty} \xi_j (\overleftrightarrow{h^{(j)}(x)} * \varphi_n(x)) \to \sum_{j=1}^{\infty} \xi_j \overleftrightarrow{g^{(j)}(x)} = 0,$$

where the arrow means strong convergence in $L_2(0, l)$. Inasmuch as

$$(\overleftrightarrow{h^{(j)}(x)} * \varphi_n(x)) * \overleftrightarrow{\psi(x)} \to g^{(j)}(x) * \overleftrightarrow{\psi(x)} \qquad (j = 1, 2, \ldots)$$

and at the same time

$$(\overleftrightarrow{h^{(j)}(x)} * \varphi_n(x)) * \overleftrightarrow{\psi(x)} = \overleftrightarrow{h^{(j)}(x)} * (\overleftrightarrow{\psi(x)} * \varphi_n(x)) \to 0,$$

we have almost everywhere $\overleftrightarrow{g^{(j)}(x)} * \overleftrightarrow{\psi(x)} = 0$. It follows from Lemma 31.2 that there does not exist any segment $[0, \epsilon]$ with $\epsilon > 0$ on which the function $\overleftrightarrow{\psi(x)}$ is equal to zero almost everywhere. Using this fact and applying Theorem 20.3, we obtain the equation $g^{(j)}(x) = 0$ $(j = 1, 2, \cdots)$.

We have proved that the intersection of the subspaces $\mathfrak{H}$ and $L_\xi^\perp$ contains only the zero vector. Accordingly, $W_\Theta(\lambda)$ $(\Theta = \mathrm{pr}_{\mathfrak{H}} \tilde{\Theta}_l)$ and $e^{i\sigma P_\xi^\perp/\lambda}$ are mutually prime functions. By Theorem 31.3 $W_\Theta(\lambda)$ is an ordered function, and therefore A is a unicellular operator.

§ 32. Decomposition of nonunicellular operators into unicellular operators

1. Nonintersecting invariant subspaces of a nonunicellular operator.

Lemma 32.1. *Suppose that A is a completely nonselfadjoint operator of the class $\Lambda^{(\exp)}$. If A is nonunicellular, then there exists a subspace $\mathfrak{H}_1$ invariant relative to A and satisfying the following conditions*: 1) *the operator A_1 induced in $\mathfrak{H}_1$ is unicellular*; 2) $\sigma[A_1] = \sigma[A]$.

Proof. In view of Theorem 8.4, we may without loss of generality suppose that the operator A is induced by the universal operator $\tilde{\mathfrak{J}}_l$ $(l = \frac{1}{2}\sigma[A])$ in a subspace $\mathfrak{H}$ of $\tilde{L}_2(0, l)$ invariant with respect to that universal operator.

By Lemma 31.3 there exists a vector $f(x) \in \mathfrak{H}$ such that $2l_f = \sigma[A]$. Denote by $\mathfrak{H}_1$ the closure of the linear envelope of the sequence $\tilde{\mathfrak{J}}_l^n f(x)$ $(n = 0, 1, \cdots)$. By Theorem 31.4 the operator A_1 induced in $\mathfrak{H}_1$ is unicellular. Applying Lemma 31.2, we obtain the equation $\sigma[A_1] = 2l_f$.

Lemma 32.2. *Suppose that A is a completely nonselfadjoint nonunicellular operator of the class $\Lambda^{(\exp)}$ operating in a space $\mathfrak{H}$, and that $\mathfrak{H}_1$ is a subspace invariant relative to A in which there is induced a unicellular operator A_1 having the property $\sigma[A_1] = \sigma[A]$. Then there exists a subspace $\mathfrak{H}_2$ invariant relative to A such that $\mathfrak{H}_1 \cap \mathfrak{H}_2 = 0$, $\mathfrak{H}_1 \tilde{\cup} \mathfrak{H}_2 = \mathfrak{H}$.*

In particular, if $\dim\{A_I\mathfrak{H}\} = n < \infty$, then for an appropriate choice of the subspace $\mathfrak{H}_2$ the inequality $\dim\{(A_2)_I\mathfrak{H}\} \le n - 1$, will be satisfied, where A_2 is the operator induced in $\mathfrak{H}_2$.

Proof. We imbed A in the node

$$\Theta = \begin{pmatrix} A & K & E \\ \mathfrak{H} & & \mathfrak{G} \end{pmatrix}$$

and put

$$W(\lambda) = W_\Theta(\lambda), \quad W_1(\lambda) = W_{\Theta_1}(\lambda), \quad (\Theta_1 = \mathrm{pr}_{\mathfrak{H}_1}\Theta).$$

By Theorem 8.3 $\sigma[W] = \sigma[A]$, $\sigma[W_1] = \sigma[A_1]$. In view of Theorem 31.3 there exists an orthoprojector P onto a one-dimensional subspace in $\mathfrak{G}$ such that $W_1(\lambda)$ and $e^{i\sigma P^\perp/\lambda}$ $(\sigma = \sigma[W],\ P^\perp = E - P)$ are mutually prime. By this same theorem the greatest common divisor $W_2(\lambda)$ of the functions $W(\lambda)$ and $e^{i\sigma P^\perp/\lambda}$ is distinct from E. Obviously the functions $W_1(\lambda)$ and $W_2(\lambda)$ are mutually prime. Accordingly, $\mathfrak{H}_1 \cap \mathfrak{H}_2 = 0$, where $\mathfrak{H}_2$ is a subspace invariant relative to A satisfying the condition

$$W_2(\lambda) = W_{\Theta_2}(\lambda) \quad (\Theta_2 = \mathrm{pr}_{\mathfrak{H}_2}\Theta).$$

We denote by P_3 the orthoprojector onto $\mathfrak{H}_3 = \mathfrak{H} \ominus \mathfrak{H}_2$ and consider the operator $A_3h = P_3Ah$ operating on elements h of $\mathfrak{H}_3$. Since

$$W(\lambda) = W_2(\lambda)\, W_3(\lambda), \quad W_3(\lambda) = W_{\Theta_3}(\lambda) \quad (\Theta_3 = \mathrm{pr}_{\mathfrak{H}_3}(\Theta),$$

by Lemma 29.3 $\tau[W_3] \le \sigma[W]$. At the same time $\sigma[W] = \sigma[W_1] = \tau[W_1]$, and therefore

$$\mathrm{tr}\,(A_3)_I \leqslant \mathrm{tr}\,(A_1)_I. \tag{32.1}$$

We introduce the operator A_3' induced by the operator A_3 in the subspace

$\mathfrak{H}_3' = (\mathfrak{H}_1 \tilde{\cup} \mathfrak{H}_2) \ominus \mathfrak{H}_2 \subseteq \mathfrak{H}_3$ which is invariant relative to it. From the last assertion of Lemma 28.3 it follows that $\operatorname{tr}(A_1)_I = \operatorname{tr}(A_3')_I$. Comparing this result with inequality (32.1) and recalling that A_3 is a completely nonselfadjoint operator, we arrive at the relation $\mathfrak{H}_1 \tilde{\cup} \mathfrak{H}_2 = \mathfrak{H}$.

In the case when $\dim\{A_I\mathfrak{H}\} = n < \infty$, the node Θ can be constructed in such a way that the space $\mathfrak{G}$ is n-dimensional. Inasmuch as $W_2(\lambda) \preccurlyeq e^{i\sigma P^\perp/\lambda}$, we have $2K^*P_2K \leq \sigma P^\perp$, where P_2 is the orthoprojector onto $\mathfrak{H}_2$. Therefore $P_2KP = 0$, and therefore

$$\dim\{(A_2)_I\,\mathfrak{H}\} = \dim\{P_2K\mathfrak{G}\} \leqslant n-1.$$

The following theorem is a consequence of Lemmas 32.1 and 32.2.

Theorem 32.1. *If A is a completely nonselfadjoint operator of the class $\Lambda^{(\exp)}$, operating in the space $\mathfrak{H}$, then in $\mathfrak{H}$ there exist subspaces $\mathfrak{H}_1$ and $\mathfrak{H}_2$ having the following properties*: 1) *they are both invariant relative to A*; 2) $\mathfrak{H}_1 \cap \mathfrak{H}_2 = 0$; 3) $\mathfrak{H}_1 \tilde{\cup} \mathfrak{H}_2 = \mathfrak{H}$; 4) *the operator A_1 induced in $\mathfrak{H}_1$ is unicellular*; 5) $\sigma[A_1] = \sigma[A]$.

2. Direct, quasidirect, and approximate sums of subspaces. We recall some definitions. If $\mathfrak{H}_1$ and $\mathfrak{H}_2$ are subspaces of Hilbert space $\mathfrak{H}$ and 1) $\mathfrak{H}_1 \tilde{\cup} \mathfrak{H}_2 = \mathfrak{H}$, 2) $\mathfrak{H}_1 \cap \mathfrak{H}_2 = 0$, then $\mathfrak{H}$ is said to be the *approximate sum* of the subspaces $\mathfrak{H}_1$ and $\mathfrak{H}_2$. An approximate sum is said to be *direct* if every vector $h \in \mathfrak{H}$ can be represented in the form $h = h_1 + h_2$ $(h_j \in \mathfrak{H}_j)$.

Suppose that $\mathfrak{H} = \mathfrak{H}_1 \oplus \mathfrak{H}_1^\perp$ and T is a bounded linear one-to-one mapping of the entire space $\mathfrak{H}_1$ into a regular part of the space $\mathfrak{H}_1^\perp$. We denote by $\mathfrak{H}_2$ the subspace consisting of all vectors of the form $f + Tf$ $(f \in \mathfrak{H}_1)$. It is easy to see that in the case when the range of the operator T is dense in $\mathfrak{H}_1^\perp$, $\mathfrak{H}$ is an approximate but not a direct sum of the subspaces $\mathfrak{H}_1$ and $\mathfrak{H}_2$.

The concepts of approximate and direct sum generalize in a natural way. Indeed, we will say that $\mathfrak{H}$ is the approximate sum of its subspaces $\mathfrak{H}_\gamma$ $(\gamma \in \Gamma)$, if

$$1)\ \tilde{\bigcup_{\gamma\in\Gamma}} \mathfrak{H}_\gamma = \mathfrak{H}, \qquad 2)\ \Big(\tilde{\bigcup_{\gamma\in\Gamma_1}} \mathfrak{H}_\gamma\Big) \cap \Big(\tilde{\bigcup_{\gamma\in\Gamma_2}} \mathfrak{H}_\gamma\Big) = 0$$

for arbitrary Γ_1, Γ_2 satisfying the conditions $\Gamma_1 \cup \Gamma_2 = \Gamma$, $\Gamma_1 \cap \Gamma_2 = 0$. Here by the symbol $\widetilde{\mathbf{U}}_{\gamma\in\Gamma}\mathfrak{H}_\gamma$ we mean the closure of the linear envelope of all the subspaces $\mathfrak{H}_\gamma$ $(\gamma\in\Gamma)$, and we are not excluding the possibility that some $\mathfrak{H}_\gamma = 0$. If in addition every vector $h\in\mathfrak{H}$ is representable in the form

$$h=h_1+h_2 \quad \left(h_j\in \widetilde{\bigcup_{\gamma\in\Gamma_j}}\mathfrak{H}_\gamma, \quad j=1,\ 2\right),$$

then $\mathfrak{H}$ will be called the direct sum of the subspaces $\mathfrak{H}_\gamma$. W

We agree to call the space $\mathfrak{H}$ a *quasidirect sum* of the subspaces $\mathfrak{H}_\gamma$ $(\gamma\in\Gamma)$ if $\widetilde{\mathbf{U}}_{\gamma\in\Gamma}\mathfrak{H}_\gamma=\mathfrak{H}$ and

$$\bigcap_{j=1}^{\omega}\widetilde{\bigcup_{\gamma\in\Gamma_j}}\mathfrak{H}_\gamma=\widetilde{\bigcup_{\gamma\in\bigcap_{j=1}^{\omega}\Gamma_j}}\mathfrak{H}_\gamma \tag{32.2}$$

for any sequence $\Gamma_j\subset\Gamma$ $(j=1,2,\cdots,\omega;\ \omega\le\infty)$.

Obviously the direct sum of subspaces is at the same time a quasidirect sum, and a quasidirect sum is an approximate sum. In the case when Γ consists of only two elements, the distinction between the concepts of quasidirect and approximate sum disappears. Therefore the example given above serves as a proof of the existence of a quasidirect sum which is not a direct sum. The following example shows that a sum of subspaces can be approximate but not quasidirect.

Suppose that $\mathfrak{H}=\mathfrak{H}_0\oplus\mathfrak{G}_1\oplus\mathfrak{G}_0\oplus\mathfrak{G}_2$, where $\mathfrak{H}_0$, $\mathfrak{G}_1$ and $\mathfrak{G}_2$ are infinite-dimensional spaces of the same dimension, and $\mathfrak{G}_0$ is one-dimensional. Consider a bounded linear operator T_1 having the following properties: 1) T_1 is defined in the entire space $\mathfrak{H}_0$; 2) its range is dense in $\mathfrak{G}_1\oplus\mathfrak{G}_0$ and does not contain the subspace $\mathfrak{G}_0$; 3) if $T_1h=0$, then $h=0$. We denote by U a unitary operator in $\mathfrak{G}_1\oplus\mathfrak{G}_0\oplus\mathfrak{G}_2$ such that: 1) $U\mathfrak{G}_1=\mathfrak{G}_2$, $U\mathfrak{G}_0=\mathfrak{G}_0$, $U\mathfrak{G}_2=\mathfrak{G}_1$; 2) the unit element is a regular point of the operator U. It is clear that the range of the operator $T_2=UT_1$ does not contain $\mathfrak{G}_0$ and is dense in $\mathfrak{G}_0\oplus\mathfrak{G}_2$.

Suppose that $\mathfrak{H}_j$ $(j=1,2)$ is a collection of vectors of the form $h+T_jh$ $(h\in\mathfrak{H}_0)$. It is not hard to see that $\mathfrak{H}$ coincides with the closure of the linear envelope of the subspaces $\mathfrak{H}_0$, $\mathfrak{H}_1$ and $\mathfrak{H}_2$.

We will show that $(\mathfrak{H}_0\,\widetilde{\cup}\,\mathfrak{H}_1)\cap\mathfrak{H}_2=0$. Assuming the contrary, we find a nonzero vector $h\in\mathfrak{H}_0$ and sequences $\{h_j\}\subset\mathfrak{H}_0$ and $\{h_j'\}\subset\mathfrak{H}_0$ such that $h_j+h_j'+T_1h_j'\to h+T_2h$. But then $T_1h_j'\to T_2h$, which means that $T_2h\in\mathfrak{G}_0$, which is impossible. Analogously one proves the equation $(\mathfrak{H}_0 \tilde{\cup} \mathfrak{H}_2) \cap \mathfrak{H}_1 = 0$.

Now we verify the fact that $(\mathfrak{H}_1 \tilde{\cup} \mathfrak{H}_2) \cap \mathfrak{H}_0 = 0$. Suppose that the sequences $\{h_j^{(1)}\} \subset \mathfrak{H}_0$ and $\{h_j^{(2)}\} \subset \mathfrak{H}_0$ are such that

$$h_j^{(1)} + T_1 h_j^{(1)} + h_j^{(2)} + T_2 h_j^{(2)} \to h_0 \qquad (h_0 \in \mathfrak{H}_0).$$

Then

$$h_j^{(1)} + h_j^{(2)} \to h_0, \qquad T_1 h_j^{(1)} + T_2 h_j^{(2)} \to 0,$$
$$(U - E)\, T_1 h_j^{(1)} = \left(T_2 h_j^{(1)} + T_2 h_j^{(2)}\right) - \left(T_1 h_j^{(1)} + T_2 h_j^{(2)}\right) \to T_2 h_0,$$

so that the sequence $\{T_1 h_j^{(1)}\}$ has a limit. If $T_1 h_j^{(1)} \to g_0$, then $T_2 h_j^{(2)} \to -g_0$, and therefore

$$g_0 \in \mathfrak{G}_0, \quad T_2 h_j^{(1)} + T_2 h_j^{(2)} \to U g_0 - g_0 \ (\in \mathfrak{G}_0).$$

On the other hand, $T_2 h_j^{(1)} + T_2 h_j^{(2)} \to T_2 h_0$, so that $T_2 h_0 \in \mathfrak{G}_0$, $h_0 = 0$.

We have proved that the space $\mathfrak{H}$ is the approximate sum of the spaces $\mathfrak{H}_0$, $\mathfrak{H}_1$ and $\mathfrak{H}_2$. At the same time it is not their quasidirect sum, inasmuch as

$$\mathfrak{H}_0 \tilde{\cup} \mathfrak{H}_1 = \mathfrak{H}_0 \oplus \mathfrak{G}_1 \oplus \mathfrak{G}_0, \quad \mathfrak{H}_0 \tilde{\cup} \mathfrak{H}_2 = \mathfrak{H}_0 \oplus \mathfrak{G}_0 \oplus \mathfrak{G}_2,$$
$$(\mathfrak{H}_0 \tilde{\cup} \mathfrak{H}_1) \cap (\mathfrak{H}_0 \tilde{\cup} \mathfrak{H}_2) = \mathfrak{H}_0 \oplus \mathfrak{G}_0.$$

Theorem 32.2. *If* $\mathfrak{H}$ *is the quasidirect sum of the subspaces* $\mathfrak{H}_j$ $(j = 1, 2, \dots)$, *and* $\mathfrak{G}_j = \mathfrak{H} \ominus \tilde{\bigcup}_{i \neq j} \mathfrak{H}_i$, *then* $\mathfrak{H}_j = \mathfrak{H} \ominus \tilde{\bigcup}_{i \neq j} \mathfrak{G}_i$ *and* $\mathfrak{H}$ *is the quasidirect sum of the subspaces* $\mathfrak{G}_j$ $(j = 1, 2, \dots)$.

Proof. By the hypothesis of the theorem

$$\bigcap_{i \neq j} \tilde{\bigcup_{k \neq i}} \mathfrak{H}_k = \mathfrak{H}_j.$$

Moreover, obviously

$$\mathfrak{H} \ominus \tilde{\bigcup_{i \neq j}} \mathfrak{G}_i = \bigcap_{i \neq j} (\mathfrak{H} \ominus \mathfrak{G}_i) = \bigcap_{i \neq j} \tilde{\bigcup_{k \neq i}} \mathfrak{H}_k.$$

Therefore $\mathfrak{H}_j = \mathfrak{H} \ominus \tilde{\bigcup}_{i \neq j} \mathfrak{G}_i$.

Suppose that N_k $(k = 1, 2, \cdots, \omega;\ \omega \leq \infty)$ is a subset of the set of natural numbers. Since

$$\tilde{\bigcup_{j \in N_k}} \mathfrak{G}_j = \mathfrak{H} \ominus \bigcap_{j \in N_k} (\mathfrak{H} \ominus \mathfrak{G}_j) = \mathfrak{H} \ominus \bigcap_{j \in N_k} \tilde{\bigcup_{i \neq j}} \mathfrak{H}_i = \mathfrak{H} \ominus \tilde{\bigcup_{j \notin N_k}} \mathfrak{H}_j,$$

$$\bigcap_{k=1}^{\omega} \tilde{\bigcup_{j \in N_k}} \mathfrak{G}_j = \mathfrak{H} \ominus \tilde{\bigcup_{k=1}^{\omega}} \left(\mathfrak{H} \ominus \tilde{\bigcup_{j \in N_k}} \mathfrak{G}_j \right) = \mathfrak{H} \ominus \tilde{\bigcup_{k=1}^{\omega}} \tilde{\bigcup_{j \notin N_k}} \mathfrak{H}_j = \mathfrak{H} \ominus \tilde{\bigcup_{j \notin N}} \mathfrak{H}_j$$

$$\left(N = \bigcap_{k=1}^{\omega} N_k \right)$$

and at the same time

$$\tilde{\bigcup_{j \in N}} \mathfrak{G}_j = \mathfrak{H} \ominus \bigcap_{j \in N} (\mathfrak{H} - \mathfrak{G}_j) = \mathfrak{H} \ominus \bigcap_{j \in N} \tilde{\bigcup_{i \neq j}} \mathfrak{H}_i = \mathfrak{H} \ominus \tilde{\bigcup_{j \notin N}} \mathfrak{H}_j,$$

it follows that

$$\bigcap_{k=1}^{\omega} \tilde{\bigcup_{j \in N_k}} \mathfrak{G}_j = \tilde{\bigcup_{j \in N}} \mathfrak{G}_j.$$

The theorem is proved.

Consider a family $\mathfrak{S}$ of subspaces of Hilbert space $\mathfrak{H}$ satisfying the following condition: if $\mathfrak{H}_\gamma \in \mathfrak{S}$ $(\gamma \in \Gamma)$, then $\tilde{\bigcup}_{\gamma \in \Gamma} \mathfrak{H}_\gamma \in \mathfrak{S}$ and $\bigcap_{\gamma \in \Gamma} \mathfrak{H}_\gamma \in \mathfrak{S}$. The family $\mathfrak{S}$ will be said to be *weighted* if there is given a numerical function τ on $\mathfrak{S}$ such that:

a) $\tau(\mathfrak{H}_0) \geq 0$ for any $\mathfrak{H}_0 \in \mathfrak{S}$, the equality sign holding if and only if $\mathfrak{H}_0 = 0$;

b) if $\mathfrak{H}_1 \subset \mathfrak{H}_2$ $(\mathfrak{H}_1, \mathfrak{H}_2 \in \mathfrak{S}, \mathfrak{H}_1 \neq \mathfrak{H}_2)$, then $\tau(\mathfrak{H}_1) < \tau(\mathfrak{H}_2)$;

c) for any two subspaces $\mathfrak{H}_1, \mathfrak{H}_2 \in \mathfrak{S}$ the equation

$$\tau(\mathfrak{H}_1) + \tau(\mathfrak{H}_2) = \tau(\mathfrak{H}_1 \tilde{\cup} \mathfrak{H}_2) + \tau(\mathfrak{H}_1 \cap \mathfrak{H}_2)$$

holds;

d) if $\mathfrak{H}_j \in \mathfrak{S}$ $(j = 1, 2, \cdots)$ and $\mathfrak{H}_j \subset \mathfrak{H}_{j+1}$, then

$$\tau\left(\tilde{\bigcup_{j=1}^{\infty}} \mathfrak{H}_j \right) = \lim_{j \to \infty} \tau(\mathfrak{H}_j);$$

e) if $\mathfrak{H}_j \in \mathfrak{S}$ $(j = 1, 2, \dots)$ and $\mathfrak{H}_{j+1} \subset \mathfrak{H}_j$, then

$$\tau\left(\bigcap_{j=1}^{\infty} \mathfrak{H}_j\right) = \lim_{j \to \infty} \tau(\mathfrak{H}_j).$$

Suppose that A is a completely nonselfadjoint operator of the class $\Lambda_0^{(\exp)}$. We assign to each subspace $\mathfrak{H}_0$ invariant relative to A the number $\tau(\mathfrak{H}_0) = 2\,\mathrm{tr}\,(A_0)_I$, where A_0 is the operator induced in $\mathfrak{H}_0$. In view of formulas (28.8) and (28.4) the collection of all subspaces invariant relative to the operator A is a weighted family.

Theorem 32.3. *Suppose that $\mathfrak{S}$ is a weighted family of subspaces of Hilbert space $\mathfrak{H}$. If $\mathfrak{H}_j \in \mathfrak{S}$ $(j = 1, 2, \dots)$ and*

$$\mathfrak{H}_k \cap \left(\tilde{\bigcup_{j>k}} \mathfrak{H}_j\right) = 0 \qquad (k = 1,\ 2,\ \dots), \tag{32.3}$$

then $\tilde{\mathfrak{H}} = \tilde{\bigcup}_{j=1}^{\infty} \mathfrak{H}_j$ is the quasidirect sum of the subspaces $\mathfrak{H}_j$ and

$$\tau(\tilde{\mathfrak{H}}) = \sum_{j=1}^{\infty} \tau(\mathfrak{H}_j). \tag{32.4}$$

Proof. Applying (32.3) and properties c), a) and d) to any subsequence $\{\mathfrak{H}^{(j)}\}$ of the sequence $\{\mathfrak{H}_j\}$, we get

$$\tau\left(\tilde{\bigcup_{j=1}^{n}} \mathfrak{H}^{(j)}\right) = \tau(\mathfrak{H}^{(1)}) + \tau\left(\tilde{\bigcup_{j=2}^{n}} \mathfrak{H}^{(j)}\right) = \tau(\mathfrak{H}^{(1)}) + \tau(\mathfrak{H}^{(2)})$$
$$+ \tau\left(\tilde{\bigcup_{j=3}^{n}} \mathfrak{H}^{(j)}\right) = \dots = \sum_{j=1}^{n} \tau(\mathfrak{H}^{(j)}),$$
$$\tau\left(\tilde{\bigcup_{j=1}^{\infty}} \mathfrak{H}^{(j)}\right) = \lim_{n \to \infty} \tau\left(\tilde{\bigcup_{j=1}^{n}} \mathfrak{H}^{(j)}\right) = \sum_{j=1}^{\infty} \tau(\mathfrak{H}^{(j)}).$$

Thus we have proved in particular the validity of equation (32.4).

Denote by I the set of all natural numbers. For arbitrary $\Gamma_1, \Gamma_2 \subset I$

$$\left(\tilde{\bigcup_{j\in\Gamma_1}}\mathfrak{H}_j\right)\cap\left(\tilde{\bigcup_{j\in\Gamma_2}}\mathfrak{H}_j\right)\supseteq\tilde{\bigcup_{j\in\Gamma_1\cap\Gamma_2}}\mathfrak{H}_j,$$

$$\tau\left(\left(\tilde{\bigcup_{j\in\Gamma_1}}\mathfrak{H}_j\right)\cap\left(\tilde{\bigcup_{j\in\Gamma_2}}\mathfrak{H}_j\right)\right)=\tau\left(\tilde{\bigcup_{j\in\Gamma_1}}\mathfrak{H}_j\right)+\tau\left(\tilde{\bigcup_{j\in\Gamma_2}}\mathfrak{H}_j\right)-\tau\left(\tilde{\bigcup_{j\in\Gamma_1\cup\Gamma_2}}\mathfrak{H}_j\right)$$

$$=\sum_{j\in\Gamma_1}\tau(\mathfrak{H}_j)+\sum_{j\in\Gamma_2}\tau(\mathfrak{H}_j)-\sum_{j\in\Gamma_1\cup\Gamma_2}\tau(\mathfrak{H}_j)=\sum_{j\in\Gamma_1\cap\Gamma_2}\tau(\mathfrak{H}_j)=\tau\left(\tilde{\bigcup_{j\in\Gamma_1\cap\Gamma_2}}\mathfrak{H}_j\right).$$

In view of property b)

$$\left(\tilde{\bigcup_{j\in\Gamma_1}}\mathfrak{H}_j\right)\cap\left(\tilde{\bigcup_{j\in\Gamma_2}}\mathfrak{H}_j\right)=\tilde{\bigcup_{j\in\Gamma_1\cap\Gamma_2}}\mathfrak{H}_j. \tag{32.5}$$

Suppose that $N_k\subset I$ $(k=1,2,\cdots)$. It follows easily by induction from (32.5) that

$$\bigcap_{k=1}^{n}\tilde{\bigcup_{j\in N_k}}\mathfrak{H}_j=\tilde{\bigcup_{j\in\bigcap_{k=1}^{n}N_k}}\mathfrak{H}_j\quad(n=1,2,\ldots).$$

Put $M_k=I-N_k$. Inasmuch as

$$\bigcap_{k=1}^{\infty}\tilde{\bigcup_{j\in N_k}}\mathfrak{H}_j\supseteq\tilde{\bigcup_{j\in\bigcap_{k=1}^{\infty}N_k}}\mathfrak{H}_j$$

and

$$\tau\left(\bigcap_{k=1}^{\infty}\tilde{\bigcup_{j\in N_k}}\mathfrak{H}_j\right)=\lim_{n\to\infty}\tau\left(\bigcap_{k=1}^{n}\tilde{\bigcup_{j\in N_k}}\mathfrak{H}_j\right)=\lim_{n\to\infty}\tau\left(\tilde{\bigcup_{j\in\bigcap_{k=1}^{n}N_k}}\mathfrak{H}_j\right)$$

$$=\lim_{n\to\infty}\tau\left(\tilde{\bigcup_{j\in I-\bigcup_{k=1}^{n}M_k}}\mathfrak{H}_j\right)=\tau(\mathfrak{H})-\lim_{n\to\infty}\tau\left(\tilde{\bigcup_{j\in\bigcup_{k=1}^{n}M_k}}\mathfrak{H}_j\right)$$

$$=\tau(\mathfrak{H})-\tau\left(\tilde{\bigcup_{j\in\bigcup_{k=1}^{\infty}M_k}}\mathfrak{H}_j\right)=\tau\left(\tilde{\bigcup_{j\in I-\bigcup_{k=1}^{\infty}M_k}}\mathfrak{H}_j\right)=\tau\left(\tilde{\bigcup_{j\in\bigcap_{k=1}^{\infty}N_k}}\mathfrak{H}_j\right),$$

we have

$$\bigcap_{k=1}^{\infty} \tilde{\bigcup_{j \in N_k}} \mathfrak{H}_j = \tilde{\bigcup_{j \in \bigcap_{k=1}^{\infty} N_k}} \mathfrak{H}_j.$$

Corollary. *For the subspaces lying in a weighted family* $\mathfrak{S}$ *the concepts of approximate and quasidirect sum coincide.*

Theorem 32.4. *If the* $\mathfrak{H}_j$ $(j = 1, 2, \cdots)$ *lie in a weighted family* $\mathfrak{S}$ *and*

$$\tau(\tilde{\mathfrak{H}}) = \sum_{j=1}^{\infty} \tau(\mathfrak{H}_j) \quad \left(\tilde{\mathfrak{H}} = \tilde{\bigcup_{j=1}^{\infty}} \mathfrak{H}_j\right),$$

then $\tilde{\mathfrak{H}}$ *is the quasidirect sum of the subspaces* $\mathfrak{H}_j$.

Proof. Using properties c) and d) of a weighted family, we get

$$\tau\left(\tilde{\bigcup_{j=k}^{k+n}} \mathfrak{H}_j\right) = \tau\left(\mathfrak{H}_k \,\tilde{\cup}\left(\tilde{\bigcup_{j=k+1}^{k+n}} \mathfrak{H}_j\right)\right) \leqslant \tau(\mathfrak{H}_k) + \tau\left(\tilde{\bigcup_{j=k+1}^{k+n}} \mathfrak{H}_j\right) \cdots \leqslant \sum_{j=k}^{k+n} \tau(\mathfrak{H}_j),$$

$$\tau\left(\tilde{\bigcup_{j=k}^{\infty}} \mathfrak{H}_j\right) = \lim_{n\to\infty} \tau\left(\tilde{\bigcup_{j=k}^{k+n}} \mathfrak{H}_j\right) \leqslant \sum_{j=k}^{\infty} \tau(\mathfrak{H}_j).$$

Accordingly,

$$\tau(\tilde{\mathfrak{H}}) = \tau\left(\left(\tilde{\bigcup_{j=1}^{k-1}} \mathfrak{H}_j\right) \tilde{\cup} \left(\tilde{\bigcup_{j=k}^{\infty}} \mathfrak{H}_j\right)\right) \leqslant \tau\left(\tilde{\bigcup_{j=1}^{k-1}} \mathfrak{H}_j\right) + \tau\left(\tilde{\bigcup_{j=k}^{\infty}} \mathfrak{H}_j\right)$$
$$\leqslant \sum_{j=1}^{k-1} \tau(\mathfrak{H}_j) + \sum_{j=k}^{\infty} \tau(\mathfrak{H}_j) = \tau(\tilde{\mathfrak{H}}),$$

so that

$$\tau\left(\tilde{\bigcup_{j=1}^{k-1}} \mathfrak{H}_j\right) + \tau\left(\tilde{\bigcup_{j=k}^{\infty}} \mathfrak{H}_j\right) = \sum_{j=1}^{k-1} \tau(\mathfrak{H}_j) + \sum_{j=k}^{\infty} \tau(\mathfrak{H}_j).$$

Inasmuch as each term in the left side of the last equation is not larger than the corresponding term in the right side,

$$\tau\left(\tilde{\bigcup_{j=k}^{\infty}} \mathfrak{H}_j\right) = \sum_{j=k}^{\infty} \tau(\mathfrak{H}_j) \quad (k = 1, 2, \ldots),$$

$$\tau\left(\mathfrak{H}_k \cap \left(\tilde{\bigcup_{j=k+1}^{\infty}} \mathfrak{H}_j\right)\right) = \tau(\mathfrak{H}_k) + \tau\left(\tilde{\bigcup_{j=k+1}^{\infty}} \mathfrak{H}_j\right) - \tau\left(\tilde{\bigcup_{j=k}^{\infty}} \mathfrak{H}_j\right)$$

$$= \tau(\mathfrak{H}_k) + \sum_{j=k+1}^{\infty} \tau(\mathfrak{H}_j) - \sum_{j=k}^{\infty} \tau(\mathfrak{H}_j) = 0.$$

Thus $\mathfrak{H}_k \cap (\tilde{\bigcup}_{j=k+1}^{\infty} \mathfrak{H}_j) = 0$ $(k = 1, 2, \cdots)$, and it remains only to apply Theorem 32.3.

3. **Decomposition theorem.**

Theorem 32.5. *If A is nonunicellular completely nonselfadjoint operator of the class $\Lambda_0^{(\exp)}$, then the space $\mathfrak{H}$ in which it is given is the quasidirect sum of subspaces $\mathfrak{H}_j$ $(j = 1, 2, \cdots, \omega;\ \omega \leq \infty)$ such that the operators A_j $(j = 1, 2, \cdots, \omega)$ induced in them are unicellular. In addition one has the relations*

$$\operatorname{tr}(A_1)_I \geqslant \operatorname{tr}(A_2)_I \geqslant \ldots; \qquad \operatorname{tr} A_I = \sum_{j=1}^{\omega} \operatorname{tr}(A_j)_I.$$

Proof. According to Theorem 32.1 there exist subspaces $\mathfrak{H}_1$ and $\mathfrak{H}^{(1)}$ invariant relative to A and satisfying the following conditions: a_1) $\mathfrak{H}_1 \cap \mathfrak{H}^{(1)} = 0$; b_1) $\mathfrak{H}_1 \tilde{\cup} \mathfrak{H}^{(1)} = \mathfrak{H}$; c_1) the operator A_1 induced in $\mathfrak{H}_1$ is unicellular; d_1) $\sigma[A_1] = \sigma[A]$. If the operator $A^{(1)}$ induced in $\mathfrak{H}^{(1)}$ is nonunicellular, by again applying Theorem 32.1 we can find subspaces $\mathfrak{H}_2 \subset \mathfrak{H}^{(1)}$ and $\mathfrak{H}^{(2)} \subset \mathfrak{H}^{(1)}$ such that a_2) $\mathfrak{H}_2 \cap \mathfrak{H}^{(2)} = 0$; b_2) $\mathfrak{H}_2 \tilde{\cup} \mathfrak{H}^{(2)} = \mathfrak{H}^{(1)}$; c_2) the operator A_2 induced in $\mathfrak{H}_2$ is unicellular; d_2) $\sigma[A_2] = \sigma[A^{(1)}]$. We continue with this process and suppose that the operators $A^{(j)}$ $(j = 1, 2, \cdots, n-1)$ are nonunicellular, and $A^{(n)}$ is unicellular. In view of Theorem 32.3, $\mathfrak{H}$ is the quasidirect sum of the subspaces $\mathfrak{H}_1, \mathfrak{H}_2, \cdots, \mathfrak{H}_n, \mathfrak{H}^{(n)}$, while

$$\operatorname{tr} A_I = \sum_{j=1}^{n} \operatorname{tr}(A_j)_I + \operatorname{tr}\left(A^{(n)}\right)_I. \tag{32.6}$$

Moreover, in view of Theorem 30.3

$$2 \operatorname{tr}(A_j)_I = \sigma[A_j] = \sigma\left[A^{(j-1)}\right] \geqslant \sigma[A_{j+1}] = 2 \operatorname{tr}(A_{j+1})_I.$$

In the case when for any n the operator $A^{(n)}$ is nonunicellular, the subspace

$\tilde{\mathfrak{H}} = \tilde{\mathbf{U}}_{j=1}^{\infty}\mathfrak{H}_j$ is the quasidirect sum of the subspaces $\mathfrak{H}_j$ $(j = 1, 2, \cdots)$, and

$$\operatorname{tr} \tilde{A}_I = \sum_{j=1}^{\infty} \operatorname{tr} (A_j)_I, \tag{32.7}$$

where $\tilde{A}$ is the operator induced in $\tilde{\mathfrak{H}}$.

Since

$$\sigma\left[A^{(j)}\right] \leqslant \sigma\left[A^{(j-1)}\right] = 2 \operatorname{tr} (A_j)_I,$$

we see that $\lim_{j\to\infty} \sigma[A^{(j)}] = 0$. Applying Theorem 29.6′, the corollary to Theorem 8.3 and relation (32.6) we obtain

$$\bigcap_{n=1}^{\infty} \mathfrak{H}^{(n)} = 0, \quad \lim_{n\to\infty} \operatorname{tr} (A^{(n)})_I = 0,$$

$$\operatorname{tr} A_I = \sum_{j=1}^{\infty} \operatorname{tr} (A_j)_I. \tag{32.8}$$

Comparison of formulas (32.7) and (32.8) leads to the equation $\tilde{\mathfrak{H}} = \mathfrak{H}$.

The theorem is proved. In the special case when $\dim\{A_I\mathfrak{H}\} < \infty$, it admits the following sharpened statement.

Theorem 32.6. *Suppose that A is a nonunicellular completely nonselfadjoint operator of the class $\Lambda_0^{(\exp)}$ operating in the space $\mathfrak{H}$. If $\dim\{A_I\mathfrak{H}\} = n < \infty$, then $\mathfrak{H}$ may be represented as the quasidirect sum of subspaces $\mathfrak{H}_j$ $(j = 1, 2, \cdots, k;$ $k \leq n)$ invariant with respect to A and such that the operators A_j induced in them are unicellular and the dimensions r_j of the subspaces $(A_j)_I\mathfrak{H}_j$ satisfy the relations $r_j \leq n - j + 1$ $(j = 1, 2, \cdots, k)$.*

The proof is easily obtained by selecting a sequence of subspaces in which unicellular operators are induced, as in the proof of the preceding theorem, and applying at each step the last assertion of Lemma 32.2.

Suppose given in $\mathfrak{H}$ a completely nonselfadjoint operator $A \in \Lambda_0^{(\exp)}$. We will say that the nonzero subspaces $\mathfrak{H}_j \subset \mathfrak{H}$ $(j = 1, 2, \cdots, \omega;\ \omega \leq \infty)$ *reduce A to Jordan form,* if: 1) $\mathfrak{H}$ is a quasidirect sum of the subspaces $\mathfrak{H}_j$ $(j = 1, 2, \cdots, \omega)$; 2) all the $\mathfrak{H}_j$ are invariant relative to A; 3) in each of the subspaces $\mathfrak{H}_j$ a unicellular operator is induced.

Theorem 32.7. *If the subspaces $\mathfrak{H}_j$ $(j = 1, 2, \cdots, \omega;\ \omega \leq \infty)$ reduce the* *operator A to Jordan form, then the subspaces* $\mathfrak{G}_j = \mathfrak{H} \ominus \tilde{\bigcup}_{i \neq j} \mathfrak{H}_i$ $(j = 1, 2, \cdots, \omega)$ *reduce the operator* $B = -A^*$ *to Jordan form.*

Proof. By Theorem 32.2 $\mathfrak{H}$ is the quasidirect sum of the subspaces $\mathfrak{G}_j$ $(j = 1, 2, \cdots, \omega)$. It is easy to see that all the $\mathfrak{G}_j$ are nonzero. Since the $\mathfrak{G}_j$ are obviously invariant relative to B, it remains only to show that a unicellular operator is induced in each of them.

We suppose that the operator induced by B in $\mathfrak{G}_k$ is nonunicellular. Applying Theorem 32.1, we find subspaces $\mathfrak{G}'_k \neq 0$ and $\mathfrak{G}''_k \neq 0$ such that $\mathfrak{G}_k = \mathfrak{G}'_k \tilde{\cup} \mathfrak{G}''_k$ and $\mathfrak{G}'_k \cap \mathfrak{G}''_k = 0$. The subspaces

$$\mathfrak{H}'_k = \mathfrak{H} \ominus \left(\left(\tilde{\bigcup_{j \neq k}} \mathfrak{G}_j \right) \tilde{\cup}\, \mathfrak{G}''_k \right), \quad \mathfrak{H}''_k = \mathfrak{H} \ominus \left(\left(\tilde{\bigcup_{j \neq k}} \mathfrak{G}_j \right) \tilde{\cup}\, \mathfrak{G}'_k \right)$$

are invariant relative to A. Inasmuch as, in view of Theorem 32.3, $\mathfrak{H}$ is the quasidirect sum of the subspaces $\tilde{\bigcup}_{j \neq k} \mathfrak{G}_j$, $\mathfrak{G}'_k$, $\mathfrak{G}''_k$, we have $\mathfrak{H}'_k \neq 0$ and $\mathfrak{H}''_k \neq 0$. At the same time

$$\mathfrak{H}'_k \cap \mathfrak{H}''_k = \mathfrak{H} \ominus \left\{ \left[\left(\tilde{\bigcup_{j \neq k}} \mathfrak{G}_j \right) \tilde{\cup}\, \mathfrak{G}''_k \right] \tilde{\cup} \left[\left(\tilde{\bigcup_{j \neq k}} \mathfrak{G}_j \right) \tilde{\cup}\, \mathfrak{G}'_k \right] \right\} = 0,$$

$$\mathfrak{H}'_k \tilde{\cup}\, \mathfrak{H}''_k = \mathfrak{H} \ominus \left\{ \left[\left(\tilde{\bigcup_{j \neq k}} \mathfrak{G}_j \right) \tilde{\cup}\, \mathfrak{G}''_k \right) \cap \left[\left(\tilde{\bigcup_{j \neq k}} \mathfrak{G}_j \right) \tilde{\cup}\, \mathfrak{G}'_k \right] \right\} = \mathfrak{H} \ominus \tilde{\bigcup_{j \neq k}} \mathfrak{G}_j = \mathfrak{H}_k,$$

which contradicts the unicellularity of the operator induced by A in the subspace $\mathfrak{H}_k$.

§ 33. Invariants of a nonunicellular operator

The collection of subspaces reducing a given completely nonselfadjoint operator $A \in \Lambda_0^{(\exp)}$ to Jordan form is not in general uniquely defined. In the present section we study the common properties of various collections of this kind.

We imbed A in the node

$$\Theta = \begin{pmatrix} A & K & E \\ \mathfrak{H} & & \mathfrak{G} \end{pmatrix} \quad (\dim \mathfrak{G} = \infty)$$

and denote by P_n the orthoprojector onto some n-dimensional subspace in $\mathfrak{G}$. By Theorem 5.6 there exists one and only one subspace $\mathfrak{H}_{P_n}$ such that the characteristic operator-function of the projection of the node Θ onto $\mathfrak{H}_{P_n}$ is equal to the greatest common divisor $W_{P_n}(\lambda)$ of the functions $W(\lambda) = W_\Theta(\lambda)$ and $e^{i\sigma P_n^\perp/\lambda}$ ($\sigma = \sigma[W]$, $P_n^\perp = E - P_n$).

We recall that to each subspace $\mathfrak{H}_0$ invariant relative to A we have agreed to assign the number $\tau(\mathfrak{H}_0) = 2\,\mathrm{tr}(A_0)$, where A_0 is the operator induced in $\mathfrak{H}_0$.

Lemma 33.1. *Suppose that the subspaces* $\mathfrak{H}_j$ ($j = 1, 2, \cdots, \omega$; $\omega \le \infty$) *reduce the operator* A *to Jordan form. If they are enumerated in such a way that* $\tau(\mathfrak{H}_{j+1}) \le \tau(\mathfrak{H}_j)$, *then for any* P_n

$$\tau(\mathfrak{H}_{P_n}) \geqslant \sum_{j=n+1}^{\omega} \tau(\mathfrak{H}_j). \tag{33.1}$$

Proof. There exist subspaces $\mathfrak{H}_j^{(n)} \subset \mathfrak{H}_j$ ($j = 1, 2, \cdots, n$) invariant relative to A and satisfying the conditions

$$\tau(\mathfrak{H}_1^{(n)}) = \tau(\mathfrak{H}_2^{(n)}) = \ldots = \tau(\mathfrak{H}_n^{(n)}) = \tau(\mathfrak{H}_{n+1}).$$

Suppose that $W^{(n)}(\lambda)$ is the characteristic operator-function of the node Θ on the subspace

$$\left(\tilde{\bigcup_{j=1}^{n}} \mathfrak{H}_j^{(n)}\right) \tilde{\cup} \left(\tilde{\bigcup_{j=n+1}^{\omega}} \mathfrak{H}_j\right)$$

and $W_{P_n}^{(n)}(\lambda)$ is the greatest common divisor of the functions $W^{(n)}(\lambda)$ and $e^{i\sigma_n P_n^\perp/\lambda}$ ($\sigma_n = \sigma[W^{(n)}]$). By Lemma 29.3 $\tau[W^{(n)}] \le \tau[W_{P_n}^{(n)}] + n\sigma_n$. Since $\sigma_n = \tau(\mathfrak{H}_{n+1})$ and

$$\tau\left[W^{(n)}\right] = n\tau(\mathfrak{H}_{n+1}) + \sum_{j=n+1}^{\omega} \tau(\mathfrak{H}_j),$$

we have $\tau[W_{P_n}^{(n)}] \ge \Sigma_{j=n+1}^{\omega} \tau(\mathfrak{H}_j)$. We now only need to note that

$$\tau(\mathfrak{H}_{P_n}) = \tau\left[W_{P_n}\right] \geqslant \tau\left[W_{P_n}^{(n)}\right].$$

The lemma is proved.

A subspace $\mathfrak{G} \subset \mathfrak{H}$ is said to be *generating* for the operator A if the collection of vectors $A^{(n)}h$ $(n = 0, 1, \cdots;\ h \in \mathfrak{G})$ is dense in $\mathfrak{H}$.

Lemma 33.2. *Suppose that* $\mathfrak{H}_1$ *and* $\mathfrak{H}_2$ *are invariant relative to* A. *Denote by* A_j $(j = 1, 2)$ *the operator induced in* $\mathfrak{H}_j$, *and put* $A_3 h = P_3 h$ $(h \in \mathfrak{H}_3)$, *where* $\mathfrak{H}_3 = \mathfrak{H} \ominus \mathfrak{H}_1$ *and* P_3 *is the orthoprojector onto* $\mathfrak{H}_3$. *Then:* 1) *if* $\mathfrak{H}_1 \tilde{\cup} \mathfrak{H}_2 = \mathfrak{H}$ *and* $\mathfrak{G}$ *is a generating subspace of the operator* A_2, *then* $\overline{P_3\mathfrak{G}}$ *is a generating subspace of the operator* A_3; 2) *if* $\mathfrak{H}_1 \tilde{\cup} \mathfrak{H}_2 = \mathfrak{H}$ *and the operator* A_2 *is unicellular, then* A_3 *is unicellular*; 3) *if* $\mathfrak{H}_1 \cap \mathfrak{H}_2 = 0$ *and the operator* A_3 *is unicellular, then* A_2 *is unicellular.*

Proof. 1) Since $P_3 A = P_3 A P_3$, we have $P_3 A^n = (P_3 A)^n P_3 = A_3^n P_3$. For any vector $h_0 \in \mathfrak{H}_3$ orthogonal to all vectors of the form $A_3^n P_3 h$ $(n = 0, 1, \cdots;\ h \in \mathfrak{G})$ we will have

$$(A_2^n h, h_0) = (P_3 A^n h, h_0) = (A_3^n P_3 h, h_0) = 0.$$

Accordingly $h_0 \perp \mathfrak{H}_2$. Inasmuch as, moreover, $h_0 \perp \mathfrak{H}_1$ and $\mathfrak{H}_1 \tilde{\cup} \mathfrak{H}_2 = \mathfrak{H}$, we get $h_0 = 0$.

2) According to Theorem 31.1 the operator A_2 has a unicellular generating subspace. The unicellularity of the operator A_3 follows from the part of the lemma already proved and from Theorem 31.4.

3) The operator $B = -A^*$ lies in the class $\Lambda_0^{(\exp)}$ and has the invariant subspaces $\mathfrak{H}_3$ and $\mathfrak{H}_4 = \mathfrak{H} \ominus \mathfrak{H}_2$. We denote by B_3 the operator induced by the operator B in $\mathfrak{H}_3$, and introduce the operator $B_2 h = P_2 B h$ operating on elements h of $\mathfrak{H}_2$, where P_2 is the orthoprojector onto $\mathfrak{H}_2$. It is easy to verify that $B_2 = -A_2^*$ and $B_3 = -A_3^*$. Inasmuch as A_3 is unicellular and $\mathfrak{H}_1 \cap \mathfrak{H}_2 = 0$, the operator B_3 is unicellular and $\mathfrak{H}_3 \tilde{\cup} \mathfrak{H}_4 = \mathfrak{H}$. By the second part of the lemma the operator B_2, and hence A_2, are unicellular.

Lemma 33.3. *Under the hypotheses of Lemma* 33.1 *the orthoprojector* P_n *may be chosen so that the equations*

$$\mathfrak{H}_{P_n} \cap \left(\tilde{\bigcup_{j=1}^{n}} \mathfrak{H}_j \right) = 0, \quad \mathfrak{H}_{P_n} \tilde{\cup} \left(\tilde{\bigcup_{j=1}^{n}} \mathfrak{H}_j \right) = \mathfrak{H} \tag{33.2}$$

are satisfied.

Proof. We first prove that there exists an orthoprojector P_n for which the first of equations (33.2) is satisfied. For $n = 1$ this assertion follows easily from Theorem 31.3. Using induction, we suppose that there exists an orthoprojector P_{n-1} satisfying the condition

$$\mathfrak{H}_{P_{n-1}} \cap \left(\tilde{\bigcup_{j=1}^{n-1}} \mathfrak{H}_j \right) = 0.$$

If in addition

$$\mathfrak{H}_{P_{n-1}} \cap \left(\tilde{\bigcup_{j=1}^{n}} \mathfrak{H}_j \right) = 0,$$

then we choose P_n so that the inequality $P_n > P_{n-1}$ will hold. Then

$$e^{\frac{i\sigma}{\lambda} P_n^{\perp}} \ll e^{\frac{i\sigma}{\lambda} P_{n-1}^{\perp}}, \quad \mathfrak{H}_{P_n} \subset \mathfrak{H}_{P_{n-1}}$$

and therefore

$$\mathfrak{H}_{P_n} \cap \left(\tilde{\bigcup_{j=1}^{n}} \mathfrak{H}_j \right) = 0. \tag{33.3}$$

Now we consider the case when

$$\mathfrak{H}_0 = \mathfrak{H}_{P_{n-1}} \cap \left(\tilde{\bigcup_{j=1}^{n}} \mathfrak{H}_j \right) \neq 0.$$

Applying the last two assertions of the preceding lemma to the closure of the linear envelope of the subspaces $\tilde{\bigcup}_{j=1}^{n-1} \mathfrak{H}_j$ and $\mathfrak{H}_n$, we find that a unicellular operator is induced in $\mathfrak{H}_0$. Accordingly there exists an orthoprojector P_1 such that $\mathfrak{H}_0 \cap \mathfrak{H}_{P_1} = 0$. Denote by P_n the orthoprojector onto the linear envelope of the ranges of the orthoprojectors P_{n-1} and P_1. Since $\mathfrak{H}_{P_n} \subset \mathfrak{H}_{P_{n-1}}$ and $\mathfrak{H}_{P_n} \subset \mathfrak{H}_{P_1}$, we arrive once again at equation (33.3).

In view of (33.1)

$$\tau\left(\mathfrak{H}_{P_n} \tilde{\cup} \left(\tilde{\bigcup_{j=1}^{n}} \mathfrak{H}_j \right) \right) = \tau\left(\tilde{\bigcup_{j=1}^{n}} \mathfrak{H}_j \right) + \tau(\mathfrak{H}_{P_n}) \geqslant \sum_{j=1}^{n} \tau(\mathfrak{H}_j) + \sum_{j=n+1}^{\omega} \tau(\mathfrak{H}_j) = \tau(\mathfrak{H}).$$

Thus

$$\mathfrak{H}_{P_n} \tilde{\cup} \left(\tilde{\bigcup_{j=1}^{n}} \mathfrak{H}_j \right) = \mathfrak{H}.$$

Theorem 33.1. *If each of the systems*

$$\mathfrak{H}_j \ (j = 1, 2, \ldots, \omega; \ \omega \leqslant \infty; \ \tau(\mathfrak{H}_{j+1}) \leqslant \tau(\mathfrak{H}_j)),$$
$$\mathfrak{H}'_j \ (j = 1, 2, \ldots, \omega'; \ \omega' \leqslant \infty; \ \tau(\mathfrak{H}'_{j+1}) \leqslant \tau(\mathfrak{H}'_j))$$

reduces the operator A to Jordan form, then $\omega = \omega'$ and $\tau(\mathfrak{H}_j) = \tau(\mathfrak{H}'_j)$ $(j = 1, 2, \cdots, \omega)$.

Proof. By Lemma 33.2 there exists an orthoprojector P_n such that $\tau(\mathfrak{H}_{P_n}) = \sum_{j=n+1}^{\omega} \tau(\mathfrak{H}_j)$. Comparing this result with (33.1), we obtain the formula

$$\sum_{j=n+1}^{\omega} \tau(\mathfrak{H}_j) = \min_{P_n} \tau(\mathfrak{H}_{P_n}),$$

in which P_n runs through the collection of orthoprojectors onto all possible n-dimensional subspaces in $\mathfrak{G}$. Inasmuch as the analogous formula holds for the subspaces $\mathfrak{H}'_j$, we have $\sum_{j=1}^{n} \tau(\mathfrak{H}_j) = \sum_{j=1}^{n} (\mathfrak{H}'_j)$, which is equivalent to the assertion of the theorem.

Once again suppose that A is a completely nonselfadjoint operator of the class $\Lambda_0^{(\exp)}$, and that $\mathfrak{H}_j$ $(j = 1, 2, \cdots, k_A)$ are subspaces reducing A to Jordan form. We denote by p_A the smallest of the dimensions of generating subspaces of the operator A.

Theorem 33.2. *For each completely nonselfadjoint operator $A \in \Lambda_0^{(\exp)}$ the equation $k_A = p_A$ holds.*

Proof. In view of Theorem 31.4 the equation $k_A = p_A$ is true when $p_A = 1$. It follows from Theorem 31.1 that it is true when $p_A = \infty$. Applying induction, we fix some integer $n > 1$ and suppose that the theorem is true for every completely nonselfadjoint operator $B \in \Lambda_0^{(\exp)}$ satisfying $p_B < n$.

Suppose that A is a completely nonselfadjoint operator of the class $\Lambda_0^{(\exp)}$ and $p_A = n$. We choose an arbitrary basis $h_1, \cdots, h_n$ in some n-dimensional generating subspace of the operator A and denote by $\mathfrak{H}_j$ $(j = 1, \cdots, n)$ the closure of the linear envelope of vectors of the form $A^n h_j$ $(n = 0, 1, \cdots)$. The subspaces $\mathfrak{H}_j$ are invariant relative to A, while the operators A_j induced in them are unicellular.

By Theorem 29.5$'$

$$\sigma[A] = \max\{\sigma[A_1], \sigma[A_2], \ldots, \sigma[A_n]\}.$$

For definiteness we suppose that $\sigma[A] = \sigma[A_1]$. We introduce an orthoprojector P onto the subspace $\mathfrak{G} = \mathfrak{H} \ominus \mathfrak{H}_1$ and an operator B operating in $\mathfrak{G}$ by the formula $Bh = PAh$ $(h \in \mathfrak{G})$. Since $B \in \Lambda_0^{(\exp)}$ and, by the first part of Lemma 33.2, the set of vectors of the form $B^k g_j$ $(k = 0, 1, \cdots; j = 2, 3, \cdots; g_j = Ph_j)$ is dense in $\mathfrak{G}$, it follows by the induction hypothesis that $\mathfrak{G}$ is the quasidirect sum of subspaces $\mathfrak{G}_2, \mathfrak{G}_3, \cdots, \mathfrak{G}_m$ $(\mathfrak{G}_j \neq 0,\ m \leq n)$, in each of which B induces a unicellular operator.

By Lemma 32.2 there exists a subspace $\mathfrak{H}'$ invariant relative to A such that $\mathfrak{H}_1 \cap \mathfrak{H}' = 0$ and $\mathfrak{H}_1 \tilde{\cup} \mathfrak{H}' = \mathfrak{H}$. The subspaces

$$\mathfrak{H}'_j = \mathfrak{H}' \cap (\mathfrak{H}_1 \oplus \mathfrak{G}_j) \qquad (j = 2, 3, \ldots, m)$$

are invariant relative to A. In view of Lemma 33.2 the operators induced in them are unicellular. By formula (28.8)

$$\tau(\mathfrak{H}_1) + \tau(\mathfrak{H}') = \tau(\mathfrak{H}),$$
$$\tau(\mathfrak{H}_1 \oplus \mathfrak{G}_j) + \tau(\mathfrak{H}') = \tau(\mathfrak{H}'_j) + \tau(\mathfrak{H}),$$

so that

$$\tau(\mathfrak{H}_1) + \tau(\mathfrak{H}'_j) = \tau(\mathfrak{H}_1 \oplus \mathfrak{G}_j) \qquad (j = 2, 3, \ldots, m). \tag{33.4}$$

Thus all the subspaces $\mathfrak{H}'_j$ are nonzero. It follows from (33.4) that $\mathfrak{H}_1 \tilde{\cup} \mathfrak{H}'_j = \mathfrak{H}_1 \oplus \mathfrak{G}_j$ as well. Therefore $\tilde{\bigcup}_{j=1}^m \mathfrak{H}'_j = \mathfrak{H}$ $(\mathfrak{H}'_1 = \mathfrak{H}_1)$, so that $m = n$.

All that remains to be proved is that $\mathfrak{H}$ is the quasidirect sum of the subspaces $\mathfrak{H}'_j$ $(j = 1, \cdots, n)$. But this follows from Theorem 32.4, inasmuch as

$$\begin{aligned}\tau(\mathfrak{H}) &= \tau\left(\left(\mathfrak{H}_1 \oplus \tilde{\bigcup_{j=2}^{n-1}} \mathfrak{G}_j\right) \tilde{\cup} (\mathfrak{H}_1 \oplus \mathfrak{G}_n)\right)\\ &= \tau\left(\mathfrak{H}_1 \oplus \tilde{\bigcup_{j=2}^{n-1}} \mathfrak{G}_j\right) + \tau(\mathfrak{H}_1 \oplus \mathfrak{G}_n) - \tau(\mathfrak{H}_1)\\ \ldots &= \sum_{j=2}^{n} \tau(\mathfrak{H}_1 \oplus \mathfrak{G}_j) - (n-2)\tau(\mathfrak{H}_1) = \sum_{j=1}^{n} \tau(\mathfrak{H}'_j).\end{aligned}$$

Theorem 33.3. *If* $\mathfrak{G}$ *is a subspace invariant relative to the completely nonselfadjoint operator* $A \in \Lambda_0^{(\exp)}$ *and* B *is the operator induced in* $\mathfrak{G}$, *then* $k_B \le k_A$.

Proof. We need only consider the case when $k_A = n < \infty$. By Theorems 33.2 and 32.7, $P_{-A^*} = k_{-A^*} = k_A$. Suppose that $\mathfrak{E}$ is some n-dimensional generating subspace of the operator $-A^*$ and P the orthoprojector onto $\mathfrak{G}$. In view of the equation $-B^*g = P(-A^*)g$ $(g \in \mathfrak{G})$ and Lemma 33.2 the subspace $P\mathfrak{E}$ is generating for the operator $-B^*$. Accordingly, $k_B = k_{-B^*} = p_{-B^*} \le \dim\{P\mathfrak{E}\} \le n$. The theorem is proved.

Suppose that the subspaces $\mathfrak{H}_j$ $(j = 1, 2, \cdots, k_A;\ k_A \le \infty)$, reducing the completely nonselfadjoint operator $A \in \Lambda_0^{(\exp)}$ to Jordan form, are enumerated so that $\tau(\mathfrak{H}_{j+1}) \le \tau(\mathfrak{H}_j)$. For any number t lying on the segment $[0, \sigma[A]]$ we construct the subspaces $\mathfrak{H}_j(t)$ $(j = 1, \cdots, k_A)$, defined as follows. 1) If $\tau(\mathfrak{H}_j) \ge t$, then by $\mathfrak{H}_j(t)$ we mean a subspace in $\mathfrak{H}_j$ invariant relative to A such that $\tau(\mathfrak{H}_j(t)) = t$. 2) If $\tau(\mathfrak{H}_j) < t$, we put $\mathfrak{H}_j(t) = \mathfrak{H}_j$. We introduce the notation $\mathfrak{H}(t) = \widetilde{\bigcup}_{j=1}^{k_A} \mathfrak{H}_j(t)$. Obviously $\mathfrak{H}(0) = 0$, $\mathfrak{H}(\sigma) = \mathfrak{H}$ $(\sigma = \sigma[A])$, and $\mathfrak{H}(t') \subset \mathfrak{H}(t'')$ $(t' < t'')$.

A subspace $\mathfrak{H}_0$ invariant relative to A will be assigned to the class $\mathfrak{J}(t)$ $(0 \le t \le \sigma)$ if the operator A_0 induced in it satisfies the condition $\sigma[A_0] \le t$.

Theorem 33.4. *The subspace* $\mathfrak{H}(t) = \widetilde{\bigcup}_{j=1}^{k_A} \mathfrak{H}_j(t)$ $(0 \le t \le \sigma)$ *coincides with the closure of the linear envelope of all spaces of the class* $\mathfrak{J}(t)$, *and therefore does not depend on the choice of the sequence* $\mathfrak{H}_j$ $(j = 1, \cdots, k_A)$ *reducing* A *to normal form.*

Proof. We imbed A in the node

$$\Theta = \begin{pmatrix} A\,K\,E \\ \mathfrak{H} \quad \mathfrak{G} \end{pmatrix}.$$

In view of Theorems 8.2 and 29.5$'$ the characteristic operator-function $W_{\mathfrak{J}(t)}(\lambda)$ of the projection of the node Θ onto the closure of the linear envelope of all subspaces of the class $\mathfrak{J}(t)$ coincides with the greatest common divisor of the functions $W(\lambda) = W_\Theta(\lambda)$ and $e^{it/\lambda}E$. Thus it suffices to show that

$$W_{\mathfrak{J}(t)}(\lambda) = W_{\Theta_t}(\lambda) \qquad (\Theta_t = \mathrm{pr}_{\mathfrak{H}(t)}\Theta).$$

By the definition of the subspace $\mathfrak{H}(t)$ we have

$$\tau\left[W_{\Theta_t}\right] = tn(t) + \sum_{j=n(t)+1}^{k_A} \tau(\mathfrak{H}_j), \tag{33.5}$$

where $n(t)$ is found from the relations

$$\tau(\mathfrak{H}_j)\geqslant t \quad (j=1,\,2,\,\ldots,\,n(t)),\ \tau(\mathfrak{H}_{n(t)+1})<t.$$

Suppose that $P_{n(t)}$ is the orthoprojector onto an $n(t)$-dimensional subspace in $\mathfrak{G}$ and that $W_{P_{n(t)}}(\lambda)$ is the greatest common divisor of the functions $W(\lambda)$ and $e^{i\sigma P^{\perp}_{n(t)}/\tau}$ $(\sigma=\sigma[W],\ P^{\perp}_{n(t)}=E-P_{n(t)})$. Denote by $\mathfrak{H}_{P_{n(t)}}$ a subspace invariant relative to A and satisfying the condition

$$W_{P_{n(t)}}(\lambda)=W_{\Theta_{P_{n(t)}}}(\lambda) \quad \left(\Theta_{P_{n(t)}}=\mathrm{pr}_{\mathfrak{H}_{P_{n(t)}}}\Theta\right).$$

Applying Lemma 33.3, we choose $P_{n(t)}$ so that the equation

$$\tau\left[W_{P_{n(t)}}\right]=\sum_{j=n(t)+1}^{k_A}\tau(\mathfrak{H}_j) \tag{33.6}$$

is satisfied. Suppose that $W_t^{\cap}(\lambda)$ is the greatest common divisor of the functions $W_{\mathfrak{J}(t)}(\lambda)$ and $e^{itP^{\perp}_{n(t)}/\lambda}$. Inasmuch as $\sigma[W_{\mathfrak{J}(t)}]=t$, by Lemma 29.3

$$\tau[W_{\mathfrak{J}(t)}]=\tau\left[W_t^{\cap}\right]+tn(t). \tag{33.7}$$

Recalling that $W_t^{\cap}(\lambda)\ll W_{P_{n(t)}}(\lambda)$, and comparing relations (33.7), (33.6) and (33.5), we obtain

$$\tau[W_{\mathfrak{J}(t)}]\leqslant\tau\left[W_{\Theta_t}\right]. \tag{33.8}$$

Since obviously $W_{\Theta_t}(\lambda)\ll W_{\mathfrak{J}(t)}(\lambda)$, it follows from (33.8) that $W_{\mathfrak{J}(t)}(\lambda)=W_{\Theta_t}(\lambda)$.

§ 34. The rank of a spectral function of a dissipative operator of exponential type

1. Existence of a spectral function of the first rank. Suppose that A is a completely nonselfadjoint operator of the class $\Lambda_0^{(\exp)}$ and π a maximal chain belonging to it. By Theorem 15.5 the chain π is continuous. Imbed A into the node

$$\Theta=\begin{pmatrix} A & K & E\\ \mathfrak{H} & & \mathfrak{G}\end{pmatrix}$$

and denote by g_α ($\alpha = 1, 2, \cdots, \omega$; $\omega \leq \infty$) some orthonormalized basis in $\mathfrak{G}$. By Theorem 16.5 the sequence Kg_α ($\alpha = 1, 2, \cdots, \omega$) is reproducing for π. Noting that

$$\sum_{\alpha=1}^{\omega} \| Kg_\alpha \|^2 = \operatorname{tr}(K^*K) = \operatorname{tr}(KK^*) = \operatorname{tr} A_l < \infty$$

and applying Theorem 18.1, we find a spectral function $P(x)$ ($0 \leq x \leq l$, $l = \operatorname{tr} A_l$), satisfying the following conditions:

1) The range of the function $P(x)$ coincides with π.

2) $\sum_{\alpha=1}^{\omega} (K^*P(x)Kg_\alpha, g_\alpha) = x \qquad (0 \leqslant x \leqslant l).$

It follows from Lemma 23.1 that the function $P(x)$ is absolutely continuous.

Suppose that $\mathfrak{G}_0$ is the linear envelope of the vectors g_α ($\alpha = 1, 2, \cdots, \omega$). There exists a measurable set $\mathfrak{N} \subset [0, l]$ of measure l, at each point of which all the functions $(K^*P(x)Kg, g')$ ($g, g' \in \mathfrak{G}_0$) have derivatives. Since

$$\left\| K^* \frac{P(x+\Delta x) - P(x)}{\Delta x} K \right\| \leqslant \operatorname{tr}\left(K^* \frac{P(x+\Delta x) - P(x)}{\Delta x}\right) = 1,$$

at the points $x \in \mathfrak{N}$ all functions of the form $(K^*P(x)Kg, g')$ ($g, g' \in \mathfrak{G}$) have derivatives as well. In other words, for $x \in \mathfrak{N}$ the derivative of $K^*P(x)K$ exists in the sense of weak convergence. Denoting it by $Q(x)$ ($x \in \mathfrak{N}$), we get

$$(K^*P(x)Kg, g') = \int_0^x (Q(t)g, g')\,dt \qquad (g, g' \in \mathfrak{G}). \tag{34.1}$$

Inasmuch as

$$\sum_{\alpha=1}^{n} \left(K^* \frac{P(x+\Delta x) - P(x)}{\Delta x} Kg_\alpha, g_\alpha\right) \leqslant 1 \qquad (n = 1, 2, \ldots),$$

we see that $\sum_{\alpha=1}^{n} (Q(x)g_\alpha, g_\alpha) \leq 1$. Accordingly $Q(x)$ is a nuclear operator and $\operatorname{tr} Q(x) \leq 1$. On the other hand,

$$\int_0^l \operatorname{tr} Q(x)\,dx = \sum_{\alpha=1}^{\omega} \int_0^l (Q(x)g_\alpha, g_\alpha)\,dx = \sum_{\alpha=1}^{\omega} (K^*Kg_\alpha, g_\alpha) = l,$$

so that $\operatorname{tr} Q(x) = 1$ almost everywhere.

Lemma 34.1. *The rank of a maximal chain of a unicellular operator* $A \in \Lambda_0^{(\exp)}$ *is equal to unity.*

Proof. Imbed A into the node Θ. A maximal chain π of A coincides with the range of a spectral function $P(x)$ $(0 \le x \le l,\ l = \operatorname{tr} A_I)$, connected with a positive function $Q(x)$ satisfying almost everywhere the relation $\operatorname{tr} Q(x) = 1$ by the relation (34.1).

By Theorem 25.5

$$W_\Theta(\lambda) = \int_0^{l} e^{\frac{2i}{\lambda} d(K^*P(x)K)}.$$

Suppose that $0 \doteq x_0 < x_1 < \cdots < x_n = l$ is any subdivision of $[0, l]$. Inasmuch as

$$|(K^*\Delta P_j K g, g)| \leqslant \int_{x_{j-1}}^{x_j} |(Q(t)g, g)|\,dt \leqslant \|g\|^2 \int_{x_{j-1}}^{x_j} \|Q(t)\|\,dt$$
$$(\Delta P_j = P(x_j) - P(x_{j-1})),$$

and accordingly

$$\|K^*\Delta P_j K\| \leqslant \int_{x_{j-1}}^{x_j} \|Q(t)\|\,dt,$$

it follows that

$$\left\| \prod_{j=1}^{n} e^{\frac{2i}{\lambda} K^*\Delta P_j K} \right\| \leqslant \prod_{j=1}^{n} e^{\frac{2}{|\lambda|}\|K^*\Delta P_j K\|} \leqslant e^{\frac{2}{|\lambda|}\int_0^l \|Q(x)\|\,dx}.$$

Thus

$$\|W_\Theta(\lambda)\| \leqslant e^{\frac{2}{|\lambda|}\int_0^l \|Q(x)\|\,dx} \tag{34.2}$$

In view of Theorems 8.3 and 30.3 the type of the function $W_\Theta(1/\mu)$ is equal to $2l$, and it follows from inequality (34.2) that $l \le \int_0^l \|Q(x)\|\,dx$. Since $\|Q(x)\| \le \operatorname{tr} Q(x) = 1$, we have almost everywhere

$$\|Q(x)\| = \operatorname{tr} Q(x) = 1. \tag{34.3}$$

By Theorem 23.1 the rank of the spectral function $P(x)$ coincides with the general rank of the matrix

$$\frac{d}{dx}\|(P(x)Kg_\alpha,\ Kg_\beta)\|_{\alpha,\beta=1}^{\omega}=\|(Q(x)g_\alpha,\ g_\beta)\|_{\alpha,\beta=1}^{\omega},$$

where g_α $(\alpha = 1, 2, \cdots, \omega;\ \omega \leq \infty)$ is an orthonormalized basis in $\mathfrak{G}$. But the general rank of the matrix $\|(Q(x)g_\alpha,\ g_\beta)\|_{\alpha,\beta=1}^{\omega}$ is equal to unity, since in view of (34.3) the operator $Q(x)$ is almost everywhere an orthoprojector onto a one-dimensional subspace.

In the statement of the following lemma, we suppose in the case $\omega = \infty$ that the identity operator in $\mathfrak{H}$ has been adjoined to the collection (34.4).

Lemma 34.2. *Suppose that the operator A operates in a space $\mathfrak{H} = \Sigma_{j=1}^{\omega} \oplus \mathfrak{H}_j$ $(\omega \leq \infty)$, and suppose that all the subspaces $\Sigma_{j=1}^{n} \oplus \mathfrak{H}_j$ $(j = 1, 2, \cdots)$ are invariant relative to A. Denote by E_j the orthoprojector onto $\mathfrak{H}_j$ and put $A_jh = E_jAh$ $(h \in \mathfrak{H}_j)$. If for each j the operator A_j has a maximal chain π_j of first rank, then the collection π consisting of the orthoprojectors*

$$PE_1\ (P\in\pi_1),\ E_1+PE_2\ (P\in\pi_2),\ E_1+E_2+PE_3\ (P\in\pi_3),\ \ldots, \tag{34.4}$$

operating in $\mathfrak{H}$ is a maximal chain of the first rank, belonging to A.

Conversely, if A has a maximal chain π of the first rank containing

$$\sum_{j=1}^{n} E_j\ \ (n=1,\ 2,\ \ldots),$$

then the operators in $\mathfrak{H}_j$ orthoprojecting onto the subspaces

$$P\mathfrak{H}_j\ (P\in\pi)$$

constitute a maximal chain of the first rank belonging to the operator A_j.

The proof is obvious. We note only that if h_j is a reproducing vector of the chain π_j, then, for any collection of nonzero c_j which guarantee that the series

$$h=\sum_{j=1}^{\omega} c_jh_j,$$

converges, the vector h will be reproducing for the chain π.

Theorem 34.1. *If A is a completely nonselfadjoint operator of the class $\Lambda^{(\exp)}$, then the rank of at least one of the maximal chains belonging to A is equal to unity.*

Proof. Denote by $\mathfrak{H}$ the space in which the operator A is given. We assign a subspace $\mathfrak{H}_0$ invariant relative to A to the class $\mathfrak{K}$ if the operator A_0 induced in it has a maximal chain of the first rank. We will say that the subspace $\mathfrak{H}_1 \in \mathfrak{K}$ is a subspace of $\mathfrak{H}_2 \in \mathfrak{K}$, and write $\mathfrak{H}_1 \prec \mathfrak{H}_2$, if 1) $\mathfrak{H}_1 \subseteq \mathfrak{H}_2$ and 2) the operator A_2 induced in $\mathfrak{H}_2$ has a maximal chain of the first rank containing the orthoprojector onto $\mathfrak{H}_1$. Applying Lemma 34.2, we easily verify that if $\mathfrak{H}_1 \prec \mathfrak{H}_2$ and $\mathfrak{H}_2 \prec \mathfrak{H}_3$, then $\mathfrak{H}_1 \prec \mathfrak{H}_3$. Thus the relation "$\prec$" converts $\mathfrak{K}$ into a partially ordered set.

We shall show that every ordered part $\mathfrak{K}_0$ of the set $\mathfrak{K}$ has an upper bound. This is obvious if the space $\tilde{\mathfrak{H}}$ which is the closure of the linear envelope of all subsets of $\mathfrak{K}_0$ lies in $\mathfrak{K}_0$. In the contrary case $\tilde{\mathfrak{H}}$ is the closure of the linear envelope of some sequence $\mathfrak{H}_1 \subset \mathfrak{H}_2 \subset \cdots$ $(\mathfrak{H}_j \in \mathfrak{K}_0)$ of subspaces. Again applying Lemma 34.2, we find that all the subspaces of the collection $\mathfrak{K}_0$ precede $\tilde{\mathfrak{H}}$.

By Zorn's lemma there exists a subspace $\mathfrak{G} \in \mathfrak{K}$ which precedes only itself. If $\mathfrak{G} = \mathfrak{H}$, then the theorem is proved. Suppose that

$$\mathfrak{G}' = \mathfrak{H} \ominus \mathfrak{G} \neq 0,$$

and consider the operator

$$A'h = P'Ah \quad (h \in \mathfrak{G}'),$$

operating in $\mathfrak{G}'$, where P' is the orthoprojector onto $\mathfrak{G}'$. By Lemma 32.1 there exists a subspace

$$\mathfrak{G}'' \subset \mathfrak{G}' \quad (\mathfrak{G}'' \neq 0)$$

invariant relative to A' and such that the operator induced in it is unicellular. In view of Theorem 30.5′ and Lemma 34.1 the operator A'' induced by A' in $\mathfrak{G}''$ has a maximal chain of the first rank. We have arrived at a contradiction, inasmuch as

$$\mathfrak{G} \oplus \mathfrak{G}'' \in \mathfrak{K} \quad \text{and} \quad \mathfrak{G} \prec \mathfrak{G} \oplus \mathfrak{G}''.$$

2. **Multiplicative representations of functions of the class $\Omega_\omega^{(\exp)}$.**

Lemma 34.3. *Suppose that $A \in \Lambda^{(\exp)}$. There exists a maximal chain π having the following properties*:

1) *π belongs to A;*

2) *the rank of π is equal to unity;*

3) *the operators PAP $(P \in \pi,\ P \neq E)$ belong to the class $\Lambda_0^{(\exp)}$.*

Proof. We need only consider the case when $A \in \Lambda_\infty^{(\exp)}$. By Theorem 30.4 the space $\mathfrak{H}$ in which the operator A is given is representable in the form

$$\mathfrak{H} = \sum_{j=1}^{\infty} \oplus \mathfrak{H}_j,$$

where all the subspaces $\sum_{j=1}^{n} \oplus \mathfrak{H}_j$ $(n = 1, 2, \cdots)$ are invariant relative to A, and the operators induced in them are in the class $\Lambda_0^{(\exp)}$. Suppose that E_j is the orthoprojector onto $\mathfrak{H}_j$ and that $A_j h = E_j A h$ $(h \in \mathfrak{H}_j)$. By Theorem 34.1 there exist maximal chains π_j $(j = 1, 2, \cdots)$ of the first rank, belonging to the operators A_j respectively. As in Lemma 34.2, the chain π made up from the chains π_j $(j = 1, 2, \cdots)$ satisfies requirements 1), 2) and 3).

Theorem 34.2. *Suppose that the basic operator of the simple node*

$$\Theta = \begin{pmatrix} A & K & E \\ \mathfrak{H} & & \mathfrak{G} \end{pmatrix}$$

lies in the class $\Lambda^{(\exp)}$. There exists a spectral function $P(x)$ $(0 \leq x \leq l \leq \infty)$ of the first rank such that:

1) *$P(x)$ belongs to A;*

2) $\operatorname{tr}(K^* P(x) K) = x$ $(0 \leq x \leq l)$;

3) *the function $K^* P(x) K$ is absolutely continuous on each segment and almost everywhere has a weak derivative $Q(x)$ whose values are orthoprojectors onto one-dimensional subspaces.*

Proof. By Lemma 34.3 there exists a maximal chain π of the first rank belonging to A and satisfying the condition $PAP \in \Lambda_0^{(\exp)}$ $(P \in \pi,\ P \neq E)$. Repeating the arguments in the last part of the proof of Theorem 30.5, we verify that π coincides with the range of some spectral function $P(x)$ $(0 \leq x \leq l)$ such that

$$\operatorname{tr}(K^* P(x) K) = \operatorname{tr}(P(x) A_I P(x)) = x \qquad (0 \leqslant x \leqslant l).$$

In addition, $l < \infty$ if $A \in \Lambda_0^{(\exp)}$, and $l = \infty$ if $A \in \Lambda_\infty^{(\exp)}$.

Suppose that g_α $(\alpha = 1, 2, \cdots, \omega;\ \omega \leq \infty)$ is an orthonormalized basis in $\mathfrak{G}$. Inasmuch as the considerations presented at the beginning of the present section are applicable to the function $K^* P(x) K$ $(0 \leq x \leq l_0 < \infty)$, it is absolutely continuous and has almost everywhere a weak derivative $Q(x)$ whose trace is identically equal to unity. By Theorem 23.1 the minors of second order of the matrix

$\|Q(x)g_\alpha, g_\beta\|^{\omega}_{\alpha,\beta=1}$ $(0 \le x \le l_0)$ are almost everywhere equal to zero. Accordingly the values of the function $Q(x)$ are orthoprojectors onto one-dimensional subspaces.

Theorem 34.3. *Suppose that* $W(\lambda) \in \Omega^{(\exp)}_{\mathfrak{G}}$. *If* $\tau[W] < \infty$, *then*

$$W(\lambda) = \int_0^{l}\!\!\!\curvearrowright e^{\frac{2i}{\lambda}\, dH(t)} \qquad (\tau[W] = 2l),$$

where $H(t)$ *is absolutely continuous and has almost everywhere a weak derivative whose values are orthoprojectors onto one-dimensional subspaces. In the case* $\tau[W] = \infty$ *we have the formula*

$$W(\lambda) = \lim_{x \to \infty} \int_0^{x}\!\!\!\curvearrowright e^{\frac{2i}{\lambda}\, dH(t)}. \tag{34.5}$$

The passage to the limit in (34.5) *is carried out in the sense of strong convergence, and* $H(t)$ *has the properties indicated above on each finite interval.*

Proof. The function $W(\lambda)$ is characteristic for some simple node

$$\Theta = \begin{pmatrix} A & K & E \\ \mathfrak{H} & & \mathfrak{G} \end{pmatrix},$$

where $A \in \Lambda^{(\exp)}$. Consider a spectral function $P(x)$ $(0 \le x \le l,\ l \le \infty)$ of the operator A, satisfying the hypotheses of Theorem 34.2. If $\tau[W] < \infty$, then $A \in \Lambda_0^{(\exp)}$ and the first of the assertions being proved follows from Theorem 25.5. In the case $\tau[W] = \infty$ we denote by Θ_x $(x < \infty)$ the projection of the node Θ onto the subspace $P(x)\mathfrak{H}$. Since

$$W_{\Theta_x}(\lambda) = \int_0^{x}\!\!\!\curvearrowright e^{\frac{2i}{\lambda}\, d(K^*P(t)K)}$$

and

$$\lim_{x \to \infty} W_{\Theta_x}(\lambda) = W_\Theta(\lambda),$$

the convergence being strong, for each fixed nonzero λ, the second assertion of the theorem is true as well.

APPENDIX I

GENERALIZATION OF NAĬMARK'S THEOREM

Suppose that $F(x)$ $(a \le x \le b)$ is a nondecreasing function whose values are bounded nonnegative linear operators operating in Hilbert space $\mathfrak{H}$. Denote by $\mathfrak{M}$ the set of elements of the type

$$\begin{Bmatrix} f_1 & f_2 & \dots & f_m \\ \Delta_1 & \Delta_2 & \dots & \Delta_m \end{Bmatrix} \quad (f_k \in \mathfrak{H},\ \Delta_k = [x_k,\ y_k],\ a \leqslant x_k \leqslant y_k \leqslant b)$$

and define in $\mathfrak{M}$ the operations of addition and multiplication by a complex number, putting

$$\begin{Bmatrix} f_1 & f_2 & \dots & f_m \\ \Delta_1 & \Delta_2 & \dots & \Delta_m \end{Bmatrix} + \begin{Bmatrix} f'_1 & f'_2 & \dots & f'_n \\ \Delta'_1 & \Delta'_2 & \dots & \Delta'_n \end{Bmatrix}$$
$$= \begin{Bmatrix} f_1 & f_2 & \dots & f_m & f'_1 & f'_2 & \dots & f'_n \\ \Delta_1 & \Delta_2 & \dots & \Delta_m & \Delta'_1 & \Delta'_2 & \dots & \Delta'_n \end{Bmatrix},$$
$$\lambda \begin{Bmatrix} f_1 & f_2 & \dots & f_m \\ \Delta_1 & \Delta_2 & \dots & \Delta_m \end{Bmatrix} = \begin{Bmatrix} \lambda f_1 & \lambda f_2 & \dots & \lambda f_m \\ \Delta_1 & \Delta_2 & \dots & \Delta_m \end{Bmatrix}.$$

To each two elements

$$\varphi = \begin{Bmatrix} f_1 & f_2 & \dots & f_m \\ \Delta_1 & \Delta_2 & \dots & \Delta_m \end{Bmatrix} \quad \text{and} \quad \psi = \begin{Bmatrix} f'_1 & f'_2 & \dots & f'_n \\ \Delta'_1 & \Delta'_2 & \dots & \Delta'_n \end{Bmatrix}$$

we assign the number

$$(\varphi,\ \psi) = \sum_{j=1}^{n} \sum_{i=1}^{m} \left(F_{\Delta_i \cap \Delta'_j} f_i,\ f'_j \right),$$

where F_Δ $(\Delta = [x, y]$, denotes the difference $F(y) - F(x)$. Since, obviously,

$$\left.\begin{aligned} (\varphi,\ \psi) = \overline{(\psi,\ \varphi)},\ (\varphi_1 + \varphi_2,\ \psi) = (\varphi_1,\ \psi) + (\varphi_2,\ \psi), \\ (\lambda\varphi,\ \psi) = \lambda(\varphi,\ \psi),\quad (\varphi,\ \varphi) \geqslant 0, \end{aligned}\right\} \qquad (1)$$

we have

$$|(\varphi, \psi)|^2 \leqslant (\varphi, \varphi)(\psi, \psi). \tag{2}$$

We shall say that the elements ϕ and ψ are *equivalent* if $(\phi - \psi, \phi - \psi) = 0$. In view of (1) and (2) the set $\mathfrak{M}$ decomposes uniquely into pairwise nonintersecting classes such that two elements lie in the same class if and only if they are equivalent. We denote by k_ϕ the class containing the element ϕ. Putting

$$k_\varphi + k_\psi = k_{\varphi+\psi}, \quad \lambda k_\varphi = k_{\lambda\varphi},$$

we convert the set $\mathfrak{L}_F^{(0)}$ of all classes into a linear space, after which we introduce a scalar product into $\mathfrak{L}_F^{(0)}$ by putting $(k_\phi, k_\psi) = (\phi, \psi)$. Finally, completing the space $\mathfrak{L}_F^{(0)}$, we obtain a Hilbert space, which we shall denote by $\mathfrak{L}_F$.

In what follows we shall use the standard definition of orthogonal resolution of unity, not requiring, however, continuity from the left or right.

Theorem 1. *Suppose that $F(x)$ $(a \leq x \leq b)$ is a nondecreasing function whose values are bounded linear operators, operating in Hilbert space $\mathfrak{L}$. If $F(a) = 0$, then there exists a bounded linear mapping R of the space $\mathfrak{L}$ into $\mathfrak{L}_F$ and an orthogonal resolution of unity $E(x)$ $(a \leq x \leq b)$ in $\mathfrak{L}_F$, such that:*

1) $F(x) = R^* E(x) R$ $(a \leq x \leq b)$;

2) *the linear envelope of the set of vectors of the form $E(x)Rf$ $(f \in \mathfrak{L}$, $a \leq x \leq b)$ is dense in $\mathfrak{L}_F$;*

3) *if the function $F(x)$ is strongly continuous on the left (right) at the point x_0, then the function $E(x)$ is strongly continuous on the left (right);*

4) *if $F(x_1) = F(x_2)$ $(a \leq x_1 < x_2 \leq b)$, then $E(x_1) = E(x_2)$.*

Proof. 1) Assign to the vector $f \in \mathfrak{L}$ the vector $Rf = k_{\left\{ f \atop \Delta_0 \right\}}$ $(\Delta_0 = [a, b])$. Since $\left\{ f+g \atop \Delta \right\}$ and $\left\{ f \; g \atop \Delta \; \Delta \right\}$ are equivalent,

$$R(f+g) = k_{\left\{ f+g \atop \Delta_0 \right\}} = k_{\left\{ f \atop \Delta_0 \right\} + \left\{ g \atop \Delta_0 \right\}} = k_{\left\{ f \atop \Delta_0 \right\}} + k_{\left\{ g \atop \Delta_0 \right\}} = Rf + Rg.$$

Moreover,

$$R(\lambda f) = k_{\left\{ \lambda f \atop \Delta_0 \right\}} = k_{\lambda \cdot \left\{ f \atop \Delta_0 \right\}} = \lambda k_{\left\{ f \atop \Delta_0 \right\}} = \lambda Rf$$

and

$$\| Rf \|^2 = \left(k_{\left\{\begin{smallmatrix} f \\ \Delta_0 \end{smallmatrix}\right\}}, k_{\left\{\begin{smallmatrix} f \\ \Delta_0 \end{smallmatrix}\right\}}\right) = \left(\left\{\begin{matrix} f \\ \Delta_0 \end{matrix}\right\}, \left\{\begin{matrix} f \\ \Delta_0 \end{matrix}\right\}\right)$$
$$= (F_{\Delta_0} f, f) \leqslant \| F(b) \| \, \| f \|^2.$$

Thus the operator R is linear and bounded. It is easy to see that $R^* k_{\left\{\begin{smallmatrix} f \\ \Delta \end{smallmatrix}\right\}} = F_\Delta f$.

The linear envelope $\mathfrak{L}_F^{(0)}$ of the set of classes of the type $k_{\left\{\begin{smallmatrix} f \\ \Delta \end{smallmatrix}\right\}}$ is dense in $\mathfrak{L}_F$. We define a linear operator $E(x)$ $(a \leq x \leq b)$ in $\mathfrak{L}_F^{(0)}$, putting

$$E(x) k_{\left\{\begin{smallmatrix} f \\ \Delta \end{smallmatrix}\right\}} = k_{\left\{\begin{smallmatrix} f \\ \Delta \cap [a, x] \end{smallmatrix}\right\}}.$$

One shows by direct verification that

$$E(a) = 0, \quad E(b) = E, \tag{3}$$

$$E(x) E(y) = E(z) \quad (z = \min\{x, y\}) \tag{4}$$

and

$$(E(x) k_1, k_2) = (k_1, E(x) k_2) \quad (k_1, k_2 \in \mathfrak{L}_F^{(0)}).$$

Accordingly, $E(x)$ is a bounded operator, and its extension by continuity on the entire space $\mathfrak{L}_F$, which we again denote by $E(x)$, is an orthoprojector. Relations (3) and (4) mean that the function $E(x)$ $(a \leq x \leq b)$ is an orthogonal resolution of unity. It remains to be noted that

$$R^* E(x) Rf = R^* E(x) k_{\left\{\begin{smallmatrix} f \\ \Delta_0 \end{smallmatrix}\right\}} = R^* k_{\left\{\begin{smallmatrix} f \\ [a, x] \end{smallmatrix}\right\}} = F(x) f.$$

2) The linear envelope of the set of vectors of the type $E(x) Rf$ coincides with $\mathfrak{L}_F^{(0)}$, since

$$E_\Delta Rf = k_{\left\{\begin{smallmatrix} f \\ [a, y] \end{smallmatrix}\right\}} - k_{\left\{\begin{smallmatrix} f \\ [a, x] \end{smallmatrix}\right\}} = k_{\left\{\begin{smallmatrix} f & -f \\ [a, y] & [a, x] \end{smallmatrix}\right\}} = k_{\left\{\begin{smallmatrix} f \\ \Delta \end{smallmatrix}\right\}}$$
$$(\Delta = [x, y], \ E_\Delta = E(y) - E(x)).$$

3) Suppose, for example, that $F(x)$ is continuous on the left at the point x_0. Then

$$R^* (E(x_0) - E(x_0 - 0)) R = F(x_0) - F(x_0 - 0) = 0,$$

so that

$$(E(x_0) - E(x_0 - 0)) R = 0, \ (E(x_0) - E(x_0 - 0)) E(x) R = 0 \qquad (a \leqslant x \leqslant b).$$

In view of the assertion proved in the preceding point, $E(x_0) = E(x_0 - 0)$.

4) If $F(x_1) = F(x_2)$, then $R^*(E(x_2) - E(x_1))R = 0$ and therefore

$$(E(x_2) - E(x_1))E(x)R = 0 \quad (a \leqslant x \leqslant b).$$

Again using the density in $\mathfrak{L}_F$ of the set of vectors of the form $E(x)Rf$, we find that $E(x_1) = E(x_2)$.

Theorem 2. *Suppose that $F(x)$ $(a \leq x \leq b)$ is a nondecreasing function whose values are bounded linear operators operating in Hilbert space $\mathfrak{L}$. If $F(a) = 0$ and $F(b) = E$, then there exists a Hilbert space $\tilde{\mathfrak{L}} \supset \mathfrak{L}$ and an orthogonal resolution of unity, $E(x)$ $(a \leq x \leq b)$ in $\tilde{\mathfrak{L}}$, such that:*

1) *$F(x)f = PE(x)f$ $(a \leq x \leq b,\ f \in \mathfrak{L})$, where P is the orthoprojector onto $\mathfrak{L}$;*

2) *the linear envelope of the set of vectors of the form $E(x)f$ $(a \leq x \leq b,\ f \in \mathfrak{L})$ is dense in $\tilde{\mathfrak{L}}$;*

3) *if the function $F(x)$ is strongly left (right) continuous at the point x_0, then the function $E(x)$ is strongly left (right) continuous at x_0 as well;*

4) *if $F(x_1) = F(x_2)$, then $E(x_1) = E(x_2)$.*

Proof. Construct a mapping R of the space $\mathfrak{L}$ into $\tilde{\mathfrak{L}} = \mathfrak{L}_F$ and an orthogonal resolution of unity $E(x)$ $(a \leq x \leq b)$ in $\tilde{\mathfrak{L}}$ in the same way as in the preceding theorem. Since in the case at hand $R^*R = F(b) = E$, the operator R maps $\mathfrak{L}$ isometrically onto its range. We identify each vector $f \in \mathfrak{L}$ to the vector Rf. Then $Rf = f$ $(f \in \mathfrak{L})$, $R^*g = Pg$ $(g \in \tilde{\mathfrak{L}})$, where P is the orthoprojector onto $\mathfrak{L}$, and assertions 1)–4) follow from the corresponding assertions of Theorem 1.

APPENDIX II

INVERTIBLE HOLOMORPHIC OPERATOR-FUNCTIONS

Suppose given in a region G of the complex plane a holomorphic function $T(\lambda)$ whose values are bounded linear operators operating in Hilbert space $\mathfrak{H}$. We denote by G_0 the set of those points of the region G at which $T(\lambda)$ does not have a bounded inverse defined on all of $\mathfrak{H}$. Then $G - G_0$ is an open set, and the function $T^{-1}(\lambda)$ is holomorphic at each of its points.

We shall call the function $T(\lambda)$ *invertible* in G if it satisfies the following conditions:

1. The set G_0 has no limit points in G.

2. Every point $\lambda \in G_0$ is a pole of the function $T^{-1}(\lambda)$.

3. The coefficients on the negative powers of $\lambda - \lambda^{(0)}$ in the Laurent expansion of $T^{-1}(\lambda)$ are finite dimensional in the neighborhood of any pole $\lambda^{(0)}$.

Theorem 1. *Suppose that a function $K(\lambda)$ whose values are completely continuous operators in Hilbert space $\mathfrak{H}$ is holomorphic in a region G. If for some $\lambda_0 \in G$ the operator $E + K(\lambda_0)$ has a bounded inverse defined throughout $\mathfrak{H}$, then the function $T(\lambda) = E + K(\lambda)$ is invertible in G.*

Proof. Consider first the case when all vectors of the form $K(\lambda)h$ $(\lambda \in G,\ h \in \mathfrak{H})$ lie in a finite-dimensional subspace $\mathfrak{H}_0 \subset \mathfrak{H}$. Denote by $T_0(\lambda)$ the operator induced by $T(\lambda)$ in $\mathfrak{H}_0$. The operator $T_0(\lambda_0)$ has an inverse, since in the contrary case there would exist a nonzero vector $h_0 \in \mathfrak{H}_0$ annihilated by the operator $T_0(\lambda_0)$. This vector would also be annihilated by the operator $T(\lambda_0)$, which would contradict the hypothesis of the theorem. Thus $\det T_0(\lambda_0) \neq 0$, and therefore the set of zeros of the holomorphic function $\det T_0(\lambda)$ does not have limit points in G. Applying the rule for finding the elements of an inverse matrix, we find that the singular points of the function $T_0^{-1}(\lambda)$ can only be poles lying in G_0. Denoting the orthoprojectors onto $\mathfrak{H}_0$ and $\mathfrak{H}_1 = \mathfrak{H} \ominus \mathfrak{H}_0$ by P_0 and P_1 respectively, we get

$$T(\lambda)=(P_0+P_1)\,T(\lambda)(P_0+P_1)$$
$$=P_0T(\lambda)\,P_0+P_0T(\lambda)\,P_1+P_1T(\lambda)\,P_1$$
$$=T_0(\lambda)\,P_0+P_0T(\lambda)\,P_1+P_1.$$

Accordingly

$$T^{-1}(\lambda)=T_0^{-1}(\lambda)\,P_0-T_0^{-1}(\lambda)\,P_0T(\lambda)\,P_1+P_1,$$

from which it is clear that $T(\lambda)$ satisfies all the requirements of the definition of an invertible function.

Now suppose that the dimension of the subspace $\mathfrak{H}_0$ is arbitrary, but that G is the set of interior points of some disk. Without loss of generality, we may suppose that its center is at the origin of coordinates, and that the radius is unity. In the disk $|\lambda|<\rho$ $(|\lambda_0|<\rho<1)$ the function $K(\lambda)$ is representable by the series

$$K(\lambda)=K_0+\lambda K_1+\lambda^2K_2+\ \ldots,$$

which converges uniformly in norm relative to λ, so that for some integer N

$$\left\|K(\lambda)-\sum_{j=0}^{N}\lambda^jK_j\right\|<\frac{1}{4}\qquad(|\lambda|<\rho).$$

Using the fact that all the operators K_j are completely continuous, we select finite-dimensional operators R_j such that the relation

$$\|K_j-R_j\|<\frac{1}{4(N+1)}\qquad(j=0,\ 1,\ \ldots)$$

is satisfied. Putting $R(\lambda)=\Sigma_{j=0}^{N}\lambda^jR_j$, we arrive at the equation

$$K(\lambda)=\ R(\lambda)+S(\lambda),\ \text{where}\ \|S(\lambda)\|<\frac{1}{2}\ \ (|\lambda|<\rho).$$

For $|\lambda|<\rho$ the series $\Sigma_{j=0}^{\infty}(-1)^jS^j(\lambda)$ converges uniformly and is a holomorphic function equal to $(E+S(\lambda))^{-1}$. Representing $T(\lambda)$ in the form

$$T(\lambda)=E+R(\lambda)+S(\lambda)=\left[E+R(\lambda)(E+S(\lambda))^{-1}\right](E+S(\lambda))$$

and noting that, because of what was proved above, the function in square brackets is invertible in the disk $|\lambda|<\rho$, we find that in this disk $T(\lambda)$ is invertible as well. In view of the arbitrariness in the choice of ρ, $T(\lambda)$ is invertible in the entire disk $|\lambda|<1$.

In order to obtain the proof of the theorem free from all restrictions, we consider an arbitrary circumference γ whose interior points lie in the region G, and construct circumferences $\gamma_0, \gamma_1, \cdots, \gamma_n = \gamma$ having the following properties: 1) the center of the circumference γ_0 is at the point λ_0; 2) the interior points of all the circumferences γ_j lie in G; 3) the circumferences γ_j and γ_{j+1} $(j < n)$ have common interior points. Since $T(\lambda)$ is invertible inside γ_0, the operator $T^{-1}(\lambda)$ exists at some point lying inside γ_1, so that $T(\lambda)$ is invertible inside γ_1. Continuing this process, we obtain the proof of invertibility of $T(\lambda)$ inside γ. It remains only to note that the invertibility of $T(\lambda)$ inside any disk lying in G implies the invertibility of $T(\lambda)$ inside the entire region G.

NOTES ON THE LITERATURE AND ADDITIONAL REMARKS

To Chapter I

§§1–3. The characteristic function of a bounded linear operator A operating in a Hilbert space $\mathfrak{H}$ was first defined by M. S. Livšic [1] by means of the formula

$$W(\lambda) = E + 2i\,(\operatorname{sign} A_I)\,|A_I|^{1/2}(A^* - \lambda E)^{-1}|A_I|^{1/2} \quad \left(A_I = \frac{A - A^*}{2i}\right), \tag{1}$$

where we have written

$$\operatorname{sign} A_I = \int_{-\infty}^{\infty} \operatorname{sign} x\, dE(x), \qquad |A_I| = \int_{-\infty}^{\infty} |x|\, dE(x),$$

the $E(x)$ denoting the orthogonal resolution of unity corresponding to the operator A_I. The right side of (1) was considered only in the space $\overline{A_I\mathfrak{H}}$. A more general definition was proposed by M. S. Brodskiĭ [6,7,8]; see also the survey paper of Brodskiĭ and Livšic [1]. According to this definition the characteristic function for A could be any function of the type

$$W(\lambda) = E - 2iK^*(A - \lambda E)^{-1}KJ \qquad \left(J = J^*,\ J^2 = E,\ KJK^* = A_I\right), \tag{2}$$

where J and K are bounded linear operators operating respectively in some Hilbert space $\mathfrak{G}$ and from $\mathfrak{G}$ to $\mathfrak{H}$. Definition (2) made it possible to formulate the "multiplication theorem" without the previous restrictions, and naturally led to the concept of operator node. The reconstruction to the theory of characteristic functions in the terms of operator nodes was carried out in the papers of Brodskiĭ and Ju. L. Šmul'jan [1] and Brodskiĭ and G. È. Kisilevskiĭ [1]. The elements of the theory of operator nodes was presented also in the monograph of Livšic [1].

§4. In this section we generalize to the case of operator-functions the well-known theorems on integral representations of scalar analytic functions (see for example the course of Ahiezer and Glazman [1], §69).

§5. Purely analytic criteria for one of two functions of the class Ω_j to be a regular divisor of the other were presented in the paper [7] of Brodskiĭ for functions

$W(\lambda) \in \Omega_j$ entire relative to $1/\lambda$ (see §27), and in Šmul'jan's papers [1,2] for arbitrary functions given in a finite-dimensional space $\mathfrak{G}$. His results were obtained as applications of the theory of Hellinger operator integrals worked out by him in the papers [3,4]. Theorems 5.6 and 5.7 were obtained by Brodskiĭ and Šmul'jan [1] for the case when the canal operator of the node θ is completely continuous.

§6. The results of this section are due to L. E. Isaev [1]. Lemma 6.2 is equivalent to a theorem of Ju. P. Ginzburg [1] on operator functions analytic in the unit disk.

§7. The results of subsection 2, with the exclusion of Theorem 7.2, are published here for the first time.

§8. Theorems 8.1–8.4 were proved by Kisilevskiĭ [2] for the case $\dim \mathfrak{G} < \infty$, and then by Brodskiĭ and Kisilevskiĭ [1] under the hypothesis that $\lim_{\lambda\to\infty} [\lambda(W(\lambda) - E)]$ is a nuclear operator. Without the last restriction it was obtained by Isaev in [3].

§9. All the theorems of this section can be found in the papers of Livšic [1], Brodskiĭ [6], and Brodskiĭ and Livšic [1].

§10. Another proof of Theorem 10.1 is given in the monograph of I. G. Gohberg and M. G. Kreĭn [4]. The theorems in subsection 2 are published here for the first time.

§11. Theorem 11.2 was first obtained in a somewhat different formulation by V. P. Potapov [2].

§§12 and 13. The contents of these sections are borrowed from the paper of Brodskiĭ and Livšic [1].

The characteristic function (2) becomes a particularly effective tool when the operator A is close to a selfadjoint operator in the sense that A_I is completely continuous. For a bounded operator T close in one or another sense to a unitary operator, it is more convenient to use a characteristic function of another type, defined by the formula

$$\Theta(\zeta) = \left[T - \zeta J_{T^*} | E - TT^* |^{1/2} (E - \zeta T^*)^{-1} | E - T^*T |^{1/2}\right] |_{\mathfrak{R}_T} \tag{3}$$

$$(J_T = \operatorname{sign}(E - T^*T), \qquad \mathfrak{R}_T = \overline{(E - T^*T)\,\mathfrak{H}}).$$

For the case $\dim \mathfrak{R}_T = 1$ the function (3) was introduced by Livšic back in 1946. In matrix form, when $\dim \mathfrak{R}_T < \infty$, it was studied by Livšic and Potapov [1].

Using the function (3), V. T. Poljackiĭ [1] obtained a triangular functional model of the operator T, satisfying the conditions $\dim \mathfrak{R}_T = \dim \mathfrak{R}_{T^*} < \infty$. V. M. Adamjan and D. Z. Arov [1,2,3,4] established important connections between the theory of characteristic functions of the type (3) and the theory of dispersion operators, and also the theory of prediction of stationary random processes. In the recent monograph of Bela Sz.-Nagy and C. Foiaş [1] the characteristic function (3) is applied to the study of compression operators.

Characteristic functions of various classes of unbounded operators have been considered by A. V. Štraus [1,2,3], A. V. Kužel' [1,2] and È. R. Cekanovskiĭ [1,2,3].

To Chapter II

§§14 and 15. The theorem on the existence of a nontrivial invariant subspace for a completely continuous operator given in a Banach space is due to N. Aronszajn and K. T. Smith [1]. For the special case when the operator is given in Hilbert space, it was established by J. von Neumann in 1935. The idea of the proof of Theorem 15.2 arose in conversation between the author and M. S. Livšic. L. A. Sahnovič arrived at the same result by a somewhat different method [1].

§§16 and 17. Theorems 16.4 and 17.1 (M. S. Brodskiĭ [3]) were proved by Gohberg and Kreĭn [1] for the case when A_I is a Hilbert-Schmidt operator. Before that Theorem 16.4 had been obtained in M. S. Brodskiĭ [2] under the assumption that A_I is a nuclear operator. It follows from Theorem 17.1 that an operator A satisfying the hypotheses of Theorem 16.3 is a Volterra operator. Thus, in Theorems 16.3 and 16.4 operators belonging to the same class are reduced to triangular form.

§18. The process of parametrization which converts a maximal chain into a spectral function was first considered by Sahnovič [1]. Assertion 1 of Theorem 18.3 is taken from the same paper. Assertions 2 and 3 are due to Gohberg and Kreĭn [5] (Chapter V, Theorem 1.3). The proof presented by us differs in method from the proofs of these authors.

§19. Theorems 19.1 and 19.9 were proved by Gohberg and Kreĭn in [5].

§20. Theorem 20.2 follows easily from the results of the paper [1] of Livšic. Theorem 20.3 was obtained first by E. C. Titchmarsh [1]. G. K. Kalisch [2] discovered that it is a consequence of the theorem on unicellularity of the integration operator. The proof we present (see Brodskiĭ [9]) is close to that of Kalisch.

§21. Lemma 21.2 is due to Ky Fan [1]. The ideal $\mathfrak{S}_\omega$ was introduced by

V. I. Macaev [1]. The theory of the ideals $\mathfrak{S}_\pi$ was presented in the monograph [4] of Gohberg and Kreĭn.

§22. In Lemma 22.2 we reproduce with trivial changes the arguments of Gohberg and Kreĭn. Before this lemma was established it was known only that the integral $\int_0^1 P(x)H\,dP(x)$, where $P(x)$ $(0 \le x \le 1)$ is a continuous spectral function and H a completely continuous selfadjoint operator, may fail to exist (see M. S. Brodskiĭ [3]). Theorems 22.1 and 22.3 are due to Macaev [1]. The first assertion of Theorem 22.1 was obtained in preliminary form for the case $H \in \mathfrak{S}_1$ by Brodskiĭ [3] and for the case $H \in \mathfrak{S}_p$ $(1 < p < \infty)$ by Gohberg and Kreĭn.

§23. It follows from Theorem 18.1 and Lemma 23.1 that every continuous maximal chain is the range of some absolutely continuous spectral function. This fact was established by Gohberg and Kreĭn [5] (Chapter V, §1). Lemma 22.3 was established in another way in the theory of spectral types of selfadjoint operators (see A. I. Plesner [1]). Special cases of Theorem 23.1 were considered by I. S. Kac [1,2].

§24. Triangular functional models of Volterra operators with nuclear imaginary component can be obtained from models of a more general type, constructed using the theory of characteristic functions in the papers of Livšic [1] and Brodskiĭ and Livšic [1]. Another approach, based on Lemma 23.3 and the fact of existence of a spectral function, was considered by Sahnovič [1,2] and by Gohberg and Kreĭn [5].

§25. The assertions of Theorem 25.8 relating to the multiplicative integral (25.15) were obtained in a purely analytic way by Ginzburg [2,4].

§26. The proof of the uniqueness of the solution of the integral equation (26.4) is due to V. M. Brodskiĭ [1].

In Chapter II abstract triangular representations were obtained only for Volterra operators. We present here some results relating to operators of wider classes.

A bounded linear operator A operating in a separable Hilbert space $\mathfrak{H}$ will be assigned to the class $\Delta^{(R)}$ if 1) the weight of the spectrum of A lies on the real axis, and 2) $A_I \in \mathfrak{S}_\omega$. Macaev proved in [1] that an operator of the class $\Delta^{(R)}$ has a maximal chain π such that for each real t there exists an orthoprojector $P_t \in \pi$ cutting the spectrum of A at the point t. This last means that the spectra of the operators $A/P_t\mathfrak{H}$ and $(E - P_t)A/(E - P_t)\mathfrak{H}$ lie respectively in the intervals

$(-\infty, t]$ and $[t, \infty)$. In terms of the orthoprojectors of this chain the operator A may be represented in the form

$$A = \int_{\pi} \varphi(P)\, dP + 2i \int_{\pi} PA_I\, dP, \tag{4}$$

(see Brodskiĭ, [2,4]), where $\phi(P)$ is some left continuous nondecreasing spectral function. In the papers of L. de Branges [1,2] and in a paper of V. M. and M. S. Brodskiĭ [1] formula (4) was generalized to the case of all bounded operators with imaginary component from $\mathfrak{S}_\omega$.

Suppose that T is a bounded linear operator the weight of whose spectrum lies on the unit circumference. If $D_T = T^*T - E \in \mathfrak{S}_\omega$, then, as was proved by Gohberg and Kreĭn [8], there exists a maximal chain π belonging to T and a left continuous nondecreasing scalar function $\phi(P)$ $(P \in \pi,\ 0 \leq \phi(P) \leq 2\pi)$, such that

$$\left.\begin{aligned} &T = U(E+V), \\ &U = \int_{\pi} e^{i\varphi(P)}\, dP, \quad (E+V)^{-1} = E + \int_{\pi} (E - PD_TP)^{-1} PD_T\, dP. \end{aligned}\right\} \tag{5}$$

The triangular representation (5) was generalized by V. M. Brodskiĭ to the case of any bounded operator T satisfying the condition $T^*T - E \in \mathfrak{S}_\omega$.

Formulas (4) and (5) lead to multiplicative representations of analytic operator-functions of the corresponding classes (see V. M. Brodskiĭ [1], V. M. and M. S. Brodskiĭ [1], M. S. Brodskiĭ [8], and Gohberg and Kreĭn [6,8]).

To Chapter III

§27. The results of this section were published without proof in Brodskiĭ [7].

§28. There is an assertion close to Lemma 28.3 in the paper of Brodskiĭ and Kisilevskiĭ [1]. Theorem 28.5 was proved by Kisilevskiĭ under the hypothesis that A is a Volterra operator with nuclear imaginary component. Further information on nuclear operators and on operators with nuclear imaginary component may be found in the paper of Brodskiĭ [10] and in the monographs of I. M. Gel'fand and N. Ja. Vilenkin [1] and Gohberg and Kreĭn [4].

§29. Under narrower hypotheses Theorems 29.5, 29.6 and 29.7 were noted in a series of papers of Kisilevskiĭ and the author. The first two of them have a nonempty intersection with recent results of Ginzburg [4].

§30. The sufficiency of condition (30.5) was established by Brodskiĭ [12]. Its necessity for the case of finite-dimensional imaginary component was established by Kisilevskiĭ, and then in the general case $A_I \in \mathfrak{S}_1$ by Brodskiĭ and Kisilevskiĭ [1]. Theorems 30.4, 30.5 and 30.6 are due to Isaev [2].

§31. Theorem 31.4 was proved by Kisilevskiĭ [4].

§32. The results presented in subsections 1 and 3 of this section are due to Kisilevskiĭ [1,5]. The contents of subsection 2 are partially taken from the paper [2] of Brodskiĭ and Kisilevskiĭ.

§33. With the exception of Theorems 33.2 and 33.3, obtained by Brodskiĭ [13], all the theorems of this section are due to Kisilevskiĭ [1,5]. In comparison with the previous publications the exposition we give here is considerably simplified.

§34. Lemma 34.1 was proved by Brodskiĭ and Kisilevskiĭ [1]. It is valid, as was proved by Kisilevskiĭ [5], also without the requirement of unicellularity of the operator A. Theorem 34.1 and all the results of subsection 2 are due to Isaev [3]

In view of Theorems 24.4 and 34.1, every completely continuous nonselfadjoint dissipative Volterra operator A with nuclear imaginary component is unitarily equivalent to an operator of the type

$$\vec{A}f(x) = 2i\int_x^l K(x, y)\, f(y)\, dy \tag{6}$$

$$(K(x, y) = \xi(x)\,\xi^*(y),\ \xi(x) = \|\psi_1(x)\ \psi_2(x)\ \dots\|),$$

operating in $L_2(0, l)$ ($l = \operatorname{tr} A_I$). Here the $\psi_j(x)$ are functions from $L_2(0, l)$ satisfying almost everywhere the condition $\xi(x)\xi^*(x) = 1$. In the papers of Brodskiĭ [12], Brodskiĭ and Livšic [1], Kisilevskiĭ [6,7], and Sahnovič [3,4,5], restrictions were imposed on the kernel $K(x, y)$ which guaranteed the unicellularity of the operator (6). The most complete results are contained in the following two theorems of Kisilevskiĭ:

I. *If the functions* $\psi_j(x)$ *are of bounded variation on the segment* $[0, l]$, *if* $\sum_j v_j < \infty$, *where* v_j *is the variation of* $\psi_j(x)$, *and if* $\xi(x_0 - 0)\xi^*(x + 0) \neq 0$ *at each point of the interval* $(0, l)$, *then the operator* (6) *is unicellular.*

II. *If the vector-function* $\xi(x)$ *is piecewise continuous on* $[0, l]$ *and satisfies the condition* $\xi(x_0 - 0)\xi^*(x_0 + 0) \neq 0$ *at every point of discontinuity* $x_0 \in [0, l]$, *then the operator* (6) *is unicellular.*

To Appendix I

Theorem 2 is due to M. A. Naĭmark [1,2]. A detailed proof of it, which served as a model of the proof of Theorem 1 which we present, will be found in the course of Ahiezer and Glazman [1].

To Appendix II

Theorem 1 (without indications as to the finite dimensionality of the coefficients of the Laurent expansion) was obtained by Gohberg [1]. The proof which we present is due to Ginzburg.

BIBLIOGRAPHY

V. M. Adamjan and D. Z. Arov

1. *On a class of scattering operators and characteristic operator-functions of contractions*, Dokl. Akad. Nauk SSSR 160 (1965), 9–12 = Soviet Math. Dokl. 6 (1965), 1–5. MR 30 #5169.
2. *On scattering operators and contraction semigroups in Hilbert space*, Dokl. Akad. Nauk SSSR 165 (1965), 9–12 = Soviet Math. Dokl. 6 (1965), 1377–1380. MR 32 #6240.
3. *Unitary couplings of semi-unitary operators*, Mat. Issled. 1 (1966), no. 2, 3–64; English transl., Amer. Math. Soc. Transl (2) 95 (1970), 75–129. MR 34 #6528.
4. *General solution of a certain problem in the linear prediction of stationary processes*, Teor. Verojatnost. i Primenen. 13 (1968), 419–431 = Theor. Probability Appl. 13 (1968), 394–407. MR 38 #5281.

N. I. Ahiezer and I. M. Glazman

1. *Theory of linear operators in Hilbert space*, 2nd rev. ed., "Nauka", Moscow, 1966; English transl. of 1st ed., Ungar, New York, 1961. MR 34 #6527.

N. Aronszajn and K. T. Smith

1. *Invariant subspaces of completely continuous operators*, Ann. of Math. (2) 60 (1954), 345–350; Russian transl., Matematika 2 (1958), no. 1, 97–102. MR 16, 488.

S. N. Bernšteĭn

1. *Extremal properties of polynomials, and best approximation of continuous functions of a real variable*, Glaz. Redak. Obšč. Lit., Moscow, 1937. (Russian)

Louis de Branges

1. *Some Hilbert spaces of analytic functions*. II, J. Math. Anal. Appl. 11 (1965), 44–72. MR 35 #778.
2. *Some Hilbert spaces of analytic functions*. III, J. Math. Anal. Appl. 12 (1965), 149–186. MR 35 #779.

M. S. Brodskiĭ

1. *On the integral representation of dissipative completely continuous operators*, Naučn. Zap. Kafedr Mat. Fiz. Estest. Odessk. Gos. Ped. Inst. 24 (1959), no. 1, 3–7. (Russian)
2. *Integral representations of bounded non-selfadjoint operators with a real spectrum*, Dokl. Akad. Nauk SSSR 126 (1959), 1166–1169. (Russian) MR 21 #7438.
3. *On the triangular representation of completely continuous operators with one-point spectra*, Uspehi Mat. Nauk 16 (1961), no. 1 (97), 135–141; English transl., Amer. Math. Soc. Transl. (2) 47 (1965), 59–65. MR 24 #A426.
4. *Triangular representation of certain operators with completely continuous imaginary part*, Dokl. Akad. Nauk SSSR 133 (1960), 1271–1274 = Soviet Math. Dokl. 1 (1960), 952–955. MR 26 #6778.
5. *On a problem of I. M. Gel'fand*, Uspehi Mat. Nauk 12 (1957), no. 2 (74), 129–132. (Russian) MR 20 #1229.
6. *Characteristic matrix functions of linear operators*, Mat. Sb. 39 (81) (1956), 179–200. (Russian) MR 18, 220.
7. *Unicellularity criteria for Volterra operators*, Dokl. Akad. Nauk SSSR 138 (1961), 512–514 = Soviet Math. Dokl. 2 (1961), 637–639. MR 24 #A1015.
8. *A multiplicative representation of certain analytic operator-functions*, Dokl. Akad. Nauk SSSR 138 (1961), 751–754 = Soviet Math. Dokl. 2 (1961), 695–698. MR 24 #A1031.
9. *On the unicellularity of the integration operator and a theorem of Titchmarsh*, Uspehi Mat. Nauk 20 (1965), no. 5 (125), 189–192. (Russian) MR 32 #8055.
10. *Operators with nuclear imaginary components*, Acta Sci. Math. (Szeged) 27 (1966), 144–155. (Russian) MR 34 #1843.
11. *On a class of entire operator-functions*, Mat. Issled. 3 (1968), no. 1, 3–16. (Russian)
12. *On Jordan cells of infinite-dimensional operators*, Dokl. Akad. Nauk SSSR 111 (1956), 926–929. (Russian) MR 19, 48.
13. *On the question of the decomposition of a dissipative operator with nuclear imaginary component into unicellular ones*, Funkcional. Anal. i. Priložen. 2 (1968), no. 3, 81–83 = Functional Anal. Appl. 2 (1968), 254–256. MR 38 #5044.

M. S. Brodskiĭ and G. È. Kisilevskiĭ

1. *Criterion for unicellularity of dissipative Volterra operators with nuclear imaginary components*, Izv. Akad. Nauk SSSR Ser. Mat. 30 (1966), 1213–1228; English transl., Amer. Math. Soc. Transl. (2) 65 (1967), 282–296. MR 34 #3310.

2. *Quasidirect sums of subspaces*, Funkcional. Anal. i Priložen. 1 (1967), no. 4, 75–78 = Functional Anal. Appl. 1 (1967), 322–324. MR 36 #3158.

M. S. Brodskiĭ and M. S. Livšic

1. *Spectral analysis of non-selfadjoint operators and intermediate systems*, Uspehi Mat. Nauk 13 (1958), no. 1 (79), 3–85; English transl., Amer. Math. Soc. Transl. (2) 13 (1960), 265–346. MR 20 #7221; 22 #3982.

M. S. Brodskiĭ and Ju. L. Šmul'jan

1. *Invariant subspaces of a linear operator and divisors of its characteristic function*, Uspehi Mat. Nauk 19 (1964), no. 1 (115), 143–149. (Russian) MR 29 #2645.

V. M. Brodskiĭ

1. *Certain questions of the theory of triangular representations of operators and multiplicative expansions of their characteristic functions*, Dissertation, Kishinev, 1967. (Russian)
2. *On the integral with respect to a chain of orthogonal projectors*, Bul. Akad. Štiince RSS Moldoven. 1966, no. 4, 59–64. (Russian) MR 36 #1997.
3. *An operator integral equation*, Mat. Issled. 1 (1966), no. 1, 178–185. (Russian) MR 34 #4944.
4. *Multiplicative representation of the characteristic functions of contraction operators*, Dokl. Akad. Nauk SSSR 173 (1967), 256–259 = Soviet Math. Dokl. 6 (1967), 362–366. MR 36 #3154.

V. M. Brodskiĭ and M. S. Brodskiĭ

1. *The abstract triangular representation of bounded linear operators and multiplicative expansions of their eigenfunctions*, Dokl. Akad. Nauk SSSR 181 (1968), 511–514 = Soviet Math. Dokl. 9 (1968), 846–850. MR 37 #5719.

È. R. Cekanovskiĭ

1. *The real and imaginary parts of an unbounded operator*, Dokl. Akad. Nauk SSSR 139 (1961), 48–51 = Soviet Math. Dokl. 2 (1961), 881–885. MR 24 #A1619.
2. *Model elements of non-selfadjoint operators*, Dokl. Akad. Nauk SSSR 142 (1962), 1043–1046 = Soviet Math. Dokl. 3 (1962), 256–259. MR 25 #455.
3. *Characteristic functions of unbounded operators*, Trudy Har'kov. Gorm. Inst. 11 (1962), 95–100. (Russian)

W. F. Donoghue

1. *The lattice of invariant subspaces of a completely continuous quasi-nilpotent transformation*, Pacific J. Math. 7 (1957), 1031–1035. MR 19, 1066.

Ky Fan

1. *Maximum properties and inequalities for the eigenvalues of completely continuous operators*, Proc. Nat. Acad. Sci. U.S.A. 37 (1959), 760–766. MR 13, 661.

I. M. Gel'fand and N. Ja. Vilenkin

1. *Generalized functions.* Vol. 4: *Applications of harmonic analysis*, Fizmatgiz, Moscow, 1961; English transl., Academic Press, New York, 1964. MR 26 #4173; 30 #4152.

Ju. P. Ginzburg

1. *A maximum principle for J-contracting operator functions and some of its consequences*, Izv. Vysš. Učebn. Zaved. Matematika 1963, no. 1 (32), 42–53. (Russian) MR 26 #6793.
2. *On J-contractive operator functions*, Dokl. Akad. Nauk SSSR 117 (1957), 171–173. MR 20 #1203.
3. *The factorization of analytic matrix functions*, Dokl. Akad. Nauk SSSR 159 (1964), 489–492 = Soviet Math. Dokl. 5 (1964), 1510–1514. MR 30 #3228.
4. *Multiplicative representations of bounded analytic operator-functions*, Dokl. Akad. Nauk SSSR 170 (1966), 23–26 = Soviet Math. Dokl. 7 (1966), 1125–1128. MR 34 #611.

I. C. Gohberg

1. *On linear operators depending analytically on a parameter*, Dokl. Akad. Nauk SSSR 78 (1951), 629–632. (Russian) MR 13, 46.

I. C. Gohberg and M. G. Kreĭn

1. *Completely continuous operators with a spectrum concentrated at zero*, Dokl. Akad. Nauk SSSR 128 (1959), 227–230. (Russian) MR 24 #A1022.
2. *On the theory of triangular representations of non-selfadjoint operators*, Dokl. Akad. Nauk SSSR 137 (1961), 1034–1037 = Soviet Math. Dokl. 2 (1961), 392–395. MR 25 #3370.
3. *Volterra operators with imaginary component in one class or another*, Dokl. Akad. Nauk SSSR 139 (1961), 779–782 = Soviet Math. Dokl. 2 (1961), 983–986. MR 25 #3371.
4. *Introduction to the theory of linear non-selfadjoint operators in Hilbert space*, "Nauka", Moscow, 1965; English transl., Transl. Math. Monographs, vol. 18, Amer. Math. Soc., Providence, R.I., 1967. MR 36 #3137.
5. *Theory of Volterra operators in Hilbert space and its applications*, "Nauka", Moscow, 1967; English transl., Transl. Math. Monographs, vol. 24, Amer. Math. Soc., Providence, R. I., 1970. MR 36 #2007.

6. *Triangular representations of linear operators and multiplicative representations of their characteristic functions*, Dokl. Akad. Nauk SSSR 175 (1967), 272–275 = Soviet Math. Dokl. 8 (1967), 831–834. MR 35 #7157.

7. *The basic propositions on defect numbers, root numbers, and indices of linear operators*, Uspehi Mat. Nauk 12 (1957), no. 2 (74), 43–118; English transl., Amer. Math. Soc. Transl. (2) 13 (1960), 185–264. MR 20 #3459; 22 #3984.

8. *On the multiplicative representation of characteristic functions of operators close to the unitary ones*, Dokl. Akad. Nauk SSSR 164 (1965), 732–735 = Soviet Math. Dokl. 6 (1965), 1279–1283. MR 33 #571.

L. E. Isaev

1. *On the theory of eigenfunctions of operator colligations*, Ukrain. Mat. Ž. 20 (1968), 253–257. (Russian) MR 37 #2021.

2. *A certain class of operators with spectrum concentrated at zero*, Dokl. Akad. Nauk SSSR 178 (1968), 783–785 = Soviet Math. Dokl. 9 (1968), 198–200. MR 37 #6791.

3. *Spectral analysis of bounded linear operators with resolvent of exponential type*, Dissertation, Odessa, 1968. (Russian)

I. S. Kac

1. *On Hilbert spaces generated by monotone Hermitian matrix-functions*, Zap. Naučn. Har'kov. Gos. Univ. Uč. Zap. 34 = Zap. Mat. Otd. Fiz.-Mat. Fak. i Har'kov. Mat. Obšč. (4) 22 (1950), 95–113. (Russian) MR 18, 222.

2. *Spectral multiplicity of a second-order differential operator and expansion in eigenfunctions*, Izv. Akad. Nauk SSSR Ser. Mat. 27 (1963), 1081–1112. (Russian) MR 28 #3196.

G. K. Kalisch

1. *On similarity, reducing manifolds, and unitary equivalence of certain Volterra operators*, Ann. of Math. (2) 66 (1957), 481–494. MR 19, 970.

2. *A functional analysis proof of Titchmarsh's theorem of convolution*, J. Math. Anal. Appl. 5 (1962), 176–183. MR 25 #4307.

G. È. Kisilevskiĭ

1. *A generalization of the Jordan theory to a certain class of linear operators in Hilbert space*, Dokl. Akad. Nauk SSSR 176 (1967), 768–770 = Soviet Math. Dokl. 8 (1967), 1188–1190. MR 36 #6972.

2. *Conditions for unicellularity of dissipative Volterra operators*, Dissertation, Kharkov, 1965. (Russian)

3. *Conditions for unicellularity of dissipative Volterra operators with finite-dimensional imaginary component*, Dokl. Akad. Nauk SSSR 159 (1964),

505–508 = Soviet Math. Dokl. 5 (1964), 1527–1531. MR 30 #5162.

4. *Cyclic subspaces of dissipative operators*, Dokl. Akad. Nauk SSSR 173 (1967), 1006–1009 = Soviet Math. Dokl. 8 (1967), 517–520. MR 36 #2002.

5. *Invariant subspaces of Volterra dissipative operators with nuclear imaginary components*, Izv. Akad. Nauk SSSR Ser. Mat. 32 (1968), 3–23 = Math. USSR Izv. 2 (1968), 1–20. MR 36 #4375.

6. *On unicellularity of dissipative Volterra operators*, Ukrain. Mat. Ž. 16 (1964), 690–696. (Russian) MR 30 #419.

7. *Criteria for unicellularity of dissipative Volterra operators*, Funkcional. Anal. i Priložen. 1 (1967), no. 3, 90–91 = Functional Anal. Appl. 1 (1967), 247–248. MR 36 #2005.

A. V. Kužel′

1. *The reduction of unbounded non-selfadjoint operators to triangular form*, Dokl. Akad. Nauk SSSR 119 (1958), 868–871. MR 20 #6041.

2. *Spectral analysis of unbounded non-selfadjoint operators*, Dokl. Akad. Nauk SSSR 125 (1959), 35–37. (Russian) MR 22 #2905.

B. Ja. Levin

1. *Distribution of zeros of entire functions*, GITTL, Moscow, 1956; English transl., Transl. Math. Monographs, vol. 5, Amer. Math. Soc., Providence, R.I., 1964. MR 19, 402; 28 #217.

M. S. Livšic

1. *On the spectral decomposition of linear non-selfadjoint operators*, Mat. Sb. 34 (76) (1954), 145–199; English transl., Amer. Math. Soc. Transl. (2) 5 (1957), 67–114. MR 16, 48; 18, 748.

2. *Operators, oscillations, waves. Open systems*, "Nauka", Moscow, 1966. (Russian) MR 38 #1922.

3. *On a certain class of linear operators in Hilbert space*, Mat. Sb. 19 (61) (1946), 239–262. (Russian) MR 8, 588.

M. S. Livšic and V. P. Potapov

1. *A theorem on the multiplication of characteristic matrix-functions*, Dokl. Akad. Nauk SSSR 72 (1950), 625–628. (Russian) MR 11, 669.

Ju. I. Ljubič and V. I. Macaev

1. *On the spectral theory of linear operators in Banach space*, Dokl. Akad. Nauk SSSR 131 (1960), 21–23 = Soviet Math. Dokl. 1 (1960), 184–186. MR 22 #3980.

2. *Operators with separable spectrum*, Mat. Sb. 56 (98 (1962), 433–468; English transl., Amer. Math. Soc. Transl. (2) 47 (1965), 89–129. MR 25 #2450.

3. *Letter to the editors*, Mat. Sb. 71 (113) (1966), 287–288. (Russian) MR 33 #6406.

V. I. Macaev

1. *A class of completely continuous operators*, Dokl. Akad. Nauk SSSR 139 (1961), 548–551 = Soviet Math. Dokl. 2 (1961), 972–975. MR 24 #A1617.
2. *Volterra operators obtained from self-adjoint operators by perturbation*, Dokl. Akad. Nauk SSSR 139 (1961), 810–813 = Soviet Math. Dokl. 2 (1961), 1013–1016. MR 25 #457.

M. A. Naĭmark

1. *Spectral functions of a symmetric operator*, Izv. Akad. Nauk SSSR Ser. Mat. 4 (1940), 227–318. (Russian) MR 2, 105.
2. *On a representation of additive operator set functions*, Dokl. Akad. Nauk SSSR 41 (1943), 359- 361. (Russian) MR 6, 71.

N. K. Nikol'skiĭ

1. *The invariant subspaces of certain completely continuous operators*, Vestnik Leningrad. Univ. 20 (1965), no. 7, 68–77. (Russian) MR 32 #2911.
2. *Unicellularity and non-unicellularity of weighted shift operators*, Dokl. Akad. Nauk SSSR 172 (1967), 287–290 = Soviet Math. Dokl. 8 (1967) , 91–94. MR 36 #728.

A. I. Plessner

1. *Spectral theory of linear operators*, "Nauka", Moscow, 1965. (Russian) MR 33 #3106.

V. T. Poljackiĭ

1. *On the reduction of quasi-unitary operators to a triangular form*, Dokl. Akad. Nauk SSSR 113 (1957), 756–759. (Russian) MR 19, 873.

V. P. Potapov

1. *The multiplicative structure of J-contractive matrix functions*, Trudy Moskov. Mat. Obšč. 4 (1955), 125–236; English transl., Amer. Math. Soc. Transl. (2) 15 (1960), 131–243. MR 17, 958; 22 #5733.
2. *On holomorphic matrix functions bounded in the unit circle*, Dokl. Akad. Nauk SSSR 72 (1950), 849–852. (Russian) MR 13, 736.

F. Riesz and B. Sz.-Nagy

1. *Leçons d'analyse fonctionnelle*, Akad. Kiadó, Budapest, 1952; Russian transl., IL, Moscow, 1954; English transl., Ungar, New York, 1955, MR 14, 286; 17, 175.

L. A. Sahnovič

1. *The reduction of non-selfadjoint operators to triangular form*, Izv. Vysš. Učebn. Zaved. Matematika 1959, no. 1 (8), 180–186. (Russian) MR 25 #460.
2. *A study of the "triangular form" of non-selfadjoint operators*, Izv. Vysš. Učebn. Zaved. Matematika 1959, no. 4 (11), 141–149; English transl., Amer. Math. Soc. Transl. (2) 54 (1966), 75–84. MR 25 #461.
3. *The spectral analysis of Volterra operators and some inverse problems*, Dokl. Akad. Nauk SSSR 115 (1957), 666–669. (Russian) MR 19, 866.
4. *Spectral analysis of operators of the form* $Kf = \int_0^x f(t)k(x-t)dt$, Izv. Akad. Nauk SSSR Ser. Mat. 22 (1958), 299–308. (Russian) MR 20 #5409.
5. *On reduction of Volterra operators to the simplest form and on inverse problems*, Izv. Akad. Nauk SSSR Ser. Mat. 21 (1957), 235–262. (Russian) MR 19, 970.

Ju. L. Šmul'jan

1. *Some questions in the theory of operators with a finite non-Hermitian rank*, Mat. Sb. 57 (99) (1962), 105–136. (Russian) MR 27 #6131.
2. *Non-expanding operators in a finite-dimensional space with indefinite metric*, Uspehi Mat. Nauk 18 (1963), no. 6 (114), 225–230. (Russian) MR 30 #454.
3. *Hellinger's operator integral and some of its uses*, Dokl. Akad. Nauk SSSR 120 (1958), 722–725. (Russian) MR 23 #A1236.
4. *An operator Hellinger integral*, Mat. Sb. 49 (91) (1959), 381–430; English transl., Amer. Math. Soc. Transl. (2) 22 (1962), 289–337. MR 22 #12396.

A. V. Štraus

1. *Characteristic functions of linear operators*, Dokl. Akad. Nauk SSSR 126 (1959), 514–516. (Russian) MR 21 #5899.
2. *Characteristic functions of linear operators*, Izv. Akad. Nauk SSSR Ser. Mat. 24 (1960), 43–74; English transl., Amer. Math. Soc. Transl. (2) 40 (1964), 1–37. MR 25 #4363.
3. *A multiplication theorem for characteristic functions of linear operators*, Dokl. Akad. Nauk SSSR 126 (1959), 723–726. (Russian) MR 21 #5900.

B. Sz-Nagy and C. Foiaş

1. *Analyse harmonique des opérateurs de l'espace de Hilbert*, Akad. Kiadó, Budapest, and Masson, Paris, 1967. MR 37 #778.

E. C. Titchmarsh

1. *Introduction to the theory of Fourier integrals*, Clarendon Press, Oxford, 1937; Russian transl., GITTL, Moscow, 1948.